Collins

INTERNATIONAL PRIMARY MATHS

Teacher's Guide 6

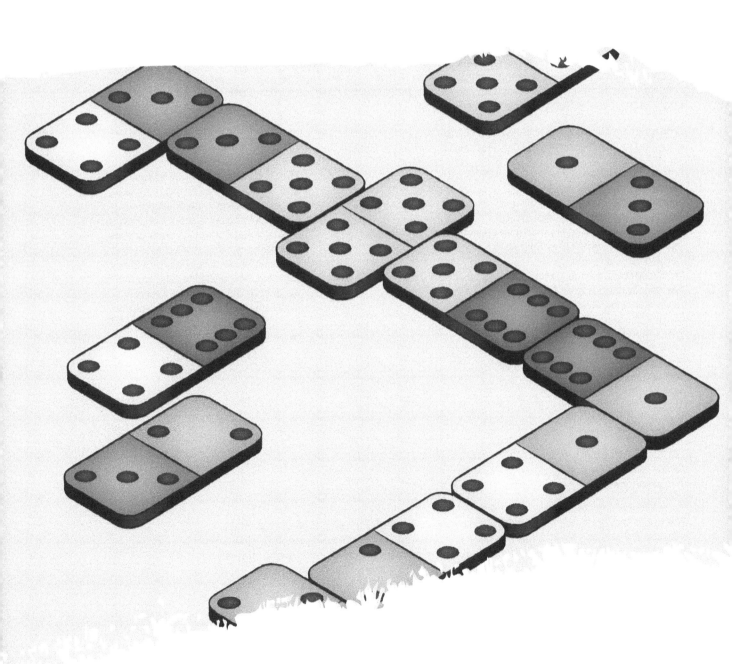

William Collins' dream of knowledge for all began with the publication of his first book in 1819. A self-educated mill worker, he not only enriched millions of lives, but also founded a flourishing publishing house. Today, staying true to this spirit, Collins books are packed with inspiration, innovation and practical expertise. They place you at the centre of a world of possibility and give you exactly what you need to explore it.

Collins. Freedom to teach.

Published by Collins
An imprint of HarperCollins*Publishers*
The News Building
1 London Bridge Street
London
SE1 9GF

HarperCollins*Publishers*
Macken House, 39/40 Mayor Street Upper
Dublin 1, D01 C9W8, Ireland

Browse the complete Collins catalogue at
www.collins.co.uk

© HarperCollins*Publishers* Limited 2021

10 9 8

ISBN 978-0-00-836956-9

British Library Cataloguing-in-Publication Data
A catalogue record for this publication is available from the British Library.

Author: Paul Hodge
Series editor: Peter Clarke
Publisher: Elaine Higgleton
Product developer: Holly Woolnough
Project manager: Mike Harman (Life Lines Editorial Services)
Development editor: Tanya Solomons
Copyeditor: Catherine Dakin
Proofreader: Catherine Dakin
Answer checker: Steven Matchett
Cover designer: Gordon MacGilp
Cover illustrator: Ann Paganuzzi
Typesetter: QBS Learning
Illustrators: Ann Paganuzzi and QBS Learning
Production controller: Lyndsey Rogers
Printed and bound in the UK by
Ashford Colour Press Ltd

With thanks to the following teachers and schools for reviewing materials in development: Antara Banerjee, Calcutta International School; Hawar International School; Melissa Brobst, International School of Budapest; Rafaella Alexandrou, Pascal Primary Lefkosia; Maria Biglikoudi, Georgia Keravnou, Sotiria Leonidou and Niki Tzorzis, Pascal Primary School Lemessos; Taman Rama Intercultural School, Bali.

MIX
Paper | Supporting responsible forestry
FSC
www.fsc.org
FSC™ C007454

This book contains FSC™ certified paper and other controlled sources to ensure responsible forest management.

For more information visit: www.harpercollins.co.uk/green

The publishers gratefully acknowledge the permission granted to reproduce the copyright material in this book. Every effort has been made to trace copyright holders and to obtain their permission for the use of copyright material. The publishers will gladly receive any information enabling them to rectify any error or omission at the first opportunity.

Cambridge International copyright material in this publication is reproduced under licence and remains the intellectual property of Cambridge Assessment International Education.

Photo acknowledgements
Every effort has been made to trace copyright holders. Any omission will be rectified at the first opportunity.
p6t Tatiana Popova/Shutterstock; p6b Studio KIWI/Shutterstock; p122l Titov Nikolai/Shutterstock; p122c UI/Shutterstock; p122r Edel/Shutterstock.

Contents

Introduction

① Key features of Collins International Primary Maths

Collins International Primary Maths places the learner at the centre of the teaching and learning of mathematics. To this end, all of the components and features of the course are aimed at helping teachers to address the distinct learning needs of *all* learners.

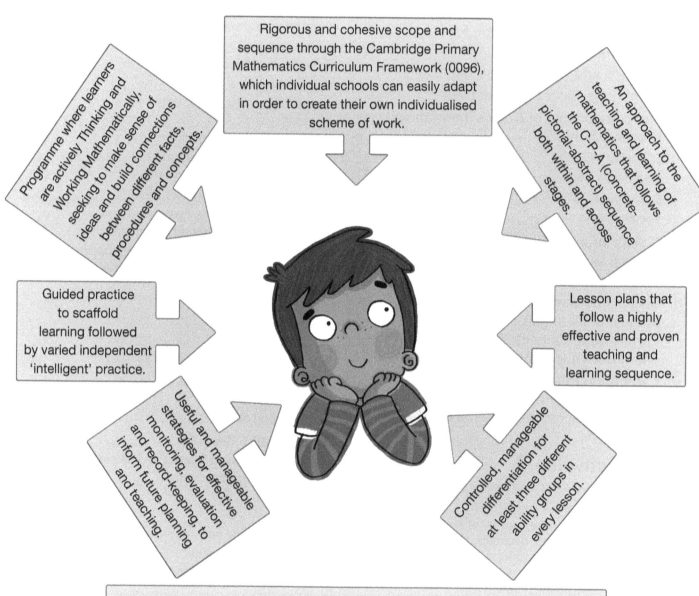

Programme where learners are actively Thinking and Working Mathematically, seeking to make sense of ideas and build connections between different facts, procedures and concepts.

Rigorous and cohesive scope and sequence through the Cambridge Primary Mathematics Curriculum Framework (0096), which individual schools can easily adapt in order to create their own individualised scheme of work.

An approach to the teaching and learning of mathematics that follows the C-P-A (concrete-pictorial-abstract) sequence both within and across stages.

Guided practice to scaffold learning followed by varied independent 'intelligent' practice.

Lesson plans that follow a highly effective and proven teaching and learning sequence.

Useful and manageable strategies for effective monitoring, evaluation and record-keeping, to inform future planning and teaching.

Controlled, manageable differentiation for at least three different ability groups in every lesson.

In addition, the course offers extensive teacher support through materials that:
- promote the most effective pedagogical methods in the teaching of mathematics
- are sufficiently detailed to aid confidence
- are rich enough to be varied and developed
- take into account issues of pace and classroom management
- give careful consideration to the key skill of appropriate and effective questioning
- provide a careful balance of teacher intervention and learner participation
- encourage communication of methods and foster mathematical rigour
- are aimed at raising levels of attainment for every learner.

② Collins International Primary Maths models of teaching and learning

Collins International Primary Maths is based on the constructivist model for teaching and learning developed by Piaget, Vygotsky, and later Bruner, and the importance of:

- starting from what learners know already and providing them with guidance that moves their thinking forwards
- students learning the fundamental principles of a subject, as well as the connections between ideas within the subject and across other subjects
- focusing on the process of learning, rather than the end product of it
- developing learners' intuitive thinking, by asking questions and providing opportunities for learners to ask questions

- learning through discovery and problem solving, which requires learners to make predictions, hypothesise, make generalisations, ask questions and discuss lines of enquiry
- using active methods that require rediscovering or reconstructing norms and truths
- using collaborative as well as individual activities, so that learners can learn from each other
- evaluating the level of each learner's development so that suitable tasks can be set.

At the core of the course are four of the major principles of Bruner's 'Theory of Instruction'* that characterise the organisation and content of Collins International Primary Maths.

Predisposition to learn

The concept of 'readiness for learning'.

A belief that any subject can be taught at any stage of development in a way that fits the learner's cognitive abilities.

Structure of knowledge

A body of knowledge can be structured so that it can be most readily grasped by the learner.

Effective sequencing

No one sequencing will fit every learner, but in general, curriculum content should be taught in increasing difficulty.

Modes of representation

Learning occurs through three modes of representation: *enactive* (action-based), *iconic* (image-based), and *symbolic* (language-based).

In particular, the teaching and learning opportunities in Collins International Primary Maths reflect Bruner's three modes of representation whereby learners develop an understanding of a concept through the three progressive steps (or representations) of concrete-pictorial-abstract, and that reinforcement of an idea or concept is achieved by going back and forth between these representations.

* Bruner, J. S. (1966) *Toward a Theory of Instruction*, Cambridge, MA: Belknap Press

Concrete Representation
The *enactive* stage

The learner is first introduced to a concept using physical objects. This 'hands on' approach is the foundation for conceptual understanding.

Pictorial Representation
The *iconic* stage

The learner has sufficiently understood the hands-on experiences and can now relate them to images, such as a picture, diagram or model of the concept.

Abstract Representation
The *symbolic* stage

The learner is now capable of using numbers, notation and mathematical symbols to represent the concept.

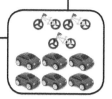

3	6
9	

$3 + 6 = 9$

The model below illustrates the six proficiencies in mathematics that Collins International Primary Maths believe learners need to command in order to become mathematically literate and achieve mastery of the subject at each stage of learning.

All of the teaching and learning units throughout the Collins International Primary Maths course aim to achieve each of these six proficiencies, in order to help teachers to establish successful mathematics learning.

⏶ Knowledge and understanding of mathematical concepts, principles, fundamental operations and procedures.

⏶ Ability to carry out procedures (or mathematical operations) flexibly, accurately, efficiently and appropriately.

⏶ Capacity for making conjectures and generalisations, and explaining and justifying conclusions.

⏶ Skill in using and applying mathematics to understand, represent, solve and interpret and evaluate problems.

⏶ Facility to communicate clearly in the language of mathematics by understanding and using precise vocabulary and symbolic and diagrammatical representations.

⏶ Process of identifying relationships and making connections by linking thoughts and ideas together.

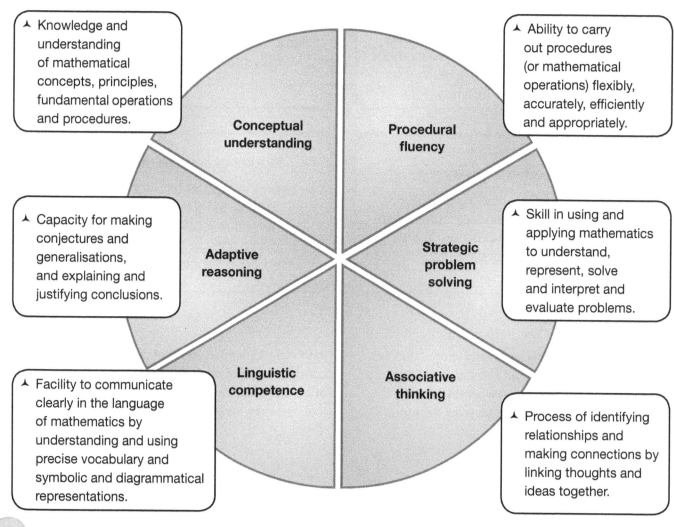

Conceptual understanding

Procedural fluency

Adaptive reasoning

Strategic problem solving

Linguistic competence

Associative thinking

③ How Collins International Primary Maths supports Cambridge Primary Mathematics

Cambridge Primary is typically for learners aged 5 to 11 years. It develops learner skills and understanding through the primary years in English, Mathematics and Science. It provides a flexible framework that can be used to tailor the curriculum to the needs of individual schools.

In Cambridge Primary Mathematics, learners:

- engage in creative mathematical thinking to generate elegant solutions
- improve numerical fluency and knowledge of key mathematical concepts to make sense of numbers, patterns, shapes, measurements and data
- develop a variety of mathematical skills, strategies and a way of thinking that will enable them to describe the world around them and play an active role in modern society
- communicate solutions and ideas logically in spoken and written language, using appropriate mathematical symbols, diagrams and representations
- understand that technology provides a powerful way of communicating mathematics, one that is particularly important in an increasingly technological and digital world.

Cambridge Primary Mathematics supports learners to become:

RESPONSIBLE	INNOVATIVE	ENGAGED
• Learners understand how principles of mathematics can be applied to real-life problems in a responsible way.	• Learners solve new and unfamiliar problems, using innovative mathematical thinking. They can select their own preferred mathematical strategies and can suggest alternative routes to develop efficient solutions.	• Learners are curious and engage intellectually to deepen their mathematical understanding. They are able to use mathematics to participate constructively in society and the economy by making informed mathematical choices.

CONFIDENT	REFLECTIVE
• Learners are confident and enthusiastic mathematical practitioners, able to use appropriate techniques without hesitation, uncertainty or fear. They are keen to ask mathematical questions in a structured, systematic, critical and analytical way. They are able to present their findings and defend their strategies and solutions as well as critique and improve the solutions of others.	• Learners reflect on the process of thinking and working mathematically as well as mastering mathematics concepts. They are keen to make conjectures by asking sophisticated questions and thus develop higher-order thinking skills.

Cambridge Primary is organised into six stages. Each stage reflects the teaching targets for a year group. Broadly speaking, Stage 1 covers the first year of primary teaching, when learners are approximately five years old. Stage 6 covers the final year of primary teaching, when learners are approximately 11 years old.

The Cambridge Primary Mathematics Curriculum Framework (0096) replaces the previous curriculum framework (0845), and is presented in three content areas (strands), with each strand divided into sub-strands. Thinking and Working Mathematically underpins all strands and sub-strands, while mental strategies are a key part of the Number content.

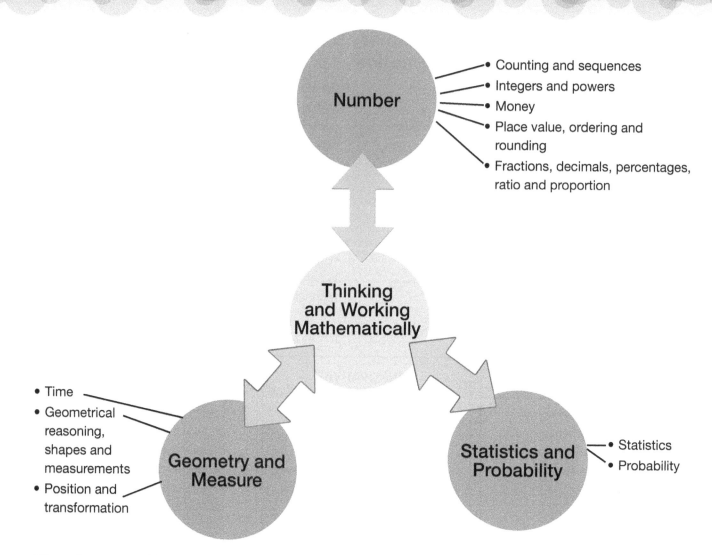

- Counting and sequences
- Integers and powers
- Money
- Place value, ordering and rounding
- Fractions, decimals, percentages, ratio and proportion

Number

Thinking and Working Mathematically

- Time
- Geometrical reasoning, shapes and measurements
- Position and transformation

Geometry and Measure

Statistics and Probability

- Statistics
- Probability

Planning, teaching and assessment

Effective planning, teaching and assessment are the three interconnected elements that contribute to promoting learning, raising learners' attainment and achieving end-of-stage expectations (mastery). All of the components of Collins International Primary Maths emphasise, and provide guidance on, the importance of this cyclical nature of teaching in order to ensure that learners reach the end-of-stage expectations of the Cambridge Primary Mathematics Curriculum Framework (0096). This teaching and learning cycle, and the important role that the teacher plays in this cycle, are at the heart of Collins International Primary Maths.

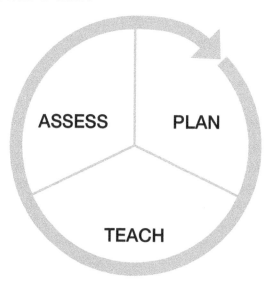

ASSESS PLAN TEACH

Collins International Primary Maths supports teachers in planning a successful mathematics programme for their unique teaching context and ensures:

- a clear understanding of learners' pre-requisite skills before they undertake particular tasks and learning new concepts
- considered progression from one lesson to another
- regular revisiting and extension of previous learning
- a judicious balance of objectives, and the time dedicated to each one
- the use of a consistent format and structure.

The elements of Collins International Primary Maths that form the basis for planning can be summarised as follows:

Long-term plans

The Cambridge Primary Mathematics Curriculum Framework (0096) constitutes the long-term plan for schools to follow at each stage across the school year. By closely reflecting the Curriculum Framework and the Cambridge Primary Mathematics Scheme of Work, the Collins International Primary Maths course embodies this long-term plan.

Medium-term plans

The Collins International Primary Maths Units and Recommended Teaching and Learning Sequence in Sections ⑦ and ⑧ (see pages 38–49) show termly/semester outlines of units of work with Cambridge Primary Mathematics Curriculum Framework (0096) references (including the Curriculum Framework codes). By using the Collins International Primary Maths Extended Teacher's Guide, these plans including curriculum coverage, delivery and timing, can be easily adapted to meet the specific needs of individual schools and teachers as well as learners' needs.

Short-term plans

Individual lesson plans and accompanying Additional practice activities represent the majority of each Teacher's Guide. The lessons provide short-term plans that can easily be followed closely, or used as a 'springboard' and varied to suit specific needs of particular classes. An editable 'Weekly Planning Grid' is also provided as Digital content, which individual teachers can fully adapt.

This includes modifying short-term planning in order to build on learners' responses to previous lessons, thereby enabling them to make greater progress in their learning.

The most important role of teaching is to promote learning and to raise learners' attainment. To best achieve these goals, Collins International Primary Maths believes in the importance of teachers:

- promoting a 'can do' attitude, where all learners can achieve success in, and enjoy, mathematics
- having high, and ambitious, expectations for *all* learners
- adopting a philosophy of equal opportunity that means *all* learners have full access to the same curriculum content
- generating high levels of engagement (*Active learning*) and commitment to learning
- offering sharply focused and timely support and intervention that matches learners' individual needs
- being *language aware* in order to understand the possible challenges and opportunities that language presents to learning
- systematically and effectively checking learners' understanding throughout lessons, anticipating where they may need to intervene, and doing so with notable impact on the quality of learning
- consistently providing high-quality marking and constructive feedback to ensure that learners make rapid gains.

To help teachers achieve these goals, Collins International Primary Maths provides:

- highly focused and clearly defined learning objectives
- examples of targeted questioning, using appropriate mathematical vocabulary, that is aimed at both encouraging and checking learner progress
- a proven lesson structure that provides clear and accurate directions, instructions and explanations
- meaningful and well-matched activities for learners, at all levels of understanding, to practise and consolidate their learning
- a balance of individual, pair, group and whole-class activities to develop both independence and collaboration and to enable learners to develop their own thinking and learn from one another
- highly effective models and images (representations) that clearly illustrate mathematical concepts, including interactive digital resources.

The lesson sequence in Collins International Primary Maths focuses on supporting learners' understanding of the learning objectives of the Cambridge Primary Mathematics Curriculum Framework (0096), as well as building their mathematical proficiency and confidence.

Based on a highly effective and proven teaching and learning sequence, each lesson is divided into six key teaching strategies that take learners on a journey of discovery. This approach is shown on the outer ring of the diagram below.

The inner ring shows the link between the six key teaching strategies and the five phases of a Collins International Primary Maths lesson plan, as well as when to use the Student's Book and Workbook.

Pedagogical cycle

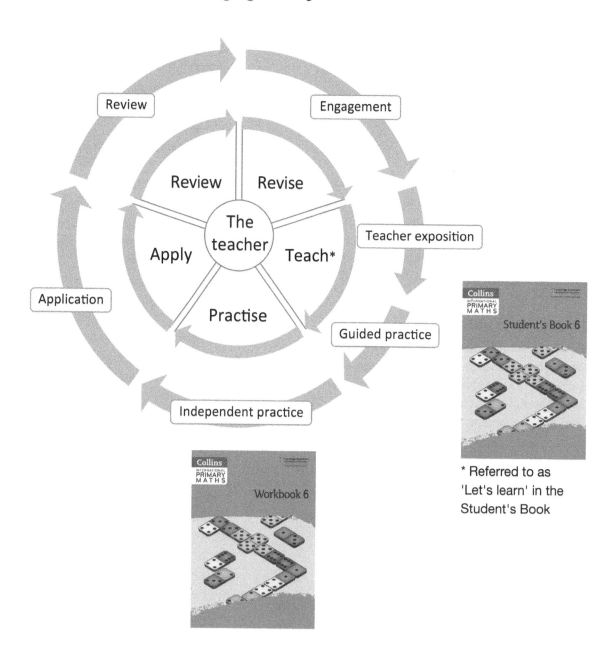

* Referred to as 'Let's learn' in the Student's Book

The chart below outlines the purposes of each phase in the Collins International Primary Maths lesson sequence, as well as the learner groupings and approximate recommended timings.

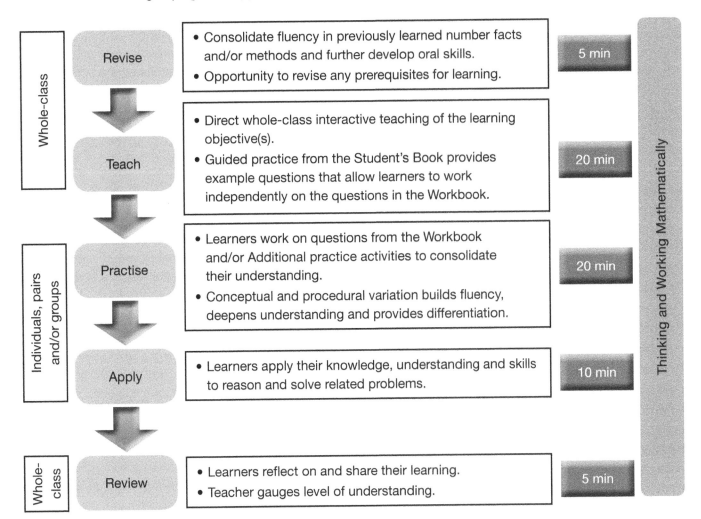

Monitoring, evaluation and feedback continue the teaching and learning cycle and are used to form the basis for adjustments to the teaching programme. Collins International Primary Maths offers meaningful, manageable and useful assessment on two of the following three levels:

Short-term 'on-going' assessment

Short-term assessments are an informal part of every lesson. A combination of carefully crafted recall, observation and thought questions is provided in each lesson of Collins International Primary Maths and these are linked to specific learning objectives.

They are designed to monitor learning and provide immediate feedback to learners and to gauge learners' progress in order to enable teachers to adapt their teaching.

Success Criteria are also provided in each unit to assist learners in identifying the steps required to achieve the unit's learning objectives.

Each unit in Collins International Primary Maths begins with a Unit introduction. One of the features of the Unit introduction is 'Common difficulties and remediation'. This feature can be used to help identify why learners do not understand, or have difficulty with, a topic or concepts and to use this information to take appropriate action to correct mistakes or misconceptions.

Medium-term 'formative' assessment

Medium-term assessments are used to review and record the progress learners make, over time, in relation to the learning objectives of the Cambridge Primary Mathematics Curriculum Framework (0096). They are used to establish whether learners have met the learning objectives or are on track to do so.

'Assessment *for* learning' is the term generally used to describe the conceptual approach to both short-term 'ongoing' assessment and medium-term 'formative' assessment.

Assessment *for* learning involves both learners and teachers finding out about the specific strengths and weaknesses of individual learners, and the class as a whole, and using this to inform future teaching and learning.

Assessment *for* learning:

- is part of the planning process
- is informed by learning objectives
- engages learners in the assessment process
- recognises the achievements of all learners
- takes account of how learners learn
- motivates learners.

In order to assist teachers with monitoring both short- and medium-term assessments, and to ensure that evidence collected is meaningful, manageable and useful, Collins International Primary Maths includes a class record-keeping document on pages 348–353.

The document helps teachers:

- identify whether learners are on track to meet end-of-stage expectations
- identify those learners working *above* and *below* end-of-stage expectations
- make long-term 'summative' assessments
- report to parents and guardians
- inform the next year's teacher about which sub-strands of the Cambridge Primary Mathematics Curriculum Framework (0096) individual learners, and the class as a whole, are exceeding, meeting or are below in expectations.

For further details on how to use the class record-keeping document, please refer to pages 30–31.

Long-term 'summative' assessment

Long-term assessment is the third level of assessment. It is used at the end of the school year in order to track progress and attainment against school and external targets, and to report to other establishments and to parents on the actual attainments of learners. By ensuring complete and thorough coverage of the Cambridge Primary Mathematics Curriculum Framework (0096), Collins International Primary Maths provides an excellent foundation for the Cambridge Primary end-of-stage tests (Cambridge Primary Progression Tests) as well as the end of primary Cambridge Primary Checkpoint.

Mental strategies and Cambridge Primary Mathematics

Mental strategies learning objectives are not included in the 2020 Cambridge Primary Mathematics Curriculum Framework (0096). However, working mentally is an important feature in the curriculum framework and is embedded not just within the Number strand but throughout all strands in the curriculum framework.

Mental strategies should be applied across all of the Cambridge Primary stages (1 to 6) and to all mathematical strands. The Cambridge Primary Mathematics Curriculum Framework (0096) is, however, less prescriptive about the specific strategies that should be learned and practised by learners at each stage. Allowing teachers greater flexibility in teaching mental strategies, and allowing learners to view mental strategies as a more personal and less formal choice, means that learners will have greater ownership over the mental strategies that they choose to use, thereby developing a deeper conceptual understanding of the number system.

In keeping with the changes that have been introduced in the Cambridge Primary Mathematics Curriculum Framework (0096), mental strategies are embedded throughout Collins International Primary Maths. Learners are given opportunities to develop and practise mental strategies, using carefully chosen numbers, and are continually encouraged to articulate their strategies verbally. This is of particular importance for the Cambridge Primary Mathematics Curriculum Framework (0096), where there are no specific learning objectives relating to mental strategies.

It is not possible to exhaustively list all of the mental strategies that can be used, and there will not be one correct strategy for any particular calculation. The most appropriate mental strategy will depend on individual learners' knowledge of mathematical facts, their working memory and their conceptual understanding of different parts of the number system. Mental strategies can be explicitly learned and practised, and doing so will enable a learner to add that strategy to their 'repertoire'. It is important therefore that learners are exposed to a wide range of strategies.

Below are some of the different mental strategies that learners may employ and that are featured throughout Collins International Primary Maths.

Addition and subtraction

- counting on and back in steps
- using known addition and subtraction number facts/number bonds/complements
- applying knowledge of place value and partitioning (i.e. compose, decompose and regroup numbers)
- compensation
- putting the larger number first and counting on (addition)
- counting back from the larger number (take away)
- counting on from the smaller number (find the difference)
- recognising that when two numbers are close together it's easier to find the difference by counting on, not counting back
- using the commutative and associative properties
- using the inverse relationship between addition and subtraction

Multiplication and division

- counting on and back in steps of constant size
- using known multiplication and division facts and related facts involving multiples of 10 and 100
- applying knowledge of place value and partitioning (i.e. compose, decompose and regroup numbers), including multiplying and dividing whole numbers and decimals by 10, 100 and 1000
- using doubling
- recognising and using factor pairs
- using the commutative, distributive and associative properties
- using the inverse relationship between multiplication and division

The use of calculators and Cambridge Primary Mathematics

When used well, calculators can assist learners in their understanding of numbers and the number system. Calculators should be used as a teaching aid to promote mental calculation and mental strategies and to explore mathematical patterns. Learners should understand when it is best to use calculators to assist calculations and when to calculate mentally or use written methods.

As Cambridge International includes calculator-based assessments at Stages 5 to 9, it is recommended that learners begin to use calculators for checking calculations from the end of Stage 3, and for performing and checking calculations from Stage 6. At Stages 5 and 6, learners should be developing effective use of calculators so that they are familiar with the functionality of a basic calculator in readiness for Stage 7 onwards.

④ Thinking and Working Mathematically

In the Cambridge Primary (0096) and Lower Secondary (0862) Mathematics Curriculum Frameworks, the problem solving strand and associated learning objectives have been replaced with four pairs of Thinking and Working Mathematically (TWM) characteristics.

The TWM characteristics represent one of the most significant changes to the Cambridge Primary Mathematics Curriculum Framework (0096). In response to this, this edition of Collins International Primary Maths has incorporated and interwoven TWM throughout all of the components; this reflects the course's most substantial change to the teaching and learning of mathematics.

Thinking and Working Mathematically is based on work by Mason, Burton and Stacey*; it places an emphasis on learners:

- actively engaging with their learning of mathematics
- talking with others, challenging ideas and providing evidence that validates conjectures and solutions
- seeking to make sense of ideas
- building connections between different facts, procedures and concepts

* Mason, J., Burton, L. and Stacey, K. (2010) *Thinking Mathematically*, 2nd edition, Harlow: Pearson

– developing higher-order thinking skills that assist them in viewing the world in a mathematical way.

This contrasts with learners simply following instructions and carrying out processes that they have been shown how to do, without appreciating why such processes work or what the results mean. Through the development of each of the TWM characteristics, learners are able to see the application of mathematics in the real world more clearly and also, crucially, to develop the skills necessary to function as citizens who are autonomous problem solvers.

If learners at any of the Cambridge International stages are to gain meaning and satisfaction from their study of mathematics, then it is vital that TWM underpins their experience of learning the subject.

The four pairs of TWM characteristics that Cambridge International identifies as fundamental to a meaningful experience of learning mathematics are represented diagrammatically and referred to as 'The Thinking and Working Mathematically Star'.

The Thinking and Working Mathematically Star

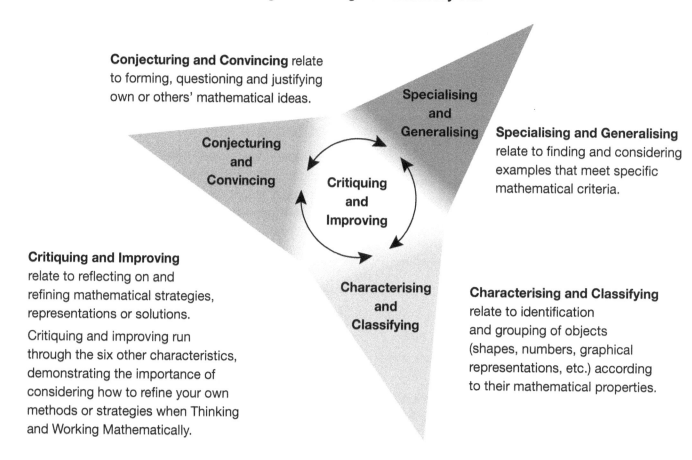

Conjecturing and Convincing relate to forming, questioning and justifying own or others' mathematical ideas.

Specialising and Generalising relate to finding and considering examples that meet specific mathematical criteria.

Critiquing and Improving relate to reflecting on and refining mathematical strategies, representations or solutions.

Critiquing and improving run through the six other characteristics, demonstrating the importance of considering how to refine your own methods or strategies when Thinking and Working Mathematically.

Characterising and Classifying relate to identification and grouping of objects (shapes, numbers, graphical representations, etc.) according to their mathematical properties.

The Thinking and Working Mathematically star, © Cambridge International, 2018

The eight characteristics of Thinking and Working Mathematically

Characteristic	Definition
TWM.01 Specialising	Choosing *an example* and checking to see if it satisfies or does not satisfy specific mathematical criteria.
TWM.02 Generalising	Recognising an underlying pattern by identifying *many* examples that satisfy the same mathematical criteria.
TWM.03 Conjecturing	Forming mathematical questions or ideas.
TWM.04 Convincing	Presenting evidence to *justify or challenge* a mathematical idea or solution.
TWM.05 Characterising	Identifying and describing the mathematical properties of an object.
TW.06 Classifying	Organising objects into groups according to their mathematical properties.
TWM.07 Critiquing	Comparing and evaluating mathematical ideas, representations or solutions to identify advantages and disadvantages.
TWM.08 Improving	Refining mathematical ideas or representations to develop a more effective approach or solution.

All eight TWM characteristics can be applied across all of the Cambridge Primary and Lower Secondary stages (1 to 9) and across all mathematical strands and sub-strands, although the prominence of different characteristics may change as learners move through the stages.

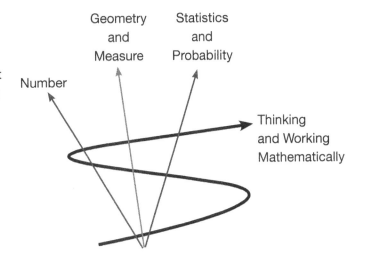

Any characteristic can be combined with any other characteristic; characteristics should be taught alongside content learning objectives and should **not** stand-alone.

The four pairs of characteristics intertwine and are interdependent, and a high-quality mathematics task may draw on one or more of them.

Thinking and Working Mathematically should **not** consist of a separate end-of-lesson or unit activity, but should be embedded throughout lessons in every unit of work. All of the characteristics identified above can be combined with most teaching topics so, when planning a unit of work, teachers should begin with one or more learning objectives and seek to draw on one or more TWM characteristics.

Thinking and Working Mathematically also enables learners' thinking to become visible, which is a crucial aspect of formative assessment.

Just as TWM is at the very heart of Cambridge Primary Mathematics, so too is this approach to the teaching and learning of mathematics a core feature of Collins International Primary Maths. Opportunities are provided in each of the 27 units for learners to develop the TWM characteristics.

Specific guidance is provided at the start of each unit as part of the Unit introduction, which highlights teaching and learning opportunities, particularly in relation to the unit's learning objectives, that promote the TWM characteristics.

At each phase of the lesson (Revise, Teach, Practise, Apply and/or Review), guidance is given in the lesson plan on how to promote the TWM characteristics.

Whenever any of the eight TWM characteristics is being promoted in a lesson plan this is shown using the initials 'TWM', followed by the Cambridge Primary Mathematics Curriculum Framework (0096) code that identifies exactly which of the eight characteristics is being developed.

Unit introduction

Promoting Thinking and Working Mathematically

Opportunities to develop all four pairs of TWM characteristics are provided throughout the unit.

In Lesson 1, learners conjecture and convince (TWM.03/04) when they suggest how a rectangle can be turned into two identical right angled triangles.

In Lesson 2, learners characterise and classify (TWM.05/06) when they investigate whether it is possible to switch the values for base and height when calculating the area of a right-angled triangle and still get the same answer.

Teach SB 🖥 [TWM.01/02]
- Display **Slide 1**. Discuss how the properties of number can be used to simplify a problem. Take learners through the example provided.
- **[T&T] [TWM.01/02]** Working in pairs, ask learners to think of a similar problem that could be simplified using the distributive property. Choose a pair of learners to explain the problem and how it could be simplified.

The **Teach** phase of this lesson aims to promote the Specialising [TWM.01] and Generalising [TWM.02] characteristics.

The **Apply** phase of this lesson aims to promote the Convincing [TWM.04] characteristic.

Apply 👥 🖥 [TWM.04]
- Display **Slide 2** and read the text to the class. Give learners time to complete the task and then ask them to discuss their ideas. Learners who finish early could investigate proof for the statement: 'Tripling the position number of a term in the square number sequence always increases the value of the term by nine times.'

Practise 📒 [TWM.01/07]
- Workbook
Title: Subtracting positive and negative integers
Pages: 22–23
- Refer to Activity 1 from the Additional practice activities.

The **Practise** phase of this lesson aims to promote the Characterising [TWM.01] and Classifying [TWM.07] characteristics.

In addition to the support provided in the Unit introductions and individual lesson plans, The Thinking and Working Mathematically Star below provides a list of prompting questions that teachers may find helpful when asking learners questions specifically aimed at developing each of the TWM characteristics.

The Thinking and Working Mathematically Star
Teacher prompting questions

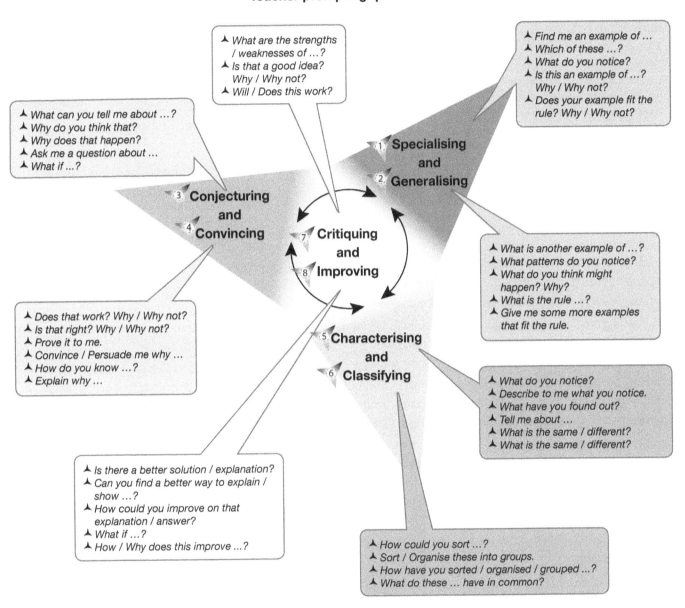

▲ What are the strengths / weaknesses of ...?
▲ Is that a good idea? Why / Why not?
▲ Will / Does this work?

▲ Find me an example of ...
▲ Which of these ...?
▲ What do you notice?
▲ Is this an example of ...? Why / Why not?
▲ Does your example fit the rule? Why / Why not?

▲ What can you tell me about ...?
▲ Why do you think that?
▲ Why does that happen?
▲ Ask me a question about ...
▲ What if ...?

3 **Conjecturing and**
4 **Convincing**

1 **Specialising and**
2 **Generalising**

7 **Critiquing and Improving**
8

▲ What is another example of ...?
▲ What patterns do you notice?
▲ What do you think might happen? Why?
▲ What is the rule ...?
▲ Give me some more examples that fit the rule.

▲ Does that work? Why / Why not?
▲ Is that right? Why / Why not?
▲ Prove it to me.
▲ Convince / Persuade me why ...
▲ How do you know ...?
▲ Explain why ...

5 **Characterising and**
6 **Classifying**

▲ What do you notice?
▲ Describe to me what you notice.
▲ What have you found out?
▲ Tell me about ...
▲ What is the same / different?
▲ What is the same / different?

▲ Is there a better solution / explanation?
▲ Can you find a better way to explain / show ...?
▲ How could you improve on that explanation / answer?
▲ What if ...?
▲ How / Why does this improve ...?

▲ How could you sort ...?
▲ Sort / Organise these into groups.
▲ How have you sorted / organised / grouped ...?
▲ What do these ... have in common?

As in the Teacher's Guide, learning opportunities aimed at developing TWM are also identified in the Stages 3 to 6 Student's Book and Workbook. Where an activity or question promotes TWM this is clearly indicated using the TWM icon and the Cambridge Primary Mathematics Curriculum Framework (0096) code that identifies exactly which of the eight characteristics is being developed.

> This paired activity from the **Student's Book** aims to promote the Convincing [TWM.04] characteristic.

  Tom increases the time he runs by the same amount each day. On Monday he ran for 5 minutes and on Saturday he ran for 45 minutes. How many minutes did Tom run for on Tuesday, Wednesday, Thursday and Friday?

Draw a diagram that explains the problem and how to solve it.

> This question from the **Workbook** aims to promote the Characterising [TWM.05] and Classifying [TWM.06] characteristics.

 5 Investigate the following statement: **A triangular number can never end in 2, 4, 7 or 9.**

Is this statement true or false? Write your working in the box below.

Similar to 'The Thinking and Working Mathematically Star – Teacher prompting questions' on page 17, the star on page 19, which is located at the back of the Stages 3 to 6 Student's Books, defines in pupil-friendly language, each of the eight TWM characteristics and numbers them 1 to 8 accordingly.

The star is aimed at helping learners think specifically about what is required when they are undertaking an activity designed to develop a specific TWM characteristic.

If used in conjunction with the Teacher's Guide, Student's Book and Workbook, which uses the initials 'TWM', followed by the Cambridge Primary Mathematics Curriculum Framework (0096) code to identify exactly which of the eight characteristics an activity, discussion prompt or practise question is developing, this star will help learners better understand the meaning and purpose of each of the eight TWM characteristics.

The star also includes some sentence stems that aim to help learners to talk with others, challenge ideas and explain their reasoning. Learners should be encouraged to use the star whenever working on an activity that develops TWM. This includes whole-class discussions and activities, group and paired activities (including those located in the Student's Book as well as Additional practice activities and Apply), and individual questions from the Workbook.

In Stages 1 and 2 a similar star is provided at the back of the Student's Book, however this star does not include the sentence stems.

The Thinking and Working Mathematically Star

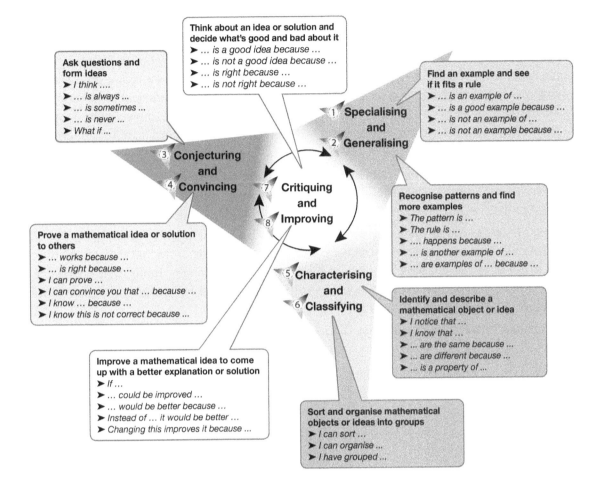

Ask questions and form ideas
- ➤ *I think ....*
- ➤ *... is always ...*
- ➤ *... is sometimes ...*
- ➤ *... is never ...*
- ➤ *What if ...*

Think about an idea or solution and decide what's good and bad about it
- ➤ *... is a good idea because ...*
- ➤ *... is not a good idea because ...*
- ➤ *... is right because ...*
- ➤ *... is not right because ...*

Find an example and see if it fits a rule
- ➤ *... is an example of ...*
- ➤ *... is a good example because ...*
- ➤ *... is not an example of ...*
- ➤ *... is not an example because ...*

3 **Conjecturing and** 4 **Convincing**

1 **Specialising and** 2 **Generalising**

7 **Critiquing and** 8 **Improving**

5 **Characterising and** 6 **Classifying**

Recognise patterns and find more examples
- ➤ *The pattern is ...*
- ➤ *The rule is ...*
- ➤ *.... happens because ...*
- ➤ *... is another example of ...*
- ➤ *... are examples of ... because ...*

Prove a mathematical idea or solution to others
- ➤ *... works because ...*
- ➤ *... is right because ...*
- ➤ *I can prove ...*
- ➤ *I can convince you that ... because ...*
- ➤ *I know ... because ...*
- ➤ *I know this is not correct because ...*

Identify and describe a mathematical object or idea
- ➤ *I notice that ...*
- ➤ *I know that ...*
- ➤ *... are the same because ...*
- ➤ *... are different because ...*
- ➤ *... is a property of ...*

Improve a mathematical idea to come up with a better explanation or solution
- ➤ *If ...*
- ➤ *... could be improved ...*
- ➤ *... would be better because ...*
- ➤ *Instead of ... it would be better ...*
- ➤ *Changing this improves it because ...*

Sort and organise mathematical objects or ideas into groups
- ➤ *I can sort ...*
- ➤ *I can organise ...*
- ➤ *I have grouped ...*

⑤ Cambridge Global Perspectives™

Cambridge Global Perspectives is a unique programme that helps learners develop outstanding transferable skills, including critical thinking, research and collaboration. The programme is available for learners aged 5-19, from Cambridge Primary through to Cambridge Advanced. For Cambridge Primary and Lower Secondary learners, the programme is made up of a series of Challenges covering a wide range of topics, using a personal, local and global perspective. The programme is available to Cambridge schools but participation in the programme is voluntary. However, whether or not your school is involved with the programme, the six skills it focuses on are relevant to **all** students in the modern world. These skills are: research, analysis, evaluation, reflection, collaboration and communication.

More information about the Cambridge Global Perspectives programme can be found on the Cambridge Assessment International Education website:
www.cambridgeinternational.org/programmes-and-qualifications/cambridge-global-perspectives.

Collins supports Cambridge Global Perspectives by including activities, tasks and projects in our Cambridge Primary and Lower Secondary courses which develop and apply these skills. Note that the content of the activities is not intended to correlate with the specific topics in the Cambridge Challenges; rather, they encourage practice and development of the Cambridge Global Perspectives to support teachers in integrating and embedding them into students' learning across all school subjects.

Activities in this book that link to the Cambridge Global Perspectives are listed at the back of this book on page 354.

⑥ The components of Collins International Primary Maths

Each of the six stages in Collins International Primary Maths consists of these four components:

Teacher's Guide

The Teacher's Guide comprises:

- a bank of **Revise activities** for the first 'warm up' phase of the lesson
- **Teaching and learning units**, which consist of Unit introductions, Lesson plans and Additional practice activities
- **Resource sheets** for use with particular lessons and activities
- **Answers** to the questions in the Workbook
- **Record-keeping document** to assist teachers with both short-term 'on-going' and medium-term 'formative' assessments.

A key aim of Collins International Primary Maths is to support teachers in planning, teaching and assessing a successful mathematics programme of work, in line with the Cambridge Primary Mathematics Curriculum Framework (0096).

To ensure complete curriculum coverage and adequate revision of the learning objectives, for each stage the learning objectives from the Cambridge Primary Mathematics Curriculum Framework (0096) have been grouped into 27 topic areas or 'units'. For a more detailed explanation of the 27 units in Collins International Primary Maths Stage 6, including a recommended teaching sequence, refer to Section ⑦ (page 38).

The charts in Section ⑧ (page 42) provide a medium-term plan, showing each of the 27 units in Collins International Primary Maths Stage 6 and which Stage 6 Cambridge Primary Mathematics Curriculum Framework (0096) strand, sub-strand and learning objectives (and codes) each of the units is teaching.

Similarly, the charts in Section ⑨ (page 50) show when each of the Stage 6 learning objectives in the Cambridge Primary Mathematics Curriculum Framework (0096) are taught in the 27 units in Collins International Primary Maths Stage 6.

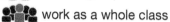 *Icons used in Collins International Primary Maths*

 work individually

 work in pairs

work in groups

 work as a whole class

[T&T] turn and talk (*Talk Partners*)

 progress check question

SB refer to the Student's Book

WB refer to the Workbook

 interactive digital resource

slide

 activity that promotes Thinking and Working Mathematically

 question/activity number typeset on a circle indicates that the question/activity is suitable for learners who require additional support with either easier questions/activities or revising pre-requisite knowledge

 question/activity number typeset on a triangle indicates that the question/activity is suitable for the majority of learners to practise and consolidate the lesson's learning objective(s)

 question/activity number typeset on a square indicates that the question/activity is suitable for learners who require enrichment and/or extension

---

A note on the use of dice

Although some activities in Collins International Primary Maths suggest the use of dice, these are not always readily available in some countries. Where dice are unavailable or the use of dice is not appropriate, we suggest using a spinner, and have provided spinners on several Resource sheets along with instructions on how to use them using a pencil and paper clip.

Revise activities

A bank of 5-minute 'warm-up' or 'starter' activities is provided for teachers to use at this first phase of the mathematics lesson. Reference is given in each lesson plan to appropriate Revise activities.

The majority of activities are for whole-class work. However, some activities may involve individual learners demonstrating something to the rest of the class, or pairs or groups working together on an activity or a game.

Strand and sub-strand

The relevant Cambridge Primary Mathematics Curriculum Framework (0096) strand and sub-strand covered is stated in the sidebar.

Title

Each activity is given a title. This is designed to help both teacher and learners identify a particular activity.

Unit number and title

2 The majority of Revise activities are suitable for the majority of learners to practise and consolidate the activity's learning objective(s).

Learning objective(s)

Each activity has clearly defined learning objective(s) to assist teachers in choosing the most appropriate activity for the concept they want the learners to practise and consolidate.

Revise

Book angles **2**

Learning objective

Code	Learning objective
6Gg.09	Classify, estimate, measure and draw angles.

Resources

book (per pair)

What to do
- Distribute books.
- Ask learners to open their books to show angles that are acute, right, obtuse and reflex.
- Play 'Simon says'. Say: **Simon says show me an acute/obtuse/right/straight/reflex angle.**
- Continue with instructions, but vary the language. For example, say: **Simon says show me an angle** that is not an obtuse angle/**Show me a reflex angle that is less than 270°.**

Variation

3 Increase the complexity of the language. Say, for example: **Simon says show me an angle** that is about halfway between an obtuse and a full 360° angle/**Show me an angle that is about halfway between 0° and a right angle.**

Estimating angles **2**

Learning objective

Code	Learning objective
6Gg.09	Classify, estimate, measure and draw angles.

What to do
- Display the Geometry set tool showing the
- Repeat for other angles in the same shape and other shapes.

rical reasoning, shapes and measurements

Classroom organisation

Icons are used to indicate whether the activity is designed to be used by the whole class working together , or for some activities for learners working in groups , pairs or individually .

Resources

To aid preparation, any resources required are listed, along with whether they are for the whole class, per group, per pair or per learner. Most of these resources are readily available in classrooms.

What to do

The activity is broken down into clear steps to support teachers in achieving the objective(s) of the activity and facilitate interactive whole-class teaching.

Variations

Where appropriate, variations are included. Variations may be designed to make the activity easier **1** or more difficult **3**, or change the focus of the activity completely. Where the variation affects the challenge level of the activity, the new challenge level is given.

Collins International Primary Maths units

There are 27 units in Collins International Primary Maths, each consisting of:

- Unit introduction
- Lesson plans
- Additional practice activities.

Unit introduction

The one-page introduction to each unit in the Teacher's Guide is designed to provide background information to help teachers plan, teach and assess that unit.

Learning objectives

The Cambridge Primary Mathematics Curriculum Framework (0096) learning objectives covered in the unit.

Collins International Primary Maths unit number and title

Unit overview

General description of the knowledge, understanding and skills taught in the unit.

Prerequisites for learning

A list of knowledge, understanding and skills that are prerequisites for learning in the unit. This list is particularly useful for diagnostic assessment.

Vocabulary

A summary is provided of key mathematical terms particularly relevant to the unit.

Common difficulties and remediation

Common errors and misconceptions, along with useful remediation hints are offered where appropriate.

Supporting language awareness

Key strategy or idea to help learners access the mathematics of the unit and overcome any barriers that the language of mathematics may present.

Collins International Primary Maths Recommended Teaching and Learning Sequence

Cambridge Primary Mathematics Curriculum Framework (0096) strand and sub-strand

Promoting Thinking and Working Mathematically (TWM)

Specific guidance on how the unit promotes the TWM characteristics.

Success criteria

Success criteria are provided to help both teachers and learners identify what learners are required to know, understand and do in order to achieve the unit's learning objective(s).

Unit introduction

Unit 1: Counting and sequences

Collins International Primary Maths Recommended Teaching and Learning Sequence: Term 1, Week 1

Learning objectives

Code	Learning objective
6Nc.01	Count on and count back in steps of constant size, including fractions and decimals, and extend beyond zero to include negative numbers.
6Nc.03	Use the relationship between repeated addition of a constant and multiplication to find and use a position-to-term rule.
6Nc.04	Use knowledge of square numbers to generate terms in a sequence, given its position.

Unit overview

In this unit, learners investigate counting on and back in decimal steps and fractions that use small denominators. They count in tenths and hundredths and examine how digits change when ones, tenths and hundredths boundaries are crossed. Counting sequences extends to beyond zero to include negative numbers.

Given a sequence of multiples, learners recognise the pattern and use their knowledge of a times table to find the term in any given position; for example, the 10th term in the sequence 4, 8, 12, 16... can be found with 4 × 10 = 40. Tables help learners see the connection between the term and its position in the sequence.

The work on square numbers in Stage 5 is extended, with learners deducing that a particular term in a square number sequence can be found by taking the square of the position.

Prerequisites for learning

Learners need to:
- understand the place value of whole numbers and decimals to two decimal places
- have experience with fractions represented on number lines
- understand that numbers extend beyond zero
- recognise, recreate and extend the patterns of square numbers, relating them to geometric sequences.

Vocabulary

counting on, counting back, ones boundary, tenths boundary, hundredths boundary, negative number, sequence, term, rule, position-to-term rule, square number

Common difficulties and remediation

Some learners find the concept of the hundredths place difficult to grasp. Use models, such as Base 10 equipment, in which the flat represents one whole, the long represents one tenth and the cube represents one hundredth, to illustrate the relationship between the numbers and help build understanding about the size of decimals.

Some learners have difficulty understanding and applying the rule for the nth term. Provide plenty

of examples to establish this link. Help learners understand that unlike a term-to-term rule, when given the rule for the nth term in a sequence of multiples, we do not need to know the previous term.

Supporting language awareness

Ensure learners understand the difference between a term and a position, as this will assist them when they are formally introduced to algebra in the Lower Secondary curriculum. Use mapping diagrams and tables to illustrate this.

Encourage frequent use of mathematical language to help embed vocabulary. For example, when learners describe the next number in a sequence, encourage them to use the more mathematically correct phrase 'next term in the sequence'.

Promoting Thinking and Working Mathematically

Opportunities to develop all four pairs of TWM characteristics are provided throughout the unit; however, because of the sequential nature of arithmetic number patterns (counting on/back sequences) and special number patterns (square numbers) particular emphasis is given to Generalising (TWM.02). Learners identify familiar sequences (multiples or square numbers) and use the position-to-term rule to determine the value of a term for a given position.

In Lessons 3 and 4, learners are given opportunities to investigate whether statements about sequences are true or not and provide convincing (TWM.04) evidence to support their conclusions. When providing proof, they are asked to compare and evaluate representations or solutions to identify advantages and disadvantages (TWM.07).

Success criteria

Learners can:
- count on or back in decimals and fractions and understand how the digits change when crossing the ones, tenths and hundredths boundaries
- identify the position-to-term rule for a sequence and use it to find any given term in the sequence
- say the value of the term when given the position in the sequence of square numbers.

Number – Counting and sequences

111

There are two different types of teaching and learning opportunities provided for each of the 27 units in Collins International Primary Maths:

• four lesson plans

• two Additional practice activities.

The lesson plans provide a clear, structured, step-by-step approach to teaching mathematics according to the learning objective(s) being covered throughout a unit. Each of the lessons has been written in a comprehensive way in order to give teachers maximum support for mixed-ability whole-class interactive teaching. It is intended, however, that the lessons will act as a model to be adapted to the particular needs of each class.

The Additional practice activities provide teachers with a bank of practical, hands-on activities that give learners opportunities for independent practice of the learning objective(s) being taught throughout a unit.

In most instances, the Additional practice activities are designed to be undertaken by pairs or small groups of learners as part of the 'Practise' phase of a Collins International Primary Maths lesson. Teachers choose which of the two Additional practice activities provided in the unit is most appropriate for the lesson's learning objective(s) and the needs of individual learners. Guidance as to which Additional practice activity consolidates a lesson's learning objective(s) is stated in each lesson plan.

The 'Practise' phase of a Collins International Primary Maths lesson also consists of written exercises found in the accompanying Workbook. Teachers need to decide how they wish to use these two different types of independent practice for individual learners or groups of learners. For example, depending on the lesson's learning objective(s), and the needs of individual learners, learners may:

- only complete the Additional practice activity

- start with the Additional practice activity and then move onto exercises in the Workbook

- start with exercises in the Workbook and then move onto the Additional practice activity

- only complete exercises in the Workbook.

It is important that the Additional practice activities are used at any time throughout a unit and therefore incorporated into each lesson as and when necessary in order to supplement, or provide an alternative to, the exercises in the Workbook. They should not be seen solely as providing teaching and learning content for the 'fifth' lesson of the week.

These two different types of teaching and learning opportunities form the weekly structure for each of the 27 units and are aimed at supporting flexibility so that the course can be tailored to meet the needs of individual classes.

Experience gained from other courses similar to Collins International Primary Maths shows that individual classes take different lengths of time to learn the content of a lesson. In light of this, rather than providing five lesson plans for each week, the decision was made to provide four core lessons which cover the unit's learning objective(s).

The intention is that as part of their weekly short-term planning, teachers make decisions as to how they will spread out the four core lessons, over the course of five days. Alternatively, as teachers progress through the week, and as part of their ongoing monitoring and evaluation, they may decide to alter their short-term planning and spend more time on teaching a particular lesson, or provide additional teaching and learning opportunities (including the Additional practice activities) in order to ensure that the class are developing a secure understanding of the unit's learning objective(s).

Lesson plan

Collins International Primary Maths unit number and title

Reference to accompanying Student's Book page and Workbook pages

Lesson number and title

Cambridge Primary Mathematics Curriculum Framework (0096) strand and sub-strand

Lesson objective(s)

The Cambridge Primary Mathematics Curriculum Framework (0096) learning objective(s) covered in the lesson.

Resources

To aid preparation, all the resources necessary to teach the lesson are listed. Each resource clearly states whether it is for the whole class, per group, per pair or per learner. Icons are displayed within the lesson plan to indicate any digital resources used in the lesson.

Number – Integers and powers

Unit **3** Addition and subtraction of whole numbers (B)

Student's Book page 14
Workbook pages 22–23

Lesson 1: **Subtracting positive and negative integers**

Learning objective

Code	Learning objective
6Ni.01	Estimate[, add] and subtract integers.

Resources

calculator (per learner); paper or mini whiteboard and pen (per learner)

Revise

Use the activity *Hot or cold* or *Temperature stories* from Unit 3: *Addition and subtraction of whole numbers (B)* in the Revise activities.

Teach 📖 💻 📊 **[TWM.08]**

- Display **Slide 1**. Read the first problem. **[T&T]** Ask: **How would you use the number line to calculate the difference between negative 7 and 5?** Take responses. Invite a learner to the board to mark the number line to demonstrate the calculation. Expect them to circle –7 and count on in ones to 5 (a difference of 12). Say: **There are a lot of ones to count. Is there a way to split the count into two stages?** Take responses. Elicit that you can bridge to and from zero. Demonstrate a count from –7 to 0 (7) and a count from 0 to 5 (5). The sum 7 + 5 gives the difference, 12.
- Referring to questions **b–e** on **Slide 1**, test learners' estimation skills by asking: **Which of the four problems, b–e, would you expect to give an answer smaller than 20? More than 20? How do you know?** Take responses. **[TWM.08]** Ask: **What do you notice about the difference between the two numbers and the values of the bridging stages? Is there a quick way to calculate the difference between a negative and positive number?** Elicit that the difference between the two numbers is equal to the sum of the two bridging stages.
- With learners helping, work out the answers to problems **b** to **e**. Use the number lines to count on.
- Display the **Thermometer tool**, scaled from –20 to 30°C. Set the pointer to 14°C. Say: **The temperature begins at 14°C and falls to –12°C. What is the difference between the two temperatures?** Take responses. Choose a learner to come to the board to demonstrate how to work out the difference by bridging to and from zero: 12 (–12 to 0) + 14 (0 to 14) = 26 degrees.
- Say: **Sally drops a metal ball into the sea from a height of 38 metres above sea level. If it drops to a depth of –87 metres, how far has the ball fallen?** Remind learners to estimate the answer before calculation. Ask learners to check their answer with a calculator. (125 m)
- Discuss the Let's learn section in the Student's Book.
- 📖 Say: **A lift rises from basement level –7 to upper level 16. How many floors has the lift risen?** (23)

- Introduce the paired activity in the Student's Book, asking learners to find the difference between daily temperatures for Canada and the UK. Choose a pair of learners to explain the strategy they used to find the differences. Ask: **Did anyone calculate the differences instead of counting on? How did you do it?**
- Discuss the Guided practice example in the Student's Book.

Practise 📝 **[TWM.01/07]**

- Workbook

Title: Subtracting positive and negative integers
Pages: 22–23

- Refer to Activity 1 from the Additional practice activities.

Apply 👥 💻

- Display **Slide 2** and read the text to the class. Give learners time to work on the problems. Then ask them which strategy they used to find the distance dropped by the ball.

Review

- Ask learners to work in pairs. They take turns to say a 'target' number between 10 and 30. They race each other to write five pairs of positive and negative numbers that have a difference equal to the target number.
- Learners swap papers or whiteboards and check whether each pair is correct. The winner is the learner who writes the most correct calculations in the shortest time.

Assessment for learning

- How would you use a number line to show that the difference between –15 and 6 is 21?
- Give me a pair of numbers, one positive and one negative, that have a difference of 23.

Same day intervention
Enrichment

- Provide problems with three-digit numbers and ask learners to use any calculation strategy they prefer, for example: find the difference between –127 and 132 (259).

Revise

Recommended teaching time: 5 min

A bank of Revise activities can be found on pages 55–109. Revise activities are designed to consolidate fluency in number facts and/or provide an opportunity to revise any prerequisites for learning.

Revise

Use the activity *Hot or cold* or *Temperature stories* from Unit 3: *Addition and subtraction of whole numbers (B)* in the Revise activities.

Teach [TWM.03]
- Display **Slide 1**. Remind learners of the terminology of subtraction. On the board, write: 15 − 7 = 8. Ask: **Which number is the subtrahend?** (7) **The minuend?** (15) **The difference?** (8)
- Read the first problem on **Slide 2**. **[T&T]** Ask: **How would you use the number line to calculate the difference between −7 and −2?** Take responses. Invite a learner to the board to mark the number line to demonstrate the calculation. Expect them to circle −7 and count on in ones to −2 (a difference of 5).
- Referring to questions b–e on **Slide 2**, test learners' estimation skills by asking: **Which of the four problems, b–e, would you expect to give an answer greater than 10? How do you know?** (c and e)
- With learners helping, work out the answers to problems b–e.
- Display the **Thermometer tool**, scaled from −20 to 30°C. Set the pointer to −19°C. **[TWM.03]** Ask: **If the temperature starts at −19°C and reaches −5°C, is that a rise or fall in temperature? How do you know?** Establish that numbers towards the top of the scale are larger and therefore, the temperature is higher (warmer) towards the right of the scale. A change from −19°C to −5°C is therefore a rise in temperature. Ask: **How much does the temperature rise from −19°C to −5°C? Use the temperature scale to show me.** Choose a learner to demonstrate the difference between the two temperatures (14 degrees).
- Ask: **What is the difference between −58 and −133?** Remind learners to estimate the answer before calculation. Ask learners to check their answer with a calculator. (75)
- 🔁 Ask: **A lift moves down from basement level −2 to basement level −13. How many floors is that?** (11)
- Discuss the Let's learn section in the Student's Book.
- Introduce the paired activity in the Student's Book, asking learners to find the amount spent by each

person. Choose a pair of learners to explain the strategy they used to find the differences. Ask: **Did anyone calculate the differences? How did you do it?** Ask learners to comment on the effectiveness of the strategies proposed.
- Discuss the Guided practice example in the Student's Book.

Practise [TWM.04]
- Workbook
Title: Subtracting two negative integers
Pages: 24–25
- Refer to Activity 1 (Variation) from the Additional practice activities.

Apply
- Display **Slide 3** and read the text to the class. Give learners time to work out how many degrees the temperature rises each day. Choose a pair to demonstrate how they worked out the difference between the two temperatures.

Review
- Display the **Thermometer tool**. Ask learners to work in threes. One learner says a negative temperature. A second learner states a second negative temperature. The third learner must work out the difference between the two temperatures. The group confirms the answer and then they swap roles.

Assessment for learning
- How would you use a thermometer scale to show that a change in temperature of −16°C to −3°C is a difference of 13 degrees?
- Tilly's bank account balance is −$5. She buys a gift and her account drops to −$23. What was the price of the gift? ($18)

Same day intervention
Enrichment
- Provide problems with three-digit numbers and ask learners to use any calculation strategy you prefer, for example: find the difference between −143 and −277 (134).

Number – Integers and powe

125

Teach

Recommended teaching time: 20 min

The main teaching activity is broken down into clear steps to support teachers in achieving the lesson objective(s) and facilitate interaction with the whole class. Suggested statements and questions are provided to support the teacher. During this phase of the lesson teachers also draw learners' attention to the 'Let's learn', Paired/TWM activity, and 'Guided practice' features in the Student's Book.

Other features of Teach include:

T&T: Turn and talk – Using *Talk Partners* helps to create a positive learning environment. Many learners feel more confident discussing with a partner before giving an answer to the whole class, and learners get opportunities to work with different students.

🔁: Progress check questions – These questions are designed to obtain an overview of learners' prior experiences before introducing a new concept or topic, to provide immediate feedback to learners and to gauge learner progress in order to adapt teaching.

TWM: Teaching and learning opportunities aimed at learners developing specific Thinking and Working Mathematically characteristics.

NOTE: Timings are approximate recommendations only.

Practise and/or and/or

Recommended teaching time: 20 min

Teach is followed by independent practice and consolidation, which provides an opportunity for all learners to focus on their newly acquired knowledge. Practice and consolidation consists of both written exercises and practical hands-on activities, with reference to the relevant Workbook pages and bank of Additional practice activities.

All of the tasks are differentiated into three different ability levels:

1 question/activity number typeset on a circle indicates that the question/activity is suitable for learners who require additional support with either easier questions/activities or revising pre-requisite knowledge

2 question/activity number typeset on a triangle indicates that the question/activity is suitable for the majority of learners to practise and consolidate the lesson's learning objective(s)

3 question/activity number typeset on a square indicates that the question/activity is suitable for learners who require enrichment and/or extension.

Introduction

Revise

Use the activity *Hot or cold* or *Temperature stories* from Unit 3: *Addition and subtraction of whole numbers (B)* in the Revise activities.

Teach 🔲 💻 📊 [TWM.03]

- Display **Slide 1**. Remind learners of the terminology of subtraction. On the board, write: $15 - 7 = 8$. Ask: **Which number is the subtrahend? (7) The minuend? (15) The difference? (8)**
- Read the first problem on **Slide 2**. **[T&T]** Ask: **How would you use the number line to calculate the difference between −7 and −2?** Take responses. Invite a learner to the board to mark the number line to demonstrate the calculation. Expect them to circle −7 and count on in ones to −2 (a difference of 5).
- Referring to questions b–e on **Slide 2**, test learners' estimation skills by asking: **Which of the four problems, b–e, would you expect to give an answer greater than 10? How do you know?** (c and e)
- With learners helping, work out the answers to problems b–e.
- Display the **Thermometer tool**, scaled from −20 to 30°C. Set the pointer to −19°C. **[TWM.03]** Ask: **If the temperature starts at −19°C and reaches −5°C, is that a rise or fall in temperature? How do you know?** Establish that numbers towards the top of the scale are larger and therefore, the temperature is higher (warmer) towards the right of the scale. A change from −19°C to −5°C is therefore a rise in temperature. Ask: **How much does the temperature rise from −19°C to −5°C? Use the temperature scale to show me.** Choose a learner to demonstrate the difference between the two temperatures (14 degrees).
- Ask: **What is the difference between −58 and −133?** Remind learners to estimate the answer before calculation. Ask learners to check their answer with a calculator. (75)
- 🔂 Ask: **A lift moves down from basement level −2 to basement level −13. How many floors is that?** (11)
- Discuss the Let's learn section in the Student's Book.
- Introduce the paired activity in the Student's Book, asking learners to find the amount spent by each

person. Choose a pair of learners to explain the strategy they used to find the differences. Ask: **Did anyone calculate the differences? How did you do it?** Ask learners to comment on the effectiveness of the strategies proposed.
- Discuss the Guided practice example in the Student's Book.

Practise 🔲 [TWM.04]

- Workbook

Title: Subtracting two negative integers

Pages: 24–25

- Refer to Activity 1 (Variation) from the Additional practice activities.

Apply 👥 💻

- Display **Slide 3** and read the text to the class. Give learners time to work out how many degrees the temperature rises each day. Choose a pair to demonstrate how they worked out the difference between the two temperatures.

Review 📊

- Display the **Thermometer tool**. Ask learners to work in threes. One learner says a negative temperature. A second learner states a second negative temperature. The third learner must work out the difference between the two temperatures. The group confirms the answer and then they swap roles.

Assessment for learning
- How would you use a thermometer scale to show that a change in temperature of −16°C to −3°C is a difference of 13 degrees?
- Tilly's bank account balance is −$5. She buys a gift and her account drops to −$23. What was the price of the gift? ($18)

Same day intervention
Enrichment
- Provide problems with three-digit numbers and ask learners to use any calculation strategy you prefer, for example: find the difference between −143 and −277 (134).

125

Apply 👤 and/or 👥 and/or 👪

Recommended teaching time: 10 min

Apply is an investigation, problem, puzzle or cross-curricular application where the learner uses and applies knowledge, understanding and skills in an applied context. The content of Apply is located on a slide 💻 (See Note* below).

Review 👪

Recommended teaching time: 5 min

The all-important conclusion to the lesson offers an opportunity for learners to make reflective comments about their learning, as well as to discuss misconceptions and common errors, and summarise what they have learned.

Same day intervention

Support and/or Enrichment

Offers same day intervention suggestions so that teachers can effectively provide either support or Enrichment where appropriate.

The aim is to provide guidance to teachers to ensure that:

- all learners reach a certain level of understanding by the end of the day, preventing an achievement gap from forming
- the needs of all learners are being met with respect to the lesson's learning objective(s).

Assessment for learning

Specific questions designed to assist teachers in checking learners' understanding of the lesson objective(s). These questions can be used at any time throughout the lesson.

NOTE*:

As learners will inevitably complete the 'Practise' activities at different times, it is recommended that teachers introduce the 'Apply' activity to the whole class at the end of 'Teach' and before learners work independently on the 'Practise' activities (Workbook page and/or Additional practice activities).

There should also be no expectation that *every* learner in *every* lesson should complete the 'Apply' activity. However, opportunities should be given, either during or

outside the maths lesson, for learners to work on 'Apply'. This could include (but not always) part of the teaching and learning content for the 'fifth' lesson of the week.

Finally, it is also recommended that as part of the whole class 'Review' phase of the lesson, learners are given the opportunity to share with the rest of the class the work they carried out as part of the 'Apply' activity. This includes not just providing the solution, answer or result (if there is one), but also the different methods and strategies they used, and what they learned from the activity.

Additional practice activities

As well as four lesson plans, each unit in Collins International Primary Maths provides two Additional practice activities.

Strand and Sub-strand

The relevant Cambridge Primary Mathematics Curriculum Framework (0096) strand and sub-strand covered is stated in the sidebar.

Collins International Primary Maths unit number and title

Unit **4** Multiples, factors, divisibility, squares and cubes

Additional practice activities

Activity 1 👥 **2**

Number – Integers

• Learning objective
  • Understand common multiples and common factors.

Resources
Resource sheet 4: Gameboard (per learner); counters (per group); two 1–6 dice or Resource sheet 5: 1–6 spinner (per learner); pencil and paper clip, for the spinner (per learner); Resource sheet 6: 10–100 number cards (per learner) (for variation)

What to do
  • Arrange the learners into groups of three or four and provide each learner with a gameboard and two dice or a 1–6 spinner, pencil and paper clip.
  • Each learner rolls both dice or spins their

  • They cover these numbers on their gameboard with counters.
  • The winner is the player who has the most numbers covered on their gameboard after five rounds.

Variation
  2 Each learner takes two cards from a pile of 10–100 number cards.
They find the common factors for their pair of numbers.
Each player scores points equivalent to the number of common factor found.

Learning objective(s)

Each activity has clearly defined learning objective(s) to assist teachers in choosing the most appropriate activity for the concept they want the learners to practise and consolidate.

Resources

To aid preparation, any resources required are listed, along with whether they are per group, per pair or per learner. Most of these resources are readily available in classrooms.

What to do

The activity is broken down into clear steps to support teachers in explaining the activity to the learners.

Variations

Where appropriate, variations are included. Variations may be designed to make the activity easier **1** or more difficult **3**, or change the focus of the activity completely. Where the variation affects the challenge level of the activity, the new challenge level is given.

Classroom organisation

Icons are used to indicate whether the activity is designed to be used by learners working in groups 👥, pairs 👤 or individually 👤.

Challenge level

The challenge level for each activity is given:

1 suitable for learners who require additional support

2 suitable for the majority of learners to practise and consolidate the lesson's learning objective(s)

3 suitable for learners who require enrichment and/or extension.

Resource sheets

Where specific paper-based resources are needed for individual lesson plans or Additional practice activities, these are provided as Resource sheets in the Teacher's Guide. Use of Resource sheets is indicated in the resources list of the relevant lesson plan or Additional practice activity.

Some Resource sheets use a spinner to generate numbers or other forms of data.

To use a spinner, hold a paper clip in the centre of the spinner using a pencil and gently flick the paper clip with your finger to make it spin.

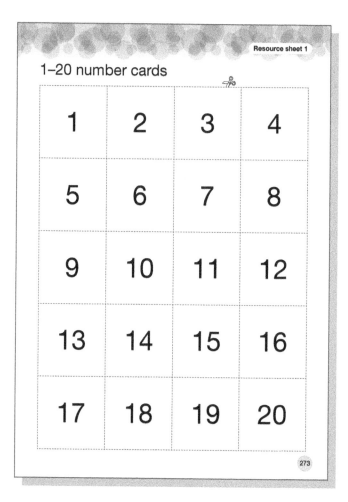

Resource sheet 1

1–20 number cards

1	2	3	4
5	6	7	8
9	10	11	12
13	14	15	16
17	18	19	20

273

Answers

Answers are provided for all the Workbook pages.

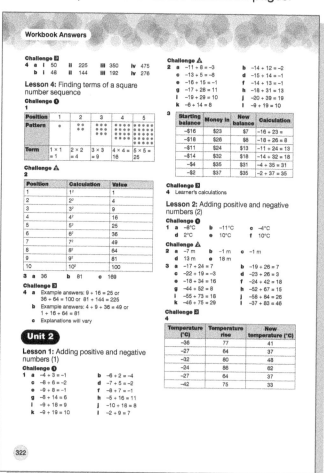

Class record-keeping document

In order to assist teachers with making both short-term 'ongoing' and medium-term 'formative' assessments manageable, meaningful and useful, Collins International Primary Maths includes a class record-keeping document on pages 348–353.

It is intended to be a working document that teachers start at the beginning of the academic year and continually update and amend throughout the course of the year.

Teachers use their own professional judgement of each learner's level of mastery in each of the sub-strands, taking into account:

- mastery of the learning objectives associated with each particular sub-strand
- performance in whole-class discussions
- participation in group work
- work presented in the Workbook
- any other evidence.

Once a decision has been made regarding the degree of mastery achieved by a learner in the particular sub-strand, teachers then write the learner's name (or initials) in the appropriate column:

A: Exceeding expectations in this sub-strand

B: Meeting expectations in this sub-strand

C: Below expectations in this sub-strand.

Given that this is a working document intended to be used throughout the entire academic year, teachers may decide to write (T1), (T2) or (T3) after the learner's name to indicate in which term/semester the judgement was made. This will also help to show the progress (or regress) that learners make during the course of a year.

Schools and/or individual teachers may decide to use a photocopy of the document at the back of this Teacher's Guide or printout the Word version from the Digital download (either enlarged to A3 if deemed appropriate). It is recommended that whichever option is taken, throughout the year teachers use pencil to fill out the document, and then at the end of the academic year complete the document in pen.

As an alternative to using a photocopy or printout, schools and/or teachers may decide to complete the document electronically using the Digital download version.

An additional bonus of enlarging the document to A3 or completing it electronically, is that it will provide additional space in each of the three columns linked to each sub-strand for teachers to provide more qualitative data should they wish to do so. Teachers can write specific comments that they feel are appropriate for individual learners related not only to the entire sub-strand, but also for specific learning objectives within the sub-strand.

Finally, the class record-keeping document should be seen as an extremely useful document as it can be used to:

- identify those learners who are working *above* and *below* expectations, thereby helping teachers to better plan for the needs of individual learners
- report to parents and guardians
- inform the next year's teacher about which sub-strands of the Cambridge Primary Mathematics Curriculum Framework (0096) individual learners, and the class as a whole, are *exceeding* (A), *meeting* (B) or are *below* (C) in expectations
- assist senior managers within the school in determining whether individual learners, and the class as a whole, are on track to meet end-of-stage expectations.

Cambridge Primary Mathematics Curriculum Framework (0096) strand and sub-strand

Cambridge Primary Mathematics Curriculum Framework (0096) learning objectives (and codes)

Overall level of mastery in the sub-strand

The degree of mastery achieved by a learner in each sub-strand is shown by writing the learner's name (or initials) in the appropriate column:

A: Exceeding expectations in this sub-strand

B: Meeting expectations in this sub-strand

C: Below expectations in this sub-strand.

Year group **Class and academic year reference**

Stage 6 Record-keeping

Class: _____ Year: _____

KEY

A: Exceeding expectations in this sub-strand	B: Meeting expectations in this sub-strand	C: Below expectations in this sub-strand

Strand: Number
Sub-strand: Counting and sequences

Code	Learning objectives
6Nc.01	Count on and count back in steps of constant size, including fractions and decimals, and extend beyond zero to include negative numbers.
6Nc.02	Recognise the use of letters to represent quantities that vary in addition and subtraction calculations.
6Nc.03	Use the relationship between repeated addition of a constant and multiplication to find and use a position-to-term rule.
6Nc.04	Use knowledge of square numbers to generate terms in a sequence, given its position.

A	B	C

348

Student's Book

There is one Student's Book for each stage in Collins International Primary Maths with one page provided for each lesson plan.

The content provided in the Student's Book is designed to be used during the 'Teach' phase of a typical Collins International Primary Maths lesson.

However, it is recommended that during the 'Practise' phase of a lesson, if appropriate, learners also use the page in the Student's Book to help them answer the questions on the accompanying Workbook pages.

On page 5 of the Student's Book is a guidance page referred to as 'How to use this book', which explains to the learners the features of the book.

The back of the Student's Book includes the TWM Star which defines, in pupil-friendly language, the eight Thinking and Working Mathematically characteristics. It also includes some sentence stems to help learners to talk with others, challenge ideas and explain their reasoning.

Reference to accompanying Workbook pages

Collins International Primary Maths unit number and title

Cambridge Primary Mathematics Curriculum Framework (0096) strand

Lesson number and title

Lesson objective(s)

The Cambridge Primary Mathematics Curriculum Framework (0096) learning objective(s) covered in the lesson, written in language appropriate for learners.

Paired activity

A short paired activity or question to discuss encourages learners to explore the key mathematical idea together.

TWM Where appropriate, indicates that the Paired activity promotes Thinking and Working Mathematically.

Key words

A list of key mathematical terms particularly relevant to the lesson.

Let's learn

Content that presents the key mathematical idea of the lesson being taught by the teacher in the 'Teach' phase.

Guided practice

Worked example(s) designed to prepare learners to work independently on the questions in the Workbook.

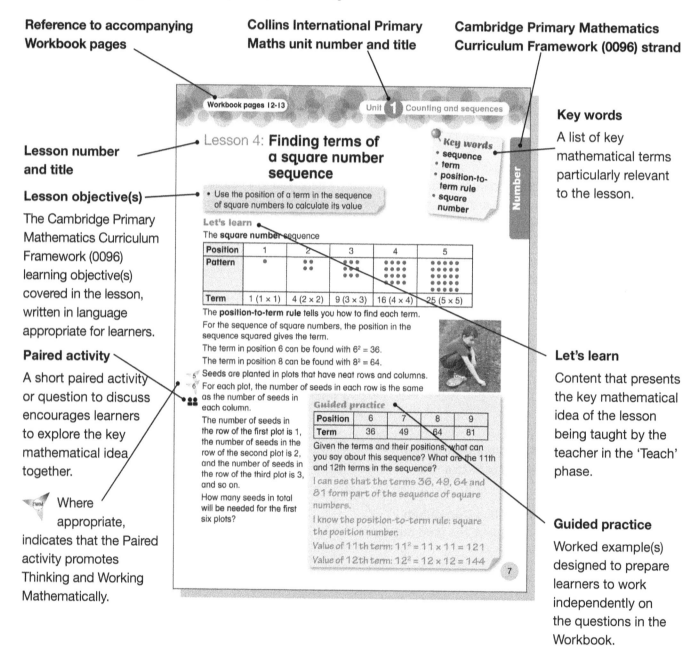

Workbook

All Workbook page exercises reinforce and build upon the main teaching points and learning objective(s) of a particular lesson in the Teacher's Guide. The work is intended to allow all learners in the class to practise and consolidate their newly acquired knowledge, understanding and skills.

The content provided in the Workbook is designed to be used during the 'Practise' phase of a typical Collins International Primary Maths lesson.

On pages 4 and 5 of the Workbook is a guidance page referred to as 'How to use this book', which explains to the learners the features of the book.

In Stage 6, two Workbook pages are provided for each lesson plan. There is no Workbook page for the Additional practice activities.

Each double page spread has three levels of challenge designed to cater, not only for the different abilities that occur in a mixed-ability or mixed-aged class, but also to assist those schools who 'set' or 'stream' their learners into ability groups. The three different levels of challenge are identified as follows:

1 Question number typeset on a circle indicates that the question is suitable for learners who require additional support with either easier questions or revising pre-requisite knowledge.

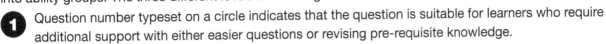 Question number typeset on a triangle indicates that the question is suitable for the majority of learners to practise and consolidate the lesson's learning objective(s).

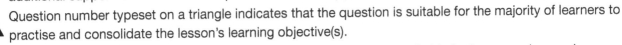 Question number typeset on a square indicates that the question is suitable for learners who require enrichment and/or extension.

Teachers should think carefully about which of the three different levels of challenge individual learners are asked to complete. There should be no expectation that *every* learner must always answer the questions in *all* three challenge levels.

It is therefore good practice to look carefully at the questions in the Workbook and assign specific questions to specific groups of learners. An effective way of doing this is to tell learners which questions you would like them to answer, and for them to circle those question numbers on their Workbook page.

When appropriate, learners should also be encouraged to work in pairs to answer some or all of the questions. Not only does this help learners learn from each other, thereby reinforcing learners' knowledge, understanding and skills, but it also encourages discussion and mathematical talk, helps create positive self-esteem, and removes the frustrations and feelings of intellectual isolation which can so often be associated with learners working alone.

Teacher's may also on occasion decide to work with a group of learners to complete some or all of the questions in the Workbook.

During the 'Review' phase of the lesson, when the whole class is back working together, teachers may decide to complete and/or mark some or all of the questions – perhaps using a different coloured pencil to differentiate those questions that learners answered independently from those they answered with some assistance.

Finally, it is important to be aware that the Workbook is *not* designed for assessment purposes, nor as a record of what learners can or can't do, nor as proof of what has been taught / learned. Its purpose is for learners to practise and consolidate the mathematical ideas that have been taught during the lesson.

Introduction

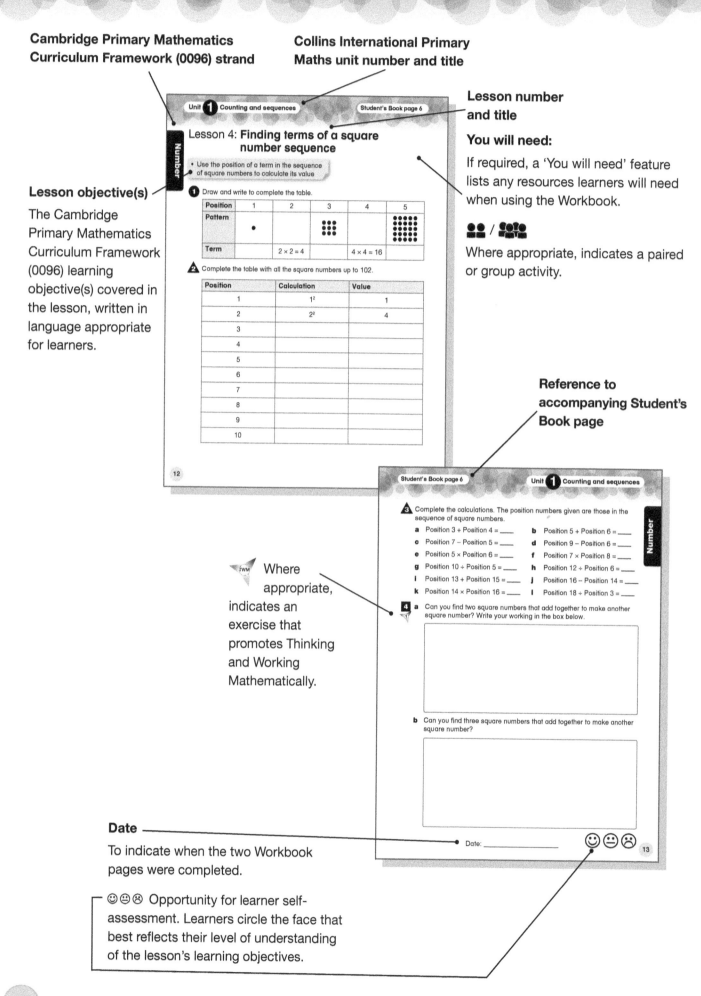

Cambridge Primary Mathematics Curriculum Framework (0096) strand

Collins International Primary Maths unit number and title

Lesson number and title

You will need:

If required, a 'You will need' feature lists any resources learners will need when using the Workbook.

Where appropriate, indicates a paired or group activity.

Lesson objective(s)

The Cambridge Primary Mathematics Curriculum Framework (0096) learning objective(s) covered in the lesson, written in language appropriate for learners.

Reference to accompanying Student's Book page

Where appropriate, indicates an exercise that promotes Thinking and Working Mathematically.

Date

To indicate when the two Workbook pages were completed.

☺☺☹ Opportunity for learner self-assessment. Learners circle the face that best reflects their level of understanding of the lesson's learning objectives.

Digital content

Collins International Primary Maths also includes a comprehensive set of digital tools and resources, designed to support teachers and learners. The digital content is organised into three sections: Teach, Interact and Support, and is available as the Extended Teacher's Guide.

Teach

The Teach section contains all of the teaching content from the Teacher's Guide, organised into units. This includes:

- Unit introductions
- Lesson plans
- Additional practice activities
- Resource sheets
- Answers

In addition to the above, the Teach section also contains:

- Weekly planning grids

  Editable short-term planning grids provide a synopsis of the teaching and learning opportunities in each of the 27 weekly units. Each Weekly planning grid highlights the content of the five phases for each of the four lessons in the unit (the 5th lesson of the week is left empty for teachers to complete) as well as providing background information, assessment opportunities and teacher and learner evaluation. The intention is for teachers to adapt the grid in order to create a bespoke weekly planning overview for the specific needs of their class.

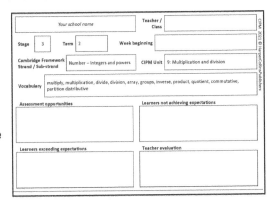

- Slideshows

  Slideshows are provided as visual aids to be shown to the whole class at various phases of a lesson (Revise, Teach, Apply and/or Review), as directed in the lesson plan.

- Interactive whiteboard mathematical tools

  Flexible interactive whiteboard (IWB) teaching tools provide additional visual representations to display to the whole class at various phases of a lesson (Revise, Teach, and/or Review), as directed in the lesson plan.

  These 41 highly adaptable teaching tools are particularly useful in generating specific examples and questions. By doing this, it enables teachers to individualise the content displayed to the class, thereby creating teaching and learning opportunities that better meet the needs of their class.

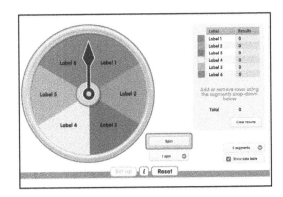

Introduction

A brief description of the functionality of each of the 40 interactive whiteboard mathematical tools is provided below.

Strand	Tool name	Description of functionality
Number	Counting tool	Use the counting tool to assist with counting from 1–20.
	Place value	Explore how multiple-digit numbers are made up of millions, hundreds of thousands, tens of thousands, thousands, hundreds, tens, ones, tenths, hundredths and thousandths.
	Base 10	Demonstrate the relationship between ones, tens, hundreds and thousands.
	Place value counters	Similar in functionality to the Base ten tool above, i.e. demonstrate the relationship between ones, tens, hundreds, thousands (and millions and decimals – tenths, hundredths and thousandths.). However this tool uses place value counters and not Dienes, as is used in the Base ten tool.
	Number line	Use the number line to assist with counting, calculations and exploring decimals.
	Fractions	Demonstrate fractions visually and display the accompanying numerical fraction, decimal, ratio or percentage alongside the pictorial fraction.
	Fraction wall	Demonstrate fractions visually with the fully customisable fraction wall.
	Snake fraction tool	Demonstrate fractions visually with the fully customisable fraction snake.
	Spinner	Demonstrate the concept of probability and making a calculated estimate.
	Bead sticks	Assist with counting, and show the relationship between thousands, hundreds, tens, ones, tenths, hundredths and thousandths.
	Number cards	Display sets of numbers and calculations on movable number cards.
	Number square	Demonstrate and explore counting and number patterns.
	Multiplication square	Demonstrate and explore multiplication and number patterns.
	Function machine	Demonstrate one- and two-step calculations with this animated tool.
	Tree tool	Use real-life objects to practise addition and subtraction.
	Dice tool	Demonstrate the concept of probability or use the tool in conjunction with activities which require dice.
	Ten frame	Use a ten frame template to demonstrate calculations.
Geometry and Measure	Coordinates	Create and interpret labelled coordinate grids, in one, two or four quadrants.
	Geoboard	Join the dots to depict shapes, routes between two points, to draw nets or to make patterns.
	Geometry set	Use the ruler, protractor or set square to measure and draw lines and angles.
	Rotate and reflect	Demonstrate the rotation and reflection across a horizontal, vertical and diagonal mirror line of a range of 2D shapes.
	Symmetry	Demonstrate lines of horizontal, vertical and diagonal symmetry on a range of 2D shapes.
	Pattern tool	Create patterns and sequences of shapes and design a jumper.

Strand	Tool name	Description of functionality
Geometry and Measure	Beads and laces	Create repeating patterns with the beads and laces tool.
	Shape set	Use the shape set to compare 2D and 3D shapes.
	Nets	Explore the nets of 3D shapes such as prisms, pyramids, tetrahedrons, cubes, and cuboids.
	Money	Practise counting money, or solving calculations involving money.
	Clock	Demonstrate the features of analogue and digital clocks, and explore time.
	Thermometer	Demonstrate how to measure and record temperature, using either Fahrenheit and Celsius scales.
	Capacity	Demonstrate how to measure capacity with this range of animated water containers.
	Weighing	Demonstrate mass by weighing objects of different masses on a range of animated weighing scales.
Statistics and Probability	Bar charter	Demonstrate the creation and interpretation of bar graphs.
	Pie charter	Demonstrate how to create and interpret pie charts with this dynamic data tool.
	Pictogram	Demonstrate the creation and interpretation of pictograms.
	Line grapher	Demonstrate the plotting and interpretation of line graphs with this dynamic data tool.
	Carroll diagram	Demonstrate how to classify and group data with this sorting tool.
	Venn diagram	Demonstrate how to classify and group data with this sorting tool. Import images to give a real-life context.
	Waffle diagram	Demonstrate the creation and interpretation of waffle diagrams.
	Frequency diagram	Demonstrate the creation and interpretation of frequency diagrams for continuous data.
	Scatter graph	Demonstrate the creation and interpretation of scatter graphs.
	Dot plot	Demonstrate the creation and interpretation of dot plot diagrams.

Within Teach, the planning tool allows schools and individual teachers to customise the sequence of units in Collins International Primary Maths within and across all stages. This allows schools and individual teachers to develop their own unique scheme of work.

Interact

The Interact section contains 16 interactive mathematical games. The audio glossary of terms for all stages is also located here.

Support

The Support section contains useful documents for the teacher, such as the medium-term plan, Record-keeping documents described on page 30 and the introduction to the Teacher's Guide and the Collins International Primary Maths Training Package.

Ebook

Ebooks are available for all of the components: Teacher's Guide, Student's Book and Workbook. These enable greater teacher-learner interaction during the whole-class 'Teach' phase of the lesson and also assist teachers in explaining activities and questions to learners as well as in discussing results, solutions and answers once learners have completed an activity or set of questions.

The ebooks can be used in a reader view on computer screens and are also designed to be used with interactive whiteboards (IWB) and if available, iPads and tablets.

Each ebook has standard functionality such as scrolling, zooming, an interactive Contents page and the ability to make notes and highlight sections digitally.

⑦ Collins International Primary Maths Stage 6 Units and Recommended Teaching and Learning Sequence

The Stage 6 learning objectives from the Cambridge Primary Mathematics Curriculum Framework (0096) have been grouped into the following 27 topic areas or 'units':

The Thinking and Working Mathematically characteristics are developed throughout each unit.

Cambridge Primary Mathematics Curriculum Framework (0096)		Collins International Primary Maths	
Strand	Sub-strand	Unit number	Topic
Number	Counting and sequences	1	Counting and sequences
	Integers and powers	2	Addition and subtraction of whole numbers (A)
		3	Addition and subtraction of whole numbers (B)
		4	Multiples, factors, divisibility, squares and cubes
		5	Whole number calculations
		6	Multiplication of whole numbers
		7	Division of whole numbers (A)
		8	Division of whole numbers (B)
	Place value, ordering and rounding	9	Place value and ordering decimals
		10	Place value, ordering and rounding decimals
	Fractions, decimals, percentages, ratio and proportion	11	Fractions (A)
		12	Fractions (B)
		13	Percentages
		14	Addition and subtraction of decimals
		15	Multiplication of decimals
		16	Division of decimals
		17	Proportion and ratio
Geometry and Measure	Geometrical reasoning, shapes and measurements	18	2D shapes and symmetry
		19	3D shapes
		20	Angles
		21	Measurements, including time
		22	Area and surface area
	Position and transformation	23	Coordinates
		24	Translation, reflection and rotation
Statistics and Probability	Statistics	25	Statistics (A)
		26	Statistics (B)
	Probability	27	Probability

STRAND:	Number	Geometry and Measure	Statistics and Probability

The Cambridge Primary Mathematics Scheme of Work offers an approach to organising the learning objectives of the Stage 6 curriculum. An overview of this can be seen below.

Unit 6.1   Number
Unit 6.2   2D and 3D shape
Unit 6.3   Calculation
Unit 6.4   Statistical methods
Unit 6.5   Fractions, percentages, decimals and proportion
Unit 6.6   The coordinate grid
Unit 6.7   Probability

The table below shows how the 27 units of Collins International Primary Maths Stage 6 link to the Cambridge Primary Mathematics Stage 6 Scheme of Work units. Please note that while the units in the Cambridge Primary Mathematics Scheme of Work may differ from that of Collins International Primary Maths, guidance from Cambridge states that there is no requirement for endorsed resources to follow the teaching order suggested in the Cambridge scheme of work. If a resource is endorsed, schools can be confident that all the learning objectives are covered.

Cambridge Primary Mathematics Stage 6 Scheme of Work units		Collins International Primary Maths Stage 6 units	
Unit number		Unit number	Topic
6.1	Number	1	Counting and sequences
		9	Place value and ordering decimals
		10	Place value, ordering and rounding decimals
6.2	2D and 3D shape	18	2D shapes and symmetry
		19	3D shapes
		20	Angles
		21	Measurements, including time
		22	Area and surface area
6.3	Calculation	2	Addition and subtraction of whole numbers (A)
		3	Addition and subtraction of whole numbers (B)
		4	Multiples, factors, divisibility, squares and cubes
		5	Whole number calculations
		6	Multiplication of whole numbers
		7	Division of whole numbers (A)
		8	Division of whole numbers (B)
6.4	Statistical methods	25	Statistics (A)
		26	Statistics (B)
6.5	Fractions, percentages, decimals and proportion	11	Fractions (A)
		12	Fractions (B)
		13	Percentages
		14	Addition and subtraction of decimals
		15	Multiplication of decimals
		16	Division of decimals
		17	Proportion and ratio
6.6	The coordinate grid	23	Coordinates
		24	Translation, reflection and rotation
6.7	Probability	27	Probability

STRAND: | Number | Geometry and Measure | Statistics and Probability |

The table below shows a recommended teaching and learning sequence (often referred to as a 'medium-term plan') for the 27 units in Collins International Primary Maths Stage 6.

However, as with the Cambridge Primary Mathematics Scheme of Work, schools and individual teachers are free to teach the learning objectives in any order to best meet the needs of individual schools, teachers and learners.

It is important to note that in order to allow for greater flexibility, the 27 units in each stage in Collins International Primary Maths are **not** ordered according to the recommended teaching sequence. Instead, they are in numerical order: Units 1 to 27, according to how the Strands and Sub-strands are arranged in the Cambridge Primary Mathematics Curriculum Framework (0096).

In other words, progression through the components in Collins International Primary Maths does not start at the beginning of the Teacher's Guide, Student's Book and Workbook and end at the back of the Guide / Book. Rather, units are covered as and when is appropriate, according to the Recommended Teaching and Learning Sequence provided by Collins International Primary Maths, or your school's specific scheme of work.

As a note of caution, the Collins International Primary Maths Recommended Teaching and Learning Sequence has been carefully written to ensure continuity and progression both *within* the units at a particular stage and also *across* Stages 1 to 6 and onwards into Lower Secondary.

This is extremely important in ensuring that learners have the pre-requisite knowledge, understanding and skills they require in order to successfully engage with new mathematical ideas at a deeper level and in different contexts.

Learners need to develop mastery of the learning objectives of the Strands and Sub-strands *within* a stage before they are able to apply and transfer their newly acquired knowledge, understanding and skills *across* other Strands and Sub-strands and into later stages. It is for this reason that in the Collins International Primary Maths Recommended Teaching and Learning Sequence, terms/semesters begin with units from the Number Strand, so that learners develop knowledge and skills in number that they can then apply to the other strands of mathematics (Geometry and Measure, and Statistics and Probability) and to their own lives.

Therefore, schools need to think extremely carefully when altering the Collins International Primary Maths Recommended Teaching and Learning Sequence in order to ensure that new learning builds on learners' prior knowledge, understanding and skills. In order to be confident with making such amendments, it is important that teachers are extremely familiar with the lines of progression both *within* and *across* stages of the Cambridge Primary and Lower Secondary Mathematics Curriculum Frameworks.

As with the Cambridge Primary Mathematics Scheme of Work, Collins International Primary Maths has assumed an academic year of three terms/semesters, each of 10 weeks duration. This is the minimum length of a school year and thereby allows flexibility for schools to add in more teaching time as necessary to meet the needs of the learners, and also to comfortably cover the content of the curriculum into an individual school's specific term/semesters times.

Collins International Primary Maths Stage 6 Recommended Teaching and Learning Sequence

	Term 1	Term 2	Term 3
Week 1	Unit 1: Counting and sequences	Unit 6: Multiplication of whole numbers	Unit 9: Place value and ordering decimals
Week 2	Unit 2: Addition and subtraction of whole numbers (A)	Unit 7: Division of whole numbers (A)	Unit 10: Place value, ordering and rounding decimals
Week 3	Unit 3: Addition and subtraction of whole numbers (B)	Unit 8: Division of whole numbers (B)	Unit 14: Addition and subtraction of decimals
Week 4	Unit 4: Multiples, factors, divisibility, squares and cubes	Unit 11: Fractions (A)	Revision and Cambridge Assessment *
Week 5	Unit 5: Whole number calculations	Unit 12: Fractions (B)	Unit 15: Multiplication of decimals
Week 6	Unit 18: 2D shapes and symmetry	Unit 13: Percentages	Unit 16: Division of decimals
Week 7	Unit 20: Angles	Unit 25: Statistics (A)	Unit 17: Proportion and ratio
Week 8	Unit 19: 3D shapes	Unit 23: Coordinates	Unit 21: Measurements, including time
Week 9	Unit 22: Area and surface area	Unit 24: Translation, reflection and rotation	Unit 26: Statistics (B)
Week 10	Revision	Revision	Unit 27: Probability

STRAND:	Number	Geometry and Measure	Statistics and Probability

* Timing of the Cambridge Assessment varies between countries.

No material is provided in Collins International Primary Maths for the three Revision weeks each term/semester. Individual teachers will decide the content to cover during these weeks, based on monitoring and evaluation made over the course of the term/semester, and learners' levels of achievement on the topics covered throughout the term/semester.

Teachers may decide to revisit certain topics and provide further practice of various concepts that have been taught during the term/semester, or they may use this week to catch up if lessons or units have taken longer than expected.

There is also no expectation that a Revision week will only take place at the end of each term/semester. If individual teachers feel that better use can be made of this week at another time throughout the term/semester, then they should feel free to do so.

⑧ Collins International Primary Maths Stage 6 units match to Cambridge Primary Mathematics Curriculum Framework (0096) Stage 6

The recommended teaching time for each unit is 1 week.

The Thinking and Working Mathematically characteristics are developed throughout each unit. Square brackets within objectives indicate parts of the objective that are not covered in this unit, but are covered elsewhere.

These learning objectives are reproduced from the Cambridge Primary Mathematics curriculum framework (0096) from 2020. This Cambridge International copyright material is reproduced under licence and remains the intellectual property of Cambridge Assessment International Education.

Unit 1 – Counting and sequences			
Cambridge Primary Mathematics Curriculum Framework (0096)			
Strand	**Sub-strand**	**Code**	**Learning objectives**
Number	Counting and sequences	6Nc.01	Count on and count back in steps of constant size, including fractions and decimals, and extend beyond zero to include negative numbers.
		6Nc.03	Use the relationship between repeated addition of a constant and multiplication to find and use a position-to-term rule.
		6Nc.04	Use knowledge of square numbers to generate terms in a sequence, given its position.
		Collins International Primary Maths Recommended Teaching and Learning Sequence:	**Term 1 Week 1**

Unit 2 – Addition and subtraction of whole numbers (A)			
Cambridge Primary Mathematics Curriculum Framework (0096)			
Strand	**Sub-strand**	**Code**	**Learning objectives**
Number	Integers and powers	6Ni.01	Estimate, add [and subtract] integers.
	Counting and sequences	6Nc.02	Recognise the use of letters to represent quantities that vary in addition and subtraction calculations.
		Collins International Primary Maths Recommended Teaching and Learning Sequence:	**Term 1 Week 2**

Unit 3 – Addition and subtraction of whole numbers (B)			
Cambridge Primary Mathematics Curriculum Framework (0096)			
Strand	**Sub-strand**	**Code**	**Learning objectives**
Number	Integers and powers	6Ni.01	Estimate[, add] and subtract integers.
	Counting and sequences	6Nc.02	Recognise the use of letters to represent quantities that vary in addition and subtraction calculations.
		Collins International Primary Maths Recommended Teaching and Learning Sequence:	**Term 1 Week 3**

Unit 4 – Multiples, factors, divisibility, squares and cubes

Cambridge Primary Mathematics Curriculum Framework Framework (0096)

Strand	Sub-strand	Code	Learning objectives
Number	Integers and powers	6Ni.06	Understand common multiples and common factors.
		6Ni.07	Use knowledge of factors and multiples to understand tests of divisibility by 3, 6 and 9.
		6Ni.08	Use knowledge of multiplication and square numbers to recognise cube numbers (from 1 to 125).
		Collins International Primary Maths Recommended Teaching and Learning Sequence:	**Term 1 Week 4**

Unit 5 – Whole numbers calculations

Cambridge Primary Mathematics Curriculum Framework (0096)

Strand	Sub-strand	Code	Learning objectives
Number	Integers and powers	6Ni.02	Use knowledge of laws of arithmetic and order of operations to simplify calculations.
		6Ni.03	Understand that brackets can be used to alter the order of operations.
		Collins International Primary Maths Recommended Teaching and Learning Sequence:	**Term 1 Week 5**

Unit 6 – Multiplication of whole numbers

Cambridge Primary Mathematics Curriculum Framework (0096)

Strand	Sub-strand	Code	Learning objectives
Number	Integers and powers	6Ni.04	Estimate and multiply whole numbers up to 10 000 by 1-digit or 2-digit whole numbers.
		Collins International Primary Maths Recommended Teaching and Learning Sequence:	**Term 2 Week 1**

Unit 7 – Division of whole numbers (A)

Cambridge Primary Mathematics Curriculum Framework (0096)

Strand	Sub-strand	Code	Learning objectives
Number	Integers and powers	6Ni.05	Estimate and divide whole numbers up to 1000 by 1-digit [or 2-digit] whole numbers.
		Collins International Primary Maths Recommended Teaching and Learning Sequence:	**Term 2 Week 2**

Unit 8 – Division of whole numbers (B)

Cambridge Primary Mathematics Curriculum Framework (0096)

Strand	Sub-strand	Code	Learning objectives
Number	Integers and powers	6Ni.05	Estimate and divide whole numbers up to 1000 by [1-digit or] 2-digit whole numbers.
		Collins International Primary Maths Recommended Teaching and Learning Sequence:	**Term 2 Week 3**

Unit 9 – Place value and ordering decimals

Cambridge Primary Mathematics Curriculum Framework (0096)

Strand	Sub-strand	Code	Learning objectives
Number	Counting and sequences	6Nc.01	Count on and count back in steps of constant size, including [fractions and] decimals, and extend beyond zero to include negative numbers.
	Place value, ordering and rounding	6Np.01	Understand and explain the value of each digit in decimals (tenths, hundredths and thousandths).
		6Np.03	Compose, decompose and regroup numbers, including decimals (tenths, hundredths and thousandths).
	Fractions, decimals, percentages, ratio and proportion	6Nf.08	Understand the relative size of quantities to compare and order numbers with one or two decimal places, [proper fractions with different denominators and percentages,] using the symbols =, > and <.

Collins International Primary Maths Recommended Teaching and Learning Sequence:	Term 3 Week 1

Unit 10 – Place value, ordering and rounding decimals

Cambridge Primary Mathematics Curriculum Framework (0096)

Strand	Sub-strand	Code	Learning objectives
Number	Place value, ordering and rounding	6Np.02	Use knowledge of place value to multiply and divide whole numbers and decimals by 10, 100 and 1000.
		6Np.04	Round numbers with 2 decimal places to the nearest tenth or whole number.

Collins International Primary Maths Recommended Teaching and Learning Sequence:	Term 3 Week 2

Unit 11 – Fractions (A)

Cambridge Primary Mathematics Curriculum Framework (0096)

Strand	Sub-strand	Code	Learning objectives
Number	Fractions, decimals, percentages, ratio and proportion	6Nf.01	Understand that a fraction can be represented as a division of the numerator by the denominator (proper and improper fractions).
		6Nf.03	Use knowledge of equivalence to write fractions in their simplest form.
		6Nf.04	Recognise that fractions, decimals (one or two decimal places) [and percentages] can have equivalent values
		6Nf.08	Understand the relative size of quantities to compare and order [numbers with one or two decimal places,] proper fractions with different denominators [and percentages,] using the symbols =, > and <.

Collins International Primary Maths Recommended Teaching and Learning Sequence:	Term 2 Week 4

Unit 12 – Fractions (B)

Cambridge Primary Mathematics Curriculum Framework (0096)

Strand	Sub-strand	Code	Learning objectives
Number	Fractions, decimals, percentages, ratio and proportion	6Nf.02	Understand that proper and improper fractions can act as operators.
		6Nf.05	Estimate, add and subtract fractions with different denominators.
		6Nf.06	Estimate, multiply and divide proper fractions by whole numbers.
		Collins International Primary Maths Recommended Teaching and Learning Sequence:	**Term 2 Week 5**

Unit 13 – Percentages

Cambridge Primary Mathematics Curriculum Framework (0096)

Strand	Sub-strand	Code	Learning objectives
Number	Fractions, decimals, percentages, ratio and proportion	6Nf.04	Recognise that fractions, decimals (one or two decimal places) and percentages can have equivalent values.
		6Nf.07	Recognise percentages (1%, and multiples of 5% up to 100%) of shapes and whole numbers.
		6Nf.08	Understand the relative size of quantities to compare and order numbers with one or two decimal places, proper fractions with different denominators and percentages, using the symbols =, > and <.
		Collins International Primary Maths Recommended Teaching and Learning Sequence:	**Term 2 Week 6**

Unit 14 – Addition and subtraction of decimals

Cambridge Primary Mathematics Curriculum Framework (0096)

Strand	Sub-strand	Code	Learning objectives
Number	Fractions, decimals, percentages, ratio and proportion	6Nf.09	Estimate, add and subtract numbers with the same or different number of decimal places.
		Collins International Primary Maths Recommended Teaching and Learning Sequence:	**Term 3 Week 3**

Unit 15 – Multiplication of decimals

Cambridge Primary Mathematics Curriculum Framework (0096)

Strand	Sub-strand	Code	Learning objectives
Number	Fractions, decimals, percentages, ratio and proportion	6Nf.10	Estimate and multiply numbers with one or two decimal places by 1-digit and 2-digit whole numbers.
		Collins International Primary Maths Recommended Teaching and Learning Sequence:	**Term 3 Week 5**

Unit 16 – Division of decimals

Cambridge Primary Mathematics Curriculum Framework (0096)

Strand	Sub-strand	Code	Learning objectives
Number	Fractions, decimals, percentages, ratio and proportion	6Nf.11	Estimate and divide numbers with one or two decimal places by whole numbers.
		Collins International Primary Maths **Recommended Teaching and Learning Sequence:**	**Term 3** **Week 6**

Unit 17 – Proportion and ratio

Cambridge Primary Mathematics Curriculum Framework (0096)

Strand	Sub-strand	Code	Learning objectives
Number	Fractions, decimals, percentages, ratio and proportion	6Nf.12	Understand the relationship between two quantities when they are in direct proportion.
		6Nf.13	Use knowledge of equivalence to understand and use equivalent ratios.
		Collins International Primary Maths **Recommended Teaching and Learning Sequence:**	**Term 3** **Week 7**

Unit 18 – 2D shapes and symmetry

Cambridge Primary Mathematics Curriculum Framework (0096)

Strand	Sub-strand	Code	Learning objectives
Geometry and Measure	Geometrical reasoning, shapes and measurements	6Gg.01	Identify, describe, classify and sketch quadrilaterals, including reference to angles, symmetrical properties, parallel sides and diagonals.
		6Gg.02	Know the parts of a circle: - centre - radius - diameter - circumference.
		6Gg.08	Identify rotational symmetry in familiar shapes, patterns or images with maximum order 4. Describe rotational symmetry as 'order x'.
		6Gg.11	Construct circles of a specified radius or diameter.
		Collins International Primary Maths **Recommended Teaching and Learning Sequence:**	**Term 1** **Week 6**

Unit 19 – 3D shapes

Cambridge Primary Mathematics Curriculum Framework (0096)

Strand	Sub-strand	Code	Learning objectives
Geometry and Measure	Geometrical reasoning, shapes and measurements	6Gg.04	Identify, describe and sketch compound 3D shapes.
		6Gg.06	Identify and sketch different nets for cubes, cuboids, prisms and pyramids.
		Collins International Primary Maths **Recommended Teaching and Learning Sequence:**	**Term 1** **Week 8**

Unit 20 – Angles

Cambridge Primary Mathematics Curriculum Framework (0096)			
Strand	Sub-strand	Code	Learning objectives
Geometry and Measure	Geometrical reasoning, shapes and measurements	6Gg.09	Classify, estimate, measure and draw angles.
		6Gg.10	Know that the sum of the angles in a triangle is 180°, and use this to calculate missing angles in a triangle.
		Collins International Primary Maths Recommended Teaching and Learning Sequence:	**Term 1 Week 7**

Unit 21 – Measurements, including time

Cambridge Primary Mathematics Curriculum Framework (0096)			
Strand	Sub-strand	Code	Learning objectives
Geometry and Measure	Time	6Gt.01	Convert between time intervals expressed as a decimal and in mixed units.
	Geometrical reasoning, shapes and measurements	6Gg.05	Understand the difference between capacity and volume.
		Collins International Primary Maths Recommended Teaching and Learning Sequence:	**Term 3 Week 8**

Unit 22 – Area and surface area

Cambridge Primary Mathematics Curriculum Framework (0096)			
Strand	Sub-strand	Code	Learning objectives
Geometry and Measure	Geometrical reasoning, shapes and measurements	6Gg.03	Use knowledge of area of rectangles to estimate and calculate the area of right-angled triangles.
		6Gg.07	Understand the relationship between area of 2D shapes and surface area of 3D shapes.
		Collins International Primary Maths Recommended Teaching and Learning Sequence:	**Term 1 Week 9**

Unit 23 – Coordinates

Cambridge Primary Mathematics Curriculum Framework (0096)			
Strand	Sub-strand	Code	Learning objectives
Geometry and Measure	Position and transformation	6Gp.01	Read and plot coordinates including integers, fractions and decimals, in all four quadrants (with the aid of a grid).
		6Gp.02	Use knowledge of 2D shapes and coordinates to plot points to form lines and shapes in all four quadrants.
		Collins International Primary Maths Recommended Teaching and Learning Sequence:	**Term 2 Week 8**

Unit 24 – Translation, reflection and rotation

Cambridge Primary Mathematics Curriculum Framework (0096)

Strand	Sub-strand	Code	Learning objectives
Geometry and Measure	Position and transformation	6Gp.03	Translate 2D shapes, identifying the corresponding points between the original and the translated image, on coordinate grids.
		6Gp.04	Reflect 2D shapes in a given mirror line (vertical, horizontal and diagonal), on square grids.
		6Gp.05	Rotate shapes 90° around a vertex (clockwise or anticlockwise).
		Collins International Primary Maths Recommended Teaching and Learning Sequence:	**Term 2 Week 9**

Unit 25 – Statistics (A)

Cambridge Primary Mathematics Curriculum Framework (0096)

Strand	Sub-strand	Code	Learning objectives
Statistics and Probability	Statistics	6Ss.01	Plan and conduct an investigation and make predictions for a set of related statistical questions, considering what data to collect (categorical, discrete and continuous data).
		6Ss.02	Record, organise and represent categorical, discrete and continuous data. Choose and explain which representation to use in a given situation: - Venn and Carroll diagrams - tally charts and frequency tables - bar charts - waffle diagrams and pie charts [- frequency diagrams for continuous data - line graphs - scatter graphs - dot plots.]
		6Ss.03	Understand that the mode, median, mean and range are ways to describe and summarise data sets. Find and interpret the mode (including bimodal data), median, mean and range, and consider their appropriateness for the context.
		6Ss.04	Interpret data, identifying patterns, within and between data sets, to answer statistical questions. Discuss conclusions, considering the sources of variation and check predictions.
		Collins International Primary Maths Recommended Teaching and Learning Sequence:	**Term 2 Week 7**

Unit 26 – Statistics (B)

Cambridge Primary Mathematics Curriculum Framework (0096)

Strand	Sub-strand	Code	Learning objectives
Statistics and Probability	Statistics	6Ss.01	Plan and conduct an investigation and make predictions for a set of related statistical questions, considering what data to collect (categorical, discrete and continuous data).
		6Ss.02	Record, organise and represent categorical, discrete and continuous data. Choose and explain which representation to use in a given situation: [- Venn and Carroll diagrams - tally charts and frequency tables - bar charts - waffle diagrams and pie charts] - frequency diagrams for continuous data - line graphs - scatter graphs - dot plots.
		6Ss.04	Interpret data, identifying patterns, within and between data sets, to answer statistical questions. Discuss conclusions, considering the sources of variation and check predictions.
		Collins International Primary Maths Recommended Teaching and Learning Sequence:	**Term 3 Week 9**

Unit 27 – Probability

Cambridge Primary Mathematics Curriculum Framework (0096)

Strand	Sub-strand	Code	Learning objectives
Statistics and Probability	Probability	6Sp.01	Use the language associated with probability and proportion to describe and compare possible outcomes.
		6Sp.02	Identify when two events can happen at the same time and when they cannot, and know that the latter are called 'mutually exclusive'.
		6Sp.03	Recognise that some probabilities can only be modelled through experiments using a large number of trials.
		6Sp.04	Conduct chance experiments or simulations, using small and large numbers of trials. Predict, analyse and describe the frequency of outcomes using the language of probability.
		Collins International Primary Maths Recommended Teaching and Learning Sequence:	**Term 3 Week 10**

⑨ Cambridge Primary Mathematics Curriculum Framework (0096) Stage 6 match to Collins International Primary Maths units

The charts below show when each of the Stage 6 learning objectives in the Cambridge Primary Mathematics Curriculum Framework (0096) are taught in the 27 units in Collins International Primary Maths Stage 6.

Cambridge Primary Mathematics Curriculum Framework (0096)				Collins International Primary Maths unit(s)
Strand	Sub-strand	Code	Learning objective	
Number	Counting and sequences	6Nc.01	Count on and count back in steps of constant size, including fractions and decimals, and extend beyond zero to include negative numbers.	1, 9
		6Nc.02	Recognise the use of letters to represent quantities that vary in addition and subtraction calculations.	2, 3
		6Nc.03	Use the relationship between repeated addition of a constant and multiplication to find and use a position-to-term rule.	1
		6Nc.04	Use knowledge of square numbers to generate terms in a sequence, given its position.	1
	Integers and powers	6Ni.01	Estimate, add and subtract integers.	2, 3
		6Ni.02	Use knowledge of laws of arithmetic and order of operations to simplify calculations.	5
		6Ni.03	Understand that brackets can be used to alter the order of operations.	5
		6Ni.04	Estimate and multiply whole numbers up to 10 000 by 1-digit or 2-digit whole numbers.	6
		6Ni.05	Estimate and divide whole numbers up to 1000 by 1-digit or 2-digit whole numbers.	7, 8
		6Ni.06	Understand common multiples and common factors.	4
		6Ni.07	Use knowledge of factors and multiples to understand tests of divisibility by 3, 6 and 9.	4
		6Ni.08	Use knowledge of multiplication and square numbers to recognise cube numbers (from 1 to 125).	4

Cambridge Primary Mathematics Curriculum Framework (0096)				Collins International Primary Maths unit(s)
Strand	**Sub-strand**	**Code**	**Learning objective**	
Number	Place value, ordering and rounding	6Np.01	Understand and explain the value of each digit in decimals (tenths, hundredths and thousandths).	9
		6Np.02	Use knowledge of place value to multiply and divide whole numbers and decimals by 10, 100 and 1000.	10
		6Np.03	Compose, decompose and regroup numbers, including decimals (tenths, hundredths and thousandths).	9
		6Np.04	Round numbers with 2 decimal places to the nearest tenth or whole number.	10
	Fractions, decimals, percentages, ratio and proportion	6Nf.01	Understand that a fraction can be represented as a division of the numerator by the denominator (proper and improper fractions).	11
		6Nf.02	Understand that proper and improper fractions can act as operators.	12
		6Nf.03	Use knowledge of equivalence to write fractions in their simplest form.	11
		6Nf.04	Recognise that fractions, decimals (one or two decimal places) and percentages can have equivalent values.	11, 13
		6Nf.05	Estimate, add and subtract fractions with different denominators.	12
		6Nf.06	Estimate, multiply and divide proper fractions by whole numbers.	12
		6Nf.07	Recognise percentages (1%, and multiples of 5% up to 100%) of shapes and whole numbers.	13
		6Nf.08	Understand the relative size of quantities to compare and order numbers with one or two decimal places, proper fractions with different denominators and percentages, using the symbols =, > and <.	9, 11, 13
		6Nf.09	Estimate, add and subtract numbers with the same or different number of decimal places.	14
		6Nf.10	Estimate and multiply numbers with one or two decimal places by 1-digit and 2-digit whole numbers.	15
		6Nf.11	Estimate and divide numbers with one or two decimal places by whole numbers.	16
		6Nf.12	Understand the relationship between two quantities when they are in direct proportion.	17
		6Nf.13	Use knowledge of equivalence to understand and use equivalent ratios.	17

Cambridge Primary Mathematics Curriculum Framework (0096)				Collins International Primary Maths unit(s)
Strand	Sub-strand	Code	Learning objective	
Geometry and Measure	Time	6Gt.01	Convert between time intervals expressed as a decimal and in mixed units.	21
	Geometrical reasoning, shapes and measurements	6Gg.01	Identify, describe, classify and sketch quadrilaterals, including reference to angles, symmetrical properties, parallel sides and diagonals.	18
		6Gg.02	Know the parts of a circle: - centre - radius - diameter - circumference.	18
		6Gg.03	Use knowledge of area of rectangles to estimate and calculate the area of right-angled triangles.	22
		6Gg.04	Identify, describe and sketch compound 3D shapes.	19
		6Gg.05	Understand the difference between capacity and volume.	21
		6Gg.06	Identify and sketch different nets for cubes, cuboids, prisms and pyramids.	19
		6Gg.07	Understand the relationship between area of 2D shapes and surface area of 3D shapes.	22
		6Gg.08	Identify rotational symmetry in familiar shapes, patterns or images with maximum order 4. Describe rotational symmetry as 'order x'.	18
		6Gg.09	Classify, estimate, measure and draw angles.	20
		6Gg.10	Know that the sum of the angles in a triangle is 180°, and use this to calculate missing angles in a triangle.	20
		6Gg.11	Construct circles of a specified radius or diameter.	18

Cambridge Primary Mathematics Curriculum Framework (0096)				Collins International Primary Maths unit(s)
Strand	**Sub-strand**	**Code**	**Learning objective**	
Geometry and Measure	Position and transformation	6Gp.01	Read and plot coordinates including integers, fractions and decimals, in all four quadrants (with the aid of a grid).	23
		6Gp.02	Use knowledge of 2D shapes and coordinates to plot points to form lines and shapes in all four quadrants.	23
		6Gp.03	Translate 2D shapes, identifying the corresponding points between the original and the translated image, on coordinate grids.	24
		6Gp.04	Reflect 2D shapes in a given mirror line (vertical, horizontal and diagonal), on square grids.	24
		6Gp.05	Rotate shapes 90° around a vertex (clockwise or anticlockwise).	24
Statistics and Probability	Statistics	6Ss.01	Plan and conduct an investigation and make predictions for a set of related statistical questions, considering what data to collect (categorical, discrete and continuous data).	25, 26
		6Ss.02	Record, organise and represent categorical, discrete and continuous data. Choose and explain which representation to use in a given situation: - Venn and Carroll diagrams - tally charts and frequency tables - bar charts - waffle diagrams and pie charts - frequency diagrams for continuous data - line graphs - scatter graphs - dot plots.	25, 26
		6Ss.03	Understand that the mode, median, mean and range are ways to describe and summarise data sets. Find and interpret the mode (including bimodal data), median, mean and range, and consider their appropriateness for the context.	25
		6Ss.04	Interpret data, identifying patterns, within and between data sets, to answer statistical questions. Discuss conclusions, considering the sources of variation and check predictions.	25, 26

Cambridge Primary Mathematics Curriculum Framework (0096)				Collins International Primary Maths unit(s)
Strand	Sub-strand	Code	Learning objective	
Statistics and Probability	Probability	6Sp.01	Use the language associated with probability and proportion to describe and compare possible outcomes.	27
		6Sp.02	Identify when two events can happen at the same time and when they cannot, and know that the latter are called 'mutually exclusive'.	27
		6Sp.03	Recognise that some probabilities can only be modelled through experiments using a large number of trials.	27
		6Sp.04	Conduct chance experiments or simulations, using small and large numbers of trials. Predict, analyse and describe the frequency of outcomes using the language of probability.	27

Revise

Decimals go round

Learning objective

Code	Learning objective
6Nc.01	Count on and count back in steps of constant size, including [fractions and] decimals, [and extend beyond zero to include negative numbers].

What to do

- Arrange learners in a circle.
- Start a forwards count from a one-place decimal in steps of 0·1.
- Ask learners to say what the next number in the sequence is.
- At any point, announce a change in the count, for example count back in steps of 0·2 or count forwards in steps of 0·3.
- **[TWM.02]** Say: **I start at 1·7 and count back in steps of 0·4. Will I say the number 0·4? How do you know? What do you notice about the digits in the tenths place for each term in the sequence?**

Variations

2 Introduce backwards sequences that extend beyond zero.

1 Allow learners to repeat the previous number, for example, if a forwards count in steps of 0·2 begins at 3·4, the first learner would say 3·4 then 3·6, the next learner would say 3·6 then 3·8, and so on.

Fractions on the line

Learning objective

Code	Learning objective
6Nc.01	Count on and count back in steps of constant size, including fractions [and decimals, and extend beyond zero to include negative numbers].

Resources

paper or mini whiteboard and pen (per learner)

What to do

- Draw a large number line on the board from 3 to 4.
- Ask learners to write down a fraction that could go on the number line.
- Invite a learner to come to the board to mark a division on the line and label it with a fraction, a third for example.
- Confirm with the class that the fraction is in the correct place.
- Invite other learners to the board to label the line with other fractions that have the same denominator.
- Choose a different interval and suggest a different denominator.

Variation

2 Use the labelled number line to count on or back in the fraction given. Ask: **What are the next three numbers in the sequence? How would you write this sequence in decimals?**

Revise

Stepping back

Learning objective

Code	Learning objective
5Nc.03	Count on and count back in steps of constant size, [including fractions and decimals,] and extend beyond zero to include negative numbers.

What to do

- Display the **Number line tool**, set to the range: –30 to 10.
- Divide the class into two teams.
- Display the **Spinner tool** with five sectors labelled 1 to 5.
- Position the snail marker at 9.
- Spin the spinner. Say: **Count back in steps of [spinner number].** Ask: **What will be the second number in the sequence? Fourth number? Sixth number?**
- The first team to give the correct answer scores a point.

- Play for seven rounds. The team with the highest score is the winner.
- [TWM.02] Say: **Count back four steps in 7s from 4 and five steps in 6s from 5. Which count gives the smaller number? How do you know?**

Variation

2 Use only three sectors on the **Spinner tool**: 7, 8 and 9.

Count forward from a negative number instead of counting back from a positive number.

Name your terms

Learning objective

Code	Learning objective
6Nc.03	Use the relationship between repeated addition of a constant and multiplication to find and use a position-to-term rule.

What to do

- On the board, write the numbers: 6, 12, 18, 24, 30, 36, 40, 48.
- [TWM.04] Ask: **Which of these numbers is incorrect?** (40) **Convince me.**
- Take responses. Invite a learner to correct the sequence and write '1st' above the first term.
- Invite other learners to continue labelling the terms in the sequence with the correct ordinal numbers.
- Model the language of term and position. Ask: **What is the term in the fourth position?** (24) **In which position is the term 42?** (7th)

- Say: **Extend the sequence to the ninth and tenth terms. What are they?** (54, 60)
- [TWM.04] Ask: **Would the numbers 117, 90 or 638 be in this sequence? How do you know?**

Variations

2 Use sequences of different multiples, for example, 7, 8 or 9.

3 Name the terms in various positions of a sequence, for example, position 4, term 36; position 8, term 72, and ask learners to name the sequence (multiples of 9).

Revise

Up above, down below

Learning objective

Code	Learning objective
5Ni.01	Estimate, add [and subtract] integers, including where one integer is negative.

What to do

- Divide the class into two teams.
- Say: **Think of an island. The trees on the island are above sea level. The fish swimming around the island are below sea level. I will tell you the height of the top of a tree, a positive number of metres, and the depth of a fish, a negative number of metres. You have to work out the distance between the top of the tree and the fish.**

- Begin the game with: **The tree is 23 metres tall; the fish is at negative 12 metres.**
- Invite the first learners to put up their hands to respond. If correct, they score a point for their team.
- Repeat with different heights above and below sea level. The team with the higher score wins.

Variation

1 Provide a number line to assist learners.

Getting warmer

Learning objective

Code	Learning objective
5Nc.02	Recognise the use of objects, shapes or symbols to represent two unknown quantities in addition and subtraction calculations.

What to do

- Display the **Spinner tool** with ten sectors labelled 0 to 9. Display the **Thermometer tool** set from −10°C to 10°C.
- Spin the spinner twice. The first spin represents a temperature below 0°C. Mark this on the thermometer scale. The second spin represents a temperature rise in degrees.
- Ask: **What is the new temperature after the rise?**
- Allow some thinking time before asking for a whole-class response.
- For example, if the original temperature was −3°C and the rise is 5 degrees, expect learners to respond with the answer, 2 degrees.
- Repeat for different pairs of temperatures and temperature increases.

Variation

3 Share temperature stories with the class. For example, 'On Monday morning, the temperature was −3°C. By the afternoon it had risen by 8 degrees. What was the temperature in the afternoon?' (5°C)

Learners calculate the new temperature without use of the **Thermometer tool**.

Number – Integers and powers

Revise

Missing numbers 👥 ▲2

Learning objective

Code	Learning objective
5Nc.02	Recognise the use of objects, shapes or symbols to represent two unknown quantities in addition and subtraction calculations.

Resources

paper or mini whiteboard and pen (per learner)

What to do [TWM.04]

- On the board, write: $46 + \square = 72$.
- Ask: **What does the square symbol mean? How would you find this missing number?** Take responses and remind learners that an inverse operation can be used to find an unknown in a number sentence. Invite a learner to write the inverse operation on the board ($\square = 72 - 46 = 26$).
- On the board, write: $\square - 38 = 49$.
- **[TWM.04] [T&T]** Ask: **Convince me that the answer can be found using addition. Explain why this works.** Take responses. (87)

- On the board, write: $\square + 56 = 98$, $\square - 88 = 56$, $67 + \square = 92$, $\square - 29 = 243$.
- Learners work in pairs to find the unknowns. Take responses and ask learners to explain their methods. (42, 144, 25, 272)
- Direct learners to write their own missing number problems for their partner to solve. They swap papers or whiteboards to check answers.

Variation

3 On the board, write: $365 - \square = 148$. **[TWM.04]** Ask: **How do you solve this problem? Do you use an inverse operation? Explain your answer.** Take responses. (217)

Going shopping 👥 ▲2

Learning objective

Code	Learning objective
5Nc.02	Recognise the use of objects, shapes or symbols to represent two unknown quantities in addition and subtraction calculations.

Resources

classroom shop with items priced between $20 and $100 (per class); paper or mini whiteboard and pen (per learner)

What to do

- Organise the class into groups of three learners.
- Two learners in each group become shopper and shopkeeper. The third learner is the 'change detective' who stands a distance away so that they cannot see or hear the transaction.
- The 'shopper' buys an item and receives change from the 'shopkeeper'.
- The third learner returns to the group and is told how much the shopper gave the shopkeeper and how much changed they received, but not the price of the item purchased.

- The shopper writes a number sentence to represent the transaction with a box to represent the unknown value (the price of the item). Next, they work out the price of the item and ask the other learners to confirm if they are correct.
- Repeat with learners swapping roles.

Variation

2 The third learner becomes the 'payment detective'. They are told the price of the item and the amount of change received, and have to work out how much money was given to the shopkeeper.

Revise

Hot or cold

Learning objective

Code	Learning objective
6Ni.01	Estimate[, add] and subtract integers.

What to do 📊

- Display the **Spinner tool** with the ten sectors labelled 0 to 9. Display the **Thermometer tool** set from 10°C to –10°C.
- Spin the spinner twice. The first spin represents a temperature above 0°C; the second spin represents a temperature below 0°C. Mark both temperatures on the thermometer scale, for example 8°C and –3°C.

- Ask: **What is the temperature difference?**
- Allow some thinking time before asking learners to respond with the answer. (11 degrees)
- Repeat for different pairs of temperatures.

Variation

3 Learners calculate the temperature difference without the **Thermometer tool**.

Temperature stories

Learning objective

Code	Learning objective
6Ni.01	Estimate[, add] and subtract integers.

Resources

paper or mini whiteboard and pen (per group)

What to do

- Organise learners into groups of three or four.
- Each group decides on a starting temperature and writes it down.
- They take turns to say describe a scenario based on a change in temperature, alternating between a temperature rise and a temperature fall. For example, with a starting temperature of –3°C, the first learner might say: 'The temperature rose by 5 degrees to 2°C.' The second learner might follow

this with: 'The temperature dropped by 6 degrees from 2°C to –4°C', and so on.
- After each sentence, the calculation and new amount/temperature are recorded.
- Encourage learners to be creative.

Variation

2 Learners repeat the activity using a different context, such as money.

Number – Integers and powers

Revise

Unknown numbers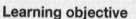

Number – Integers and powers

Learning objective

Code	Learning objective
6Nc.02	Recognise the use of letters to represent quantities that vary in addition and subtraction calculations.

What to do

- Say: **I have an unknown number. When I add 13 to my number, I get 19. What is my number?**
- Ask a learner to come to the board to write a number sentence for the problem with the unknown represented by the letter 'a'. Expect: 13 + a = 19.
- Ask: **How would you find the value of 'a'?** Take responses. Expect some learners to say that they know 13 + 6 = 19 and therefore, the unknown number is 6. Remind them that they could use their knowledge of subtraction as the inverse operation of addition to work out the unknown. Write: 19 − 13 = a, so a is 6.
- Repeat for other unknown numbers, for example: **When I subtract 8 from my number, I get 27.**

Variation

3 Introduce more difficult additions and subtraction, for example: **When I subtract 46 from my number, I get 168.**

Angles on a line

Learning objective

Code	Learning objective
5Nc.02	Recognise the use of objects, shapes or symbols to represent two unknown quantities in addition and subtraction calculations.

What to do

- Draw a horizontal line on the board. Draw a second line that intersects the first at an angle (not 90 degrees) and stops at a point on the line. Label the two angles made A and B.
- Ask: **What do you know about angles on a straight line?** (they add to 180°)
- Say angle A is 40°. Choose a learner to write an equation connecting angles A and B (40 + B = 180, or 40 = 180 − B).
- Write possible answers for A and B in the form: If A = 40°, then B = 140°.

Variation

3 Work with three of four angles on a line.

Revise

Mixed up multiples

Learning objective

Code	Learning objective
6Ni.06	Understand common multiples [and common factors].

Resources

paper or mini whiteboard and pen (per learner)

What to do

- Write the multiples of 6, 7, 8 and 9, up to the tenth multiple, randomly on the board.
- Write another four numbers that do not appear in any of these tables.
- Say: **The numbers on the board are all multiples of 6, 7, 8 or 9, except for four of the numbers. When I say 'Go!' I want you to write the numbers as a list of multiples for each table. You do not have to put them in order.**

- Give the command 'Go!'. Ask learners to identify the four numbers that are not multiples of 6, 7, 8 or 9.

Variation

Include 11th, 12th and 13th multiples.

Factor find

Learning objective

Code	Learning objective
6Ni.06	Understand common multiples and common factors.

Resources

2–10 number cards from Resource sheet 2:
0–100 number cards (one card per learner);
2–15 number from Resource sheet 2: 0–100
number cards (one card per learner) (for variation)

What to do

- Randomly distribute sets of number cards from 2 to 10 so every learner has a card.
- Ask learners to hold the card against their chest so that it can be seen.

- Say: **When I say a number, anyone holding a number that is a factor must get together with their factor partner as quickly as possible. For example, if I call out 12, numbers 2 and 6 and numbers 3 and 4 should stand together.**
- Call out any composite number from 4 to 100.
- Play the game with a different number each time.

Variation

Use number cards 2–15 and include factors of numbers to 150.

Revise

Three in a row

Learning objective

Code	Learning objective
6Ni.07	Use knowledge of factors and multiples to understand tests of divisibility by 3, 6 and 9.

Resources

paper or mini whiteboard and pen (per learner)

What to do

- Display the **Spinner tool** with sectors labelled 2 to 10.
- On the board, write: 3, 6, 9.
- Ask learners to draw a 3 by 3 grid.
- They choose a number from the board and fill the squares in the grid with multiples of that number up to ten times.

- Spin the spinner. Learners multiply the number spun by their chosen number.
- If they have the answer to the multiplication in their grid, they cross it out.
- The winner is the first learner to cross out three multiples in a row, column or a diagonal.

Variation

3 Include multiples of up to 20 times. The spinner needs to be spun twice and the sum becomes the multiplier.

Multiplication race

Learning objective

Code	Learning objective
6Ni.08	Use knowledge of multiplication and square numbers to recognise cube numbers (from 1 to 125).

What to do

- Divide the class into two teams.
- Say a number between 2 and 10.
- Learners must multiply the number by itself two times. For example, if the number 3 is called, learners must find the product of 3 x 3 x 3.
- The first team to put up their hands and answer correctly scores a point.
- Repeat for other numbers.

- The team with the higher score after seven rounds is the winner.

Variation

1 Learners must multiply the number by itself only one time, for example 4 x 4.

Number – Integers and powers

Revise

Reordering additions

Learning objective

Code	Learning objective
6Ni.02	Use knowledge of laws of arithmetic and order of operations to simplify calculations.

Resources

paper or mini whiteboard and pen (per learner)

What to do

- Write the following numbers in random positions across the board: 14, 13, 55, 28, 27, 17, 36, 42, 25.
- Ask learners to write three addition calculations using the numbers on the board. Each calculation must have three addends. Learners should arrange the numbers so that they are as easy as possible to add. For example, choosing 13 and 27 as the first two addends for the addition 13 + 27 + 25 because

13 and 27 are complementary numbers that add to a multiple of 10. (40)

- Choose learners to share their addition calculations and explain how they chose each set of three addends.

Variation

3 Increase the choice to 12 numbers: 14, 13, 55, 28, 27, 17, 36, 42, 25, 12, 35, 46, for learners to construct three addition calculations with four addends each.

Reordering multiplications

Learning objective

Code	Learning objective
6Ni.02	Use knowledge of laws of arithmetic and order of operations to simplify calculations.

Resources

paper or mini whiteboard and pen (per learner)

What to do

- Write the following numbers in random positions across the board: 6, 27, 4, 23, 25, 5, 18, 17, 50, 15, 2, 8.
- Ask learners to write four multiplication calculations, each with three multiplicands using the numbers on the board. They should arrange the numbers so that they are as easy as possible to multiply. For example, choosing 4 and 25 as the first two terms for the multiplication 4 × 25 × 17 because 4 multiplied by 25 makes 100 – an easy number to multiply by 17. (1700)

- Choose learners to share their multiplication calculations and explain how they chose each set of three terms.

Variation

1 Reduce the choice to six numbers: 6, 18, 5, 50, 17, 4.

Number – Integers and powers

Revise

Ordering operations

Learning objective

Code	Learning objective
6Ni.03	Understand that brackets can be used to alter the order of operations.

Resources

A4 sheet of paper (per learner); scissors (per learner); paper or mini whiteboard and pen (per learner)

What to do

- Each learner folds and cuts an A4 sheet of paper into six pieces. On each piece of paper, they write a different operation and a one-digit number, for example: + 7, × 5, ÷ 2, – 6, + 8, × 3.
- Working in pairs, learners combine all 12 papers, shuffle them and place them face down on the table.
- Each learner writes down their own start number.

- They then take turns to turn over three pieces of paper and lay them out in the order they were turned.
- Each learner writes them down after their start number and works out the calculation according to the order of operations.
- They then swap with their partner and check each other's calculations.

Variation

3 Learners turn over four pieces of paper.

Spinning operations

Learning objective

Code	Learning objective
6Ni.03	Understand that brackets can be used to alter the order of operations.

Resources

one spinner from Resource sheet 35: Operations spinner (per pair); two 1–6 dice or Resource sheet 5: 1–6 spinner (per pair); pencil and paper clip for the spinner (per pair); paper or mini whiteboard and pen (per learner)

What to do

- Arrange the learners into pairs. One learner spins the Operations spinner three times and both learners write down the operations.
- The other learner rolls both dice or spins the 1–6 spinner twice and uses the digits generated to make a two-digit start number.

- Learners then work independently to write a calculation using the start number and all three operations.
- The learner whose answer is closer to 20 scores a point.

Variation

3 Learners work with three-digit numbers. They roll the dice or spin the 1–6 spinner three times to create each three-digit number.

Revise

Complete the table (1)

Learning objective

Code	Learning objective
6Ni.04	Estimate and multiply whole numbers up to 10 000 by 1-digit [or 2-digit] whole numbers.

Resources

paper or mini whiteboard and pen (per learner)

What to do

- Draw a 4 by 8 grid on the board. Write the multiplication symbol (×) in the top left-hand corner.
- Write four one-digit numbers down the left-hand column. Write eight two-digit numbers they are to be multiplied by along the top row.
- Invite learners to choose a multiplication problem by selecting from the left-hand column and the top row. They should solve the problem using any preferred mental or written method. Remind them to estimate the answer first.

- If the answer is correct, they write it in the corresponding position on the grid.

Variation

2 Learners devise their own grid of numbers and swap with a partner to solve. They return their papers for marking.

Expanded written method (1)

Learning objective

Code	Learning objective
6Ni.04	Estimate and multiply whole numbers up to 10 000 by 1-digit [or 2-digit] whole numbers.

Resources

paper or mini whiteboard and pen (per learner)

What to do

- On the board, write the numbers: 5, 6, 7 and 8.
- Ask learners to arrange the numbers into a THO × O multiplication.
- Have them estimate the product and then use the expanded written method to multiply.
- They check that their answer is reasonable by comparing it with the estimate.
- Choose a learner to come to the board and demonstrate the multiplication.

- Ask learners to arrange the digits into two more THO × O multiplication problems and solve.

Variation

1 Learners calculate the product using the grid method.

Number – Integers and powers

Revise

Complete the table (2)

Learning objective

Code	Learning objective
6Ni.04	Estimate and multiply whole numbers up to 10 000 by 1-digit or 2-digit whole numbers.

Resources

paper or mini whiteboard and pen (per learner)

What to do

- Draw a 4 by 8 grid on the board. Write the multiplication symbol (×) in the top left-hand corner.
- Write four two-digit numbers down the left-hand column. Write eight two-digit numbers they are to be multiplied by along the top row.
- Invite learners to choose a multiplication problem by selecting from the left-hand column and the top row. They should solve the calculation using any preferred mental or written method. Remind them to estimate the answer first.
- If the answer is correct, they write it in the corresponding position on the grid.

Variation

2 Learners devise their own grid of numbers and swap with a partner to solve. They return their papers for marking.

Expanded written method (2)

Learning objective

Code	Learning objective
6Ni.04	Estimate and multiply whole numbers up to 10 000 by 1-digit or 2-digit whole numbers.

Resources

paper or mini whiteboard and pen (per learner)

What to do

- On the board, write the numbers: 5, 6, 7 and 8.
- Ask learners to arrange the numbers into a TO × TO multiplication.
- Have them estimate the product and then use the expanded written method to multiply.
- They check that their answer is reasonable by comparing it with the estimate.
- Choose a learner to come to the board and demonstrate the multiplication.
- Ask learners to arrange the digits into two more TO × TO multiplication calculations and solve them.

Variation

1 Learners calculate the product using the grid method.

Revise

Division wheel (1)

Learning objective

Code	Learning objective
6Ni.05	Estimate and divide whole numbers up to 1000 by 1-digit [or 2-digit] whole numbers.

Resources

paper or mini whiteboard and pen (per learner)

What to do

- Draw three concentric circles on the board with spokes radiating out from the centre circle.
- Write ÷ 6 in the centre. Write a multiple of 6 from 36 to 108 in the inner circle.

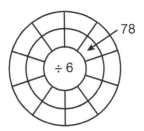

- Ask learners to divide the number by 6 mentally and record the answer.
- Confirm the quotient as a class and then record it in the appropriate space in the outer circle of the wheel.

- Add further multiples of 6 randomly, one at a time, and repeat.
- Redraw the diagram and repeat for other divisors and multiples.

Variation

3 Include divisions where there is a remainder. For example, 39 and 62 for ÷ 6. Ask learners to express the remainder as a whole number.

Division strategies (1)

Learning objective

Code	Learning objective
6Ni.05	Estimate and divide whole numbers up to 1000 by 1-digit [or 2-digit] whole numbers.

Resources

place value counters or Resource sheet 8: Place value counters (1s and 10s) (per pair); paper or mini whiteboard and pen (per pair)

What to do

- On the board, write the calculations: 72 ÷ 3 =, 95 ÷ 5 =, 81 ÷ 6 =, 90 ÷ 8 =.
- Provide place value counters. Ask learners to work in pairs to solve the four calculations.
- Encourage them to use a range of strategies from mental 'grouping', counting backwards/forwards in multiples of a number, or a concrete method where they use place value counters to group (or 'chunk') multiples of the divisor on a place value grid.

- For each calculation, choose a pair of learners to explain the strategy they used. Confirm the answers as a class (24, 19, 13 r 3, 11 r 2).

Variation

1 Include only calculations with multiples of the divisor up to ten times the divisor. No remainders.

Number – Integers and powers

Number – Integers and powers

Revise

Spot the error

Learning objective

Code	Learning objective
6Ni.05	Estimate and divide whole numbers up to 1000 by 1-digit [or 2-digit] whole numbers.

Resources

paper or mini whiteboard and pen (per learner)

What to do

- On the board, write the following:
  a) $78 \div 3 = 28$, b) $96 \div 4 = 24$, c) $85 \div 6 = 14 \, r \, 3$,
  d) $96 \div 7 = 13 \, r \, 5$, e) $99 \div 8 = 12 \, r \, 3$, f) $99 \div 8 = 12 \, r \, 7$.

- Explain to learners that some of the calculations contain errors. Ask them to check the answers using the expanded written method or short division to confirm which of the calculations are incorrect.

- For each calculation, choose a pair of learners to say whether they think the answer is correct or incorrect. Confirm this as a class. (incorrect: a, c, f)

Variation

1 Learners use any strategy they are confident with to decide whether each calculation is correct or not.

Division strategies (2)

Learning objective

Code	Learning objective
6Ni.05	Estimate and divide whole numbers up to 1000 by 1-digit [or 2-digit] whole numbers.

Resources

place value counters or Resource sheet 8: Place value counters (1s, 10s and 100s) (per pair); paper or mini whiteboard and pen (per pair)

What to do

- On the board, write the calculations: $255 \div 3 =$, $384 \div 4 =$, $615 \div 5 =$, $402 \div 6 =$, $420 \div 8 =$, $419 \div 9 =$.

- Provide place value counters. Ask learners to work in pairs to solve the six calculations.

- Encourage them to use a range of strategies from mental 'grouping', counting backward/forward in multiples of a number or a concrete method where they use place value counters to group (or 'chunk') multiples of the divisor on a place value chart.

- For each calculation, choose a pair of learners to explain the strategy they used. Confirm the answers as a class (85, 96, 123, 67, 52 r 4, 46 r 5).

Variation

1 Include only calculations with multiples of 10 as the divisor, for example: $240 \div 3 =$, $280 \div 4 =$. No remainders.

Revise

Division wheel (2)

Learning objective

Code	Learning objective
6Ni.05	Estimate and divide whole numbers up to 1000 by [1-digit or] 2-digit whole numbers.

Resources:

paper or mini whiteboard and pen (per pair)

What to do

- Draw three concentric circles on the board with spokes radiating out from the centre circle.

- Write ÷ 24 in the centre. Write a multiple of 24 from 24 to 240 in the inner circle.

- Ask learners to divide the number by 24 mentally and record the answer.

- Confirm the quotient as a class and then record it in the appropriate space in the outer circle of the wheel.

- Add further multiples of 24 randomly, one at a time, and repeat.

- Redraw the diagram and repeat for other two-digit divisors and multiples.

Variation

3 Include divisions where there is a remainder. For example, 49 and 77 for ÷ 24. Ask learners to express the remainder as a whole number.

Two-digit divisors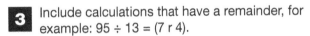

Learning objective

Code	Learning objective
6Ni.05	Estimate and divide whole numbers up to 1000 by [1-digit or] 2-digit whole numbers.

Resources:

paper or mini whiteboard and pen (per pair)

What to do

- On the board, write the calculations: 96 ÷ 16 =, 91 ÷ 13 =, 85 ÷ 17 =, 92 ÷ 23 =.

- Ask learners to work in pairs to solve the four calculations.

- Encourage them to use a range of strategies from mental 'grouping', counting backwards/forwards in multiples on a number line or a trial and improvement method, for example, 92 ÷ 23 = (3 × 23 = 69 (not 92); 4 × 23 = 92 (yes)).

- Confirm the answers as a class (6, 7, 5, 4).

Variation

3 Include calculations that have a remainder, for example: 95 ÷ 13 = (7 r 4).

Revise

Find the error (1)

Learning objective

Code	Learning objective
6Ni.05	Estimate and divide whole numbers up to 1000 by [1-digit or] 2-digit whole numbers.

Resources

paper or mini whiteboard and pen (per pair)

What to do

- On the board, write the following:
  a) $90 \div 18 = 4$, b) $76 \div 19 = 4$, c) $75 \div 24 = 3 \, r \, 3$,
  d) $91 \div 26 = 3 \, r \, 11$, e) $99 \div 28 = 3 \, r \, 15$,
  f) $97 \div 23 = 4 \, r \, 3$.

- Explain to learners that some of the calculations contain errors. Ask them to work in pairs to check the answers using the expanded written method to confirm which of the calculations are incorrect.

- For each calculation, choose a pair of learners to say whether they think the answer is correct or not. Confirm this as a class. (incorrect: a, d, f)

Variation

1 Learners use any strategy they are confident with to decide whether each calculation is correct or not.

Find the error (2)

Learning objective

Code	Learning objective
6Ni.05	Estimate and divide whole numbers up to 1000 by [1-digit or] 2-digit whole numbers.

Resources

paper or mini whiteboard and pen (per pair)

What to do

- On the board, write the following:
  a) $238 \div 34 = 7$, b) $288 \div 36 = 8$, c) $643 \div 43 = 15$,
  d) $846 \div 47 = 17$, e) $897 \div 39 = 27$, f) $783 \div 29 = 27$,
  g) $891 \div 33 = 27$, h) $988 \div 38 = 27$

- Explain to learners that some of the calculations contain errors. Ask them to work in pairs to check the answers using the expanded written method to confirm which of the calculations are incorrect.

- For each calculation, choose a pair of learners to say whether they think the answer is correct or not. Confirm this as a class. (incorrect: c, d, e, h)

Variation

1 Learners use any strategy they are confident with to decide whether each calculation is correct or not.

Revise

Decimal-go-round

Learning objectives

Code	Learning objective
6Nc.01	Count on and count back in steps of constant size, including [fractions and] decimals, and extend beyond zero to include negative numbers.
6Np.01	Understand and explain the value of each digit in decimals (tenths, hundredths and thousandths).

What to do

- Arrange learners in a circle.
- Start a forwards count from a two-place decimal in steps of 0·01, for example, 1·57.
- Ask learners to say what the next number in the sequence is. (1·58)
- At any point, announce a change in the count, for example count back in steps of 0·02 from 5·76 or count forwards in steps of 0·03 from 0·85.

Variation

- Help learners with step-counting by allowing them to repeat the previous number, for example, if a forwards count in steps of 0·02 begins at 2·74, the first learner would say 2·74 then 2·76, the next learner would say 2·76 then 2·78, and so on.

Compose and decompose

Learning objective

Code	Learning objective
6Np.03	Compose, decompose and regroup numbers, including decimals (tenths, hundredths and thousandths).

Resources:

paper or mini whiteboard and pen (per learner)

What to do

- Arrange the learners into pairs. One player is the 'composer', the other is the 'decomposer'.
- The 'composer' writes down a number with two decimal places and describes the number to their partner, with a sentence that begins: 'My number is composed of...'. For example, if the number is 2·73, the player may say: 'My number is composed of 2 ones, 7 tenths and 3 hundredths' or 'My number is composed of 27 tenths and 3 hundredths' or '... 2 ones and 73 hundredths'. The 'decomposer' tries to guess the number and scores a point if successful.

- The 'decomposer' writes down a number with two decimal places and describes it to their partner with a number sentence that begins: 'My number can be decomposed into...'. For example, if the number is 7·38 the player may say: 'My number can be decomposed into 7 ones, 3 tenths and 8 hundredths' or '... 7 ones and 38 hundredths' or '... 738 hundredths'. The 'composer' tries to guess the number and scores a point if successful.
- The winner is the player with the higher score after nine rounds.

Variation

2 Each player gets three clues to guess the number. They score three points for the first 'difficult' clue, two points for the next 'less difficult' clue, and one point for the 'easy clue'.

Revise

Regrouping race 🧑‍🤝‍🧑 ⚠2

Learning objective

Code	Learning objective
6Np.03	Compose, decompose and regroup numbers, including decimals (tenths, hundredths and thousandths).

Resources

1–9 spinner made from Resource sheet 5: 1–6 and 1–9 spinners (per pair); pencil and paper clip for the spinner (per pair); paper or mini whiteboard and pen (per learner)

What to do [TWM.03/04]

* Arrange the learners into pairs. Learners take turns to spin the spinner three times to create a three-digit number with two decimal places for example, 4·83. They record the number.
* Learners have two minutes to record as many different ways of regrouping the number as they can. On the board, write abbreviations for learners to use in their regrouping statements: o (ones), t (tenths) and h (hundredths).
* For example, if the number is 4·83, a learner might record 4·83 = 4o + 8t + 3h = 48t + 3h = 483h = 4o + 83h = 403h + 8t.

* The player with most correct regroupings statements scores a point. If players have the same number of statements, they both score a point.
* The winner is the player with the higher score after five rounds.
* **[TWM.03/04]** Ask: **What is the maximum number of regroupings you can find for a three-digit number with two decimal places?** (5) **Why?**

Variation

3 Ask learners to work with four-digit numbers with two decimal places. They create the numbers by spinning the spinner four times.

Comparing hundredths 🧑‍🤝‍🧑 ⚠2

Learning objective

Code	Learning objective
6Nf.08	Understand the relative size of quantities to compare and order numbers with one or two decimal places, [proper fractions with different denominators and percentages,] using the symbols =, > and <.

What to do

* Divide the class into two teams, A and B.
* Say: **I am going to give each team a decimal number. Compare the two decimals and indicate whether your number is greater or less than the other, by holding your hand up high or down low.**
* Start with a pair of three-digit numbers with two decimal places, for example, Team A: 7·23; Team B: 7·32. (7·23 < 7·32)
* Repeat with different pairs of decimal tenths. Look out for learners who are unable to compare numbers correctly.

Variation

⚠2 Introduce pairs of four-digit numbers with two decimal places, for example, 54·54 and 54·45.

Revise

10, 100 or 1000 times larger

Learning objective

Code	Learning objective
6Np.02	Use knowledge of place value to multiply [and divide] whole numbers and decimals by 10, 100 and 1000.

What to do

- Ask questions that require multiplying whole numbers by 10, 100 or 1000, and multiplying decimals by 10 or 100, for example:
  - **Lucas has a remote-controlled car with a mass of 867 g. What would the mass of the car be if it were 10/100/1000 times larger?**
  - **Ruby has walked a distance of 0·09 km. How far would she have walked if the distance was 10/100 times longer?**

Variation

 Introduce questions that involve money, for example: **Maisie has $2.30 in her account. How much money would she have if the amount was 10/100 times larger?**

10 or 100 times smaller

Learning objective

Code	Learning objective
6Np.02	Use knowledge of place value to [multiply and] divide whole numbers and decimals by 10, 100 and 1000.

What to do

- Ask questions that require dividing whole numbers by 10, 100 or 1000, and dividing decimals by 10 or 100, for example:
  - **Khaled has filled a bathtub with 47 litres of water. How much water would be in the bath if the amount was 10/100/1000 times smaller?**
  - **Serena drives a distance of 607 km. How far would she have driven if the distance was 10/100 times shorter?**

Variation

Introduce questions that involve money, for example: **A house is valued at $470 235. How much would cost if it was $10/$100 cheaper?**

Number – Place value, ordering and rounding

Revise

Rounding targets

Learning objective

Code	Learning objective
6Np.04	Round numbers with 2 decimal places to the nearest [tenth or] whole number.

Resources

paper or mini whiteboard and pen (per learner)

What to do

- Ask a set of questions that test knowledge of rounding. For example:

Say a number that:

- **rounds up to 60 when rounded to the nearest 10**
- **rounds down to 90 when rounded to the nearest 10**
- **rounds up to 3 when rounded to the nearest whole number**
- **rounds down to 7 when rounded to the nearest whole number.**

- Learners write the answers on paper or their whiteboards. They swap whiteboards with a partner to confirm that the answers are correct.

Variation

1 Provide learners with a number line to help them find the numbers to round.

Rounding decimals

Learning objective

Code	Learning objective
6Np.04	Round numbers with 2 decimal places to the nearest tenth or whole number.

Resources

paper or mini whiteboard and pen (per learner)

What to do

- Write a one-digit number on the board, for example, 7.
- Ask learners to write five decimals that would round to this number, for example, 7·1, 6·9.
- Check learners' answers. Record them on a number line on the board, if appropriate.

Variation

2 Use two-digit whole numbers, for example, 38. Ask learners to write five decimals that would round to this number, for example, 38·3, 37·8.

Revise

Decimal fractions

Learning objective

Code	Learning objective
6Nf.01	Understand that a fraction can be represented as a division of the numerator by the denominator (proper and improper fractions).

Resources

mini whiteboard and pen (per learner)

What to do

- On the board, write: 0·43. Ask: **How many decimal places does this number have?** (2) **How do you write the number as a fraction?** ($\frac{43}{100}$)
- Say: **I am going to call out a number with two decimal places. I want you to write the equivalent fraction.**
- Call out 0·59. Ask learners to raise their whiteboards to confirm they are correct. ($\frac{59}{100}$)
- Repeat for other decimals, for example 0·73. ($\frac{73}{100}$)

Variation

 Ask similar questions with numbers less than 10 with two decimal places, for example: **How do you write 4·67 as a fraction?** ($4\frac{67}{100}$)

Highest common factors

Learning objective

Code	Learning objective
6Nf.03	Use knowledge of equivalence to write fractions in their simplest form.

What to do

- On the board, write the numbers 16 and 24 in separate rows.
- Ask volunteer learners to come to the board to list the factors of each number.
- Expect for 16: 1, 2, 4, 8, 16; for 24: 1, 2, 3, 4, 6, 8, 12, 24.
- Invite a learner to the board to circle the common factors for both numbers. (1, 2, 4, 8)
- Ask: **What is the highest common factor for both numbers?** (8)
- Repeat for different pairs of numbers, for example 42 and 36.

Variation

 Use pairs of numbers of greater value, for example 96 and 72.

Number – Fractions, decimals, percentages, ratio and proportion

Revise

Spin and match (1)

Number – Fractions, decimals, percentages, ratio and proportion

Learning objective

Code	Learning objective
6Nf.04	Recognise that fractions, decimals (one or two decimal places) [and percentages] can have equivalent values

What to do 📊

- Display the **Spinner tool** in two windows, one dial labelled: $\frac{1}{2}$, $\frac{1}{10}$, $\frac{3}{10}$, $\frac{7}{10}$, $\frac{1}{100}$, $\frac{3}{100}$, $\frac{7}{100}$; the other dial labelled: 0·5, 0·1, 0·3, 0·7, 0·01, 0·03, 0·07.
- Ask learners to stand up.
- Say: **I am going to spin a fraction on one spinner and a decimal fraction on the other. I want you to sit down if both numbers are equivalent, for example $\frac{1}{2}$ and 0·5.**
- Ask a learner to help operate the second spinner.
- Spin both spinners. Learners sit down if both numbers spun are equivalent.
- If learners are sitting, they should stand up if the numbers spun are equivalent. Continue for ten rounds.

Variation

3 Introduce labels for the extra numbers: $\frac{1}{4}$ and 0·25, $\frac{3}{4}$ and 0·75, $\frac{1}{5}$ and 0·2.

Equivalent fractions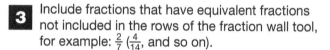

Learning objective

Code	Learning objective
6Nf.04	Recognise that fractions, decimals (one or two decimal places) [and percentages] can have equivalent values.

What to do 📊

- Display the **Fraction wall tool** showing the fraction wall with eighths on the bottom row.
- Point to a fraction, for example $\frac{2}{3}$ and ask learners to find a fraction that is equivalent. ($\frac{4}{6}$)
- Repeat for other fractions.

Variation

3 Include fractions that have equivalent fractions not included in the rows of the fraction wall tool, for example: $\frac{2}{7}$ ($\frac{4}{14}$, and so on).

Revise

Fractions of length measurements

Learning objective

Code	Learning objective
6Nf.02	Understand that proper and improper fractions can act as operators.

What to do

- Write a fraction on the board that represents a length of ribbon, for example: $\frac{1}{4}$ of 20 cm =.
- Ask: **What is a quarter of 20 cm?** (5 cm) **How did you work this out?** ($20 \div 4$ or $\frac{20}{1} \times \frac{1}{4}$)
- Remind learners that a fraction can act as an operator. Finding a unit fraction of a number or a quantity can be found in more than one way, for example, dividing the quantity by the denominator of the fraction or converting the quantity to a fraction and multiplying this number by the fraction.
- On the board, draw the table:

Amount	Fraction of amount		
	$\frac{1}{2}$	$\frac{1}{3}$	$\frac{1}{4}$
24 cm			
36 cm			

- Ask learners to complete the table.

- Choose one of the fraction questions in the table at a time and ask learners to state the answer and how they calculated it.

Answers:

Amount	Fraction of amount		
	$\frac{1}{2}$	$\frac{1}{3}$	$\frac{1}{4}$
24 cm	12 cm	8 cm	6 cm
36 cm	18 cm	12 cm	9 cm

Variation

3 Change the table so that learners are asked to find $\frac{1}{2}$, $\frac{1}{3}$, $\frac{1}{4}$ and $\frac{1}{5}$ of 240 cm and 300 cm.

The answer is...

Learning objective

Code	Learning objective
6Nf.05	Estimate, add and subtract fractions with different denominators.

What to do

- Write a proper fraction on the board, for example $\frac{5}{8}$.
- Ask: **This is the answer. What could the addition or subtraction question have been?** Take responses and confirm answers. Expect answers similar to the following: $\frac{3}{8} + \frac{2}{8}$; $\frac{7}{8} - \frac{2}{8}$.
- Repeat for different fractions, for example $\frac{3}{5}$ or $\frac{7}{10}$.
- Record the questions and the answers on the board.

Variation

2 Ask learners to suggest the fraction that is the 'answer'.

Number – Fractions, decimals, percentages, ratio and proportion

Revise

Multiplying fractions

Learning objective

Code	Learning objective
6Nf.06	Estimate, multiply and divide proper fractions by whole numbers.

What to do

- On the board, write: $\frac{1}{5} \times 4 =$.
- Ask: **How do you multiply a fraction by a whole number?** Take responses.
- Remind learners that you can either divide the whole number by the denominator of the fraction or convert the whole number to a fraction with a denominator of 1 and multiply the numerators together, and then the denominators together.
- Split the class in half. Ask one half to solve the problem on the board using the first method, and the other half of the class, using the second method. Choose learners from each group to state the answer. $(\frac{4}{5})$

- Write: $\frac{1}{8} \times 5 =$. Ask the two groups to switch methods and answer the problem. $(\frac{5}{8})$
- Repeat for other multiplication problems: $\frac{1}{2} \times 3 =$, $\frac{1}{4} \times 6$, $\frac{1}{8} \times 7 =$, $\frac{1}{10} \times 9 =$. $(\frac{3}{2}, \frac{6}{4}, \frac{7}{8}, \frac{9}{10})$

Variation

 Introduce calculations with measurements, for example $\frac{1}{10} \times 5$ kg =.

Fraction stories

Learning objective

Code	Learning objective
6Nf.06	Estimate, multiply and divide proper fractions by whole numbers.

What to do

- Describe a fraction in a story, for example: **I pour $\frac{3}{4}$ of a litre of orange juice into a container. I share the orange juice between three friends. How much orange juice does each person get?**
- Learners work in pairs to write a division calculation that gives the answer $(\frac{3}{4} \div 3 =)$.
- Repeat with a different 'fraction' story, for example: **I have a piece of wood $\frac{2}{3}$ m in length. I saw it into four equal sections. Write a division calculation that will give me the length of each section of wood.** $(\frac{2}{3} \div 4 =)$
- Repeat for other fraction stories.

Variation

Learners invent their own fraction stories for their partner to answer as a division problem.

Number – Fractions, decimals, percentages, ratio and proportion

Revise

Out of $100

Learning objective

Code	Learning objective
6Nf.07	Recognise percentages (1%, and multiples of 5% up to 100%) of shapes and whole numbers.

What to do 📊

- Display the **Number square tool** in two windows, and set up blank 100 grids.
- Explain that each square in the grid represents $1. The whole grid is $100.
- Shade 23 squares in the first grid and 21 squares in the second.
- Ask: **How much money does each set of shaded squares represent?** ($23, $21) **What are these percentages as fractions?** ($\frac{23}{100}$, $\frac{21}{100}$) **Which percentage is greater?** (23%) **What percentage of $100 does each set of shaded squares represent?** (23%, 21%)
- Repeat for different numbers of shaded squares.

Variation

1 Restrict the percentages to 25%, 50%, 75% and multiples of 10%.

Spin and match (2)

Learning objective

Code	Learning objective
6Nf.07	Recognise percentages (1%, and multiples of 5% up to 100%) of shapes and whole numbers.

What to do 📊

- Display the **Spinner tool** in two windows, one dial labelled 20%, 30%, 40%, 50%, 80% and 90%; the other dial labelled: $\frac{2}{10}$, $\frac{3}{10}$, $\frac{4}{10}$, $\frac{5}{10}$, $\frac{8}{10}$ or $\frac{9}{10}$.
- Ask learners to stand up.
- Say: **I am going to spin a fraction on one spinner and a percentage on the other. I want you to sit down if both numbers are equivalent, for example $\frac{2}{10}$ and 20%.**
- Ask a learner to help operate the second spinner.
- Spin both spinners. Learners sit down if both numbers spun are equivalent.
- If learners are sitting, they should stand up if the numbers spun are equivalent. Continue for ten rounds.

Variation

1 Reduce the number of percentage-fraction pairs from six to four.

Revise

Spin and match (3)

Learning objective

Code	Learning objective
6Nf.07	Recognise percentages (1%, and multiples of 5% up to 100%) of shapes and whole numbers.

What to do

- Display the **Spinner tool** in two windows, one dial labelled 10%, 25%, 50%, 60%, 75%, 90%; the other dial labelled: $\frac{1}{10}$, $\frac{1}{4}$, $\frac{1}{2}$, $\frac{3}{5}$, $\frac{3}{4}$, $\frac{9}{10}$.
- Ask learners to stand up.
- Say: **I am going to spin a fraction on one spinner and a percentage on the other. I want you to sit down if both numbers are equivalent, for example $\frac{3}{5}$ and 60%.**
- Ask a learner to help operate the second spinner.

- Spin both spinners. Learners sit down if both numbers spun are equivalent.
- If learners are sitting, they should stand up if the numbers spun are equivalent. Continue for ten rounds.

Variation

1. Reduce the number of percentage-fraction pairs from six to four.

Higher or lower? (1)

Learning objective

Code	Learning objective
6Nf.08	Understand the relative size of quantities to compare and order numbers with one or two decimal places, proper fractions with different denominators and percentages, using the symbols =, > and <.

What to do

- Display the **Spinner tool** in two windows, one dial labelled $\frac{1}{10}$, 30%, $\frac{1}{2}$, $\frac{7}{10}$, 90%; the other dial labelled: $\frac{1}{5}$, 40%, $\frac{3}{5}$, 80%, $\frac{10}{10}$.
- On one dial, click 'spin' and note the percentage or fraction spun.
- Ask learners to predict if the next number spun on the second dial will be higher or lower.
- Spin the next number. Choose a learner to decide if the number spun is higher or lower. Learners score a point if their prediction is correct.
- Play ten rounds. The learner with the highest score is the winner.

Variation

1. Display a chart of fraction and percentage equivalents to help learners compare fractions and percentages.

Revise

Quick sums

Learning objective

Code	Learning objective
5Nf.09	Estimate, add [and subtract] numbers with the same number of decimal places.

Resources

paper or mini whiteboard and pen (per learner)

What to do

- Divide the class into two teams, A and B.
- Display the **Spinner tool** with ten sectors labelled 0 to 9. Spin twice to create a number less than 10 with one decimal place. Record the number on the board.
- Do the same again and record the number on the board as an addition calculation with the first number, for example, 2·9 + 6·5 =.
- Learners add the numbers using any mental strategy, with jottings if required. Strategies might include using a number line, compensation (2·9 + 6·5 = 3 + 6·5 − 0·1 = 3 + 6·4 = 9·4) or decomposition (2·9 + 6·5 = 2 + 6 + 0·9 + 0·5 = 8 + 1·4 = 9·4).

- The first learner to raise their hand and answer correctly scores a point for their team.
- The winner is the team with more points after nine rounds.

Variation

3 Change the numbers to two numbers less than 100 with one decimal place, for example, 24·8 + 38·6 =.

Find the sum

Learning objective

Code	Learning objective
6Nf.09	Estimate, add [and subtract] numbers with the same or different number of decimal places.

Resources

paper or mini whiteboard and pen (per learner)

What to do

- Display the **Spinner tool** with ten sectors labelled 0 to 9. Spin twice to create a number less than 10 with one decimal place. Record the number on the board.
- Draw a column addition layout on the board.
- Write the number from the spinner in the top row of the layout.
- Spin two more times to create a second number less than 10 with one decimal place.
- Write the new number in the second row of the layout.

- Ask learners to use column addition to calculate the sum of the numbers.
- Repeat with different numbers.

Variation

3 Use column addition layouts for the sum of two numbers less than 10 with two decimal places.

Revise

Quick subtractions

Learning objective

Code	Learning objective
6Nf.09	Estimate, [add] and subtract numbers with the same or different number of decimal places.

Resources

paper or mini whiteboard and pen (per learner)

What to do

- Divide the class into two teams, A and B.
- Display the **Spinner tool** with ten sectors labelled 0 to 9. Spin twice to create a number less than 10 with one decimal place. Record the number on the board.
- Do the same again and record the number on the board as a subtraction calculation with the first number so that the larger number is the minuend, for example, 7·2 – 4·9 =.
- Learners subtract the numbers using any mental strategy, with jottings if required. Strategies might include using a number line or compensation (7·2 – 4·9 = 7·2 – 5·0 + 0·1 = 2·2 + 0·1 = 2·3).

- The first learner to raise their hand and answer correctly scores a point for their team.
- The winner is the team with more points after nine rounds.

Variation

3 Change to two numbers less than 100 with one decimal place, for example, 53·4 – 15·6 =.

Find the difference

Learning objective

Code	Learning objective
6Nf.09	Estimate, [add] and subtract numbers with the same or different number of decimal places.

Resources

paper or mini whiteboard and pen (per learner)

What to do

- Display the **Spinner tool** with ten sectors labelled 0 to 9. Spin four times to create two numbers less than 10 with one decimal place.
- Draw a column subtraction layout on the board and write the two numbers in the rows of the layout so that the larger number is on top.
- Ask learners to use column subtraction to calculate the difference between the numbers.
- Repeat with different numbers.

Variation

3 Use column subtraction layouts for the subtraction of two numbers less than 10 with two decimal places.

Number – Fractions, decimals, percentages, ratio and proportion

Revise

Multiplication function machines

Learning objective

Code	Learning objective
6Nf.10	Estimate and multiply numbers with one or two decimal places by 1-digit [and 2-digit] whole numbers.

What to do

- Display the **Spinner tool** with segments labelled 2 to 9.
- On the board, draw a function machine labelled for multiplication.
- Spin the spinner and write the number as the multiplier in the function machine box.
- Spin again for the multiplicand. This is the tenths digit of a number less than one. Write the number spun on the left of the function machine as an input.
- Learners multiply the number by the multiplier mentally and write down the output.

- Repeat for different inputs. Learners show their answer each time.
- Spin the spinner to produce a different multiplier.

Variation

 Spin twice for the multiplicand (input) to make a one-place decimal. For example, if a '3' and a '6' are spun, this makes the number 3·6.

Grid method (1)

Learning objective

Code	Learning objective
6Nf.10	Estimate and multiply numbers with one or two decimal places by 1-digit [and 2-digit] whole numbers.

Resources

paper or mini whiteboard and pen (per learner)

What to do

- Display the **Spinner tool** with ten sectors labelled 0 to 9.
- Spin once to create a one-digit multiplier.
- Spin twice to create the multiplicand, a number less than 10 with one decimal place. Record the multiplication on the board, for example, 5·3 × 6.
- Ask learners to use the grid method to calculate the product.
- Repeat with different numbers.

Variation

Spin three times to create the multiplicand, a number less than 10 with two decimal places.

Number – Fractions, decimals, percentages, ratio and proportion

Revise

Related facts 👥 🔺2

Learning objective

Code	Learning objective
6Nf.10	Estimate and multiply numbers with one or two decimal places by 1-digit [and 2-digit] whole numbers.

Resources

paper or mini whiteboard and pen (per learner)

What to do 📊

- Display the **Spinner tool** with ten sectors labelled 0 to 9.
- Spin once to create a one-digit multiplier.
- Spin twice to create the multiplicand, a number less than 10 with one decimal place. Record the multiplication on the board, for example, 3·7 × 4.
- Learners use any strategy to calculate the answer, for example, 3·7 × 4 = (3 × 4) + (0·7 × 4) = 12 + 2·8 = 14·8.

- Alongside, they write and solve two related calculations of the form O.t × T and 0.th × O using knowledge of place value, for example, 3·7 × 40 = 148 and 0·37 × 4 = 1·48.
- Learners write the answers to each of the calculations.

Variation

1 Spin once to create a one-digit multiplier and once to create the multiplicand, a one-place decimal less than 1, for example, 0·8 × 4 =.
Alongside, learners write and solve one related calculation of the form 0.th × T, for example 0·8 × 40 = 32.

Grid method (2) 👥 🔺2

Learning objective

Code	Learning objective
6Nf.10	Estimate and multiply numbers with one or two decimal places by [1-digit and] 2-digit whole numbers.

Resources

paper or mini whiteboard and pen (per learner)

What to do 📊

- Display the **Spinner tool** with ten sectors labelled 0 to 9.
- Spin twice to create a two-digit multiplier.
- Spin twice to create the multiplicand, a number less than 10 with one decimal place. Record the multiplication on the board, for example, 6·8 × 27 =.
- Ask learners to use the grid method to calculate the product.
- Repeat with different numbers.

Variation

3 Spin four times to create the multiplicand, a number less than 100 with two decimal places.

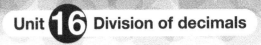

Revise

Estimate and divide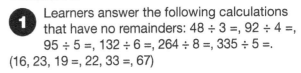

Learning objective

Code	Learning objective
6Nf.11	Estimate and divide numbers with one or two decimal places by whole numbers.

Resources

paper or mini whiteboard and pen (per learner)

What to do

- On the board, write: $54 \div 4 =$, $86 \div 5 =$, $87 \div 6 =$, $173 \div 4 =$, $261 \div 6 =$, $334 \div 8 =$.
- Ask learners to estimate the quotient of each division.
- Then working in pairs, partners race against each other to be the first to answer all the calculations correctly. They write the quotients as mixed numbers.
- Learners check their answers against the estimates and confirm the winner. ($13\frac{1}{2}$, $17\frac{1}{5}$, $14\frac{1}{2}$, $43\frac{1}{4}$, $43\frac{1}{2}$, $41\frac{3}{4}$)

Variation

1 Learners answer the following calculations that have no remainders: $48 \div 3 =$, $92 \div 4 =$, $95 \div 5 =$, $132 \div 6 =$, $264 \div 8 =$, $335 \div 5 =$.
(16, 23, 19 =, 22, 33 =, 67)

Quick division

Learning objective

Code	Learning objective
6Nf.11	Estimate and divide numbers with one or two decimal places by whole numbers.

Resources

paper or mini whiteboard and pen (per learner)

What to do

- On the board, write: $6 \cdot 6 \div 2 =$, $8 \cdot 8 \div 4 =$, $9 \cdot 9 \div 3 =$, $66 \cdot 6 \div 3 =$, $88 \cdot 8 \div 4 =$, $99 \cdot 9 \div 3 =$.
- Say: **These calculations are easily solved mentally. How quickly can you find the quotients?**
- Then working in pairs, partners race each other to be the first to record the answers to all six questions.
- The first person to raise their hand lists all of the quotients. Confirm the answers as a class. (3·3, 2·2, 3·3, 22·2, 22·2, 33·3)

Variation

3 Learners answer the following calculations: $333 \cdot 33 \div 3 =$, $888 \cdot 88 \div 4 =$, $666 \cdot 66 \div 3 =$, $555 \cdot 55 \div 5 =$, $999 \cdot 99 \div 3 =$, $444 \cdot 44 \div 2 =$. (111·11, 222·22, 222·22, 111·11, 333·33, 222·22)

Number – Fractions, decimals, percentages, ratio and proportion

Revise

Division stories

Learning objective

Code	Learning objective
6Nf.11	Estimate and divide numbers with one or two decimal places by whole numbers.

Resources

paper or mini whiteboard and pen (per learner)

What to do

- Say some 'I share…' stories where learners have to find the decimal quotient. For example:
  - **I share $9.60 between three people. How much does each person get?**
  - **I share 8·6 kg of sand equally between three buckets. How much sand is in each bucket?**
  - **I share 76·8 ml of orange juice equally between three cups. How much juice is in each cup?**
  - **I cut 248·4 cm of ribbon into six equal strips. How long is each strip?**

- Learners use any preferred strategy, including mental methods.
- Choose learners to explain how they calculated the quotient.

Variation

 Invite learners to the front of the class to read out their own 'I share…' stories.

Compare the methods

Learning objective

Code	Learning objective
6Nf.11	Estimate and divide numbers with one or two decimal places by whole numbers.

Resources

paper or mini whiteboard and pen (per learner)

What to do

- On the board, write: 6·21 ÷ 3 =, 8·32 ÷ 4 =, 48·42 ÷ 6 =, 64·56 ÷ 8 =.
- Learners work in pairs to find the quotient for each calculation, one partner uses a mental method or partitioning, and the other uses the expanded written method.
- They complete the divisions and confirm that they have the same quotients.

Variation

 Provide word problems that involve dividing decimals by whole numbers, for example: $9.27 is split equally between three people. How much does each person get? ($3.09)

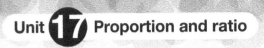

Revise

Shape proportions

Learning objective

Code	Learning objective
6Nf.12	Understand the relationship between two quantities when they are in direct proportion.

What to do

- Display the **Pattern tool**. Place five circles and three triangles on the jumper.
- Ask: **What proportion of the shapes are circles?** ($\frac{5}{8}$) **Triangles?** ($\frac{3}{8}$)
- Place seven hexagons and five pentagons.
- Ask: **What proportion of the shapes are hexagons?** ($\frac{7}{12}$) **Pentagons?** ($\frac{5}{12}$)
- Repeat for different combinations of shapes.

Variation

2 Introduce ratios of shapes that can be simplified, for example, six circles and two triangles.

Scaling triangles

Learning objective

Code	Learning objective
6Nf.12	Understand the relationship between two quantities when they are in direct proportion.

What to do

- Draw a right-angled triangle on the board with the following sides: 5 cm, 4 cm and 3 cm.
- Display the **Spinner tool**. Label sectors 2 to 10. Explain that the number spun will be a multiplier.
- Spin the spinner and ask learners to multiply the side lengths by the multiplier spun.
- Draw and label the new-sized triangle. **[TWM.05]** Ask: **What is the relationship between the side lengths of the two triangles?** Take responses. Praise learners who describe the relationship as 'directly proportional'. Establish that the two triangles are similar: their corresponding angles are identical and their corresponding sides are in a proportional relationship that can be described by the multiplier.

- Spin a second scale factor, followed by a third, and ask learners to calculate the new side lengths each time.

Variation

3 After increasing the size of the triangle, ask learners to suggest a scale factor that is a fraction that will proportionally reduce the side lengths but keep them as whole numbers.

Number – Fractions, decimals, percentages, ratio and proportion

Revise

Simplifying ratios

Learning objective

Code	Learning objective
6Nf.13	Use knowledge of equivalence to understand and use equivalent ratios.

Resources

1–20 number cards from Resource sheet 2: 0–100 number cards (per pair)

What to do

- Learners shuffle the pack of cards.
- They take turns to turn over two cards.
- They arrange and read the numbers as a ratio. If possible, they read the ratio in its simplest form.
- For example, if they reveal numbers 2 and 8, they read the ratio as '2 for every 8' then simplify to '1 for every 4'.

Variation

 Learners simplify the ratio if possible, then express the ratio in a form double, triple and four times the size.

Equivalent fraction chains

Learning objective

Code	Learning objective
6Nf.13	Use knowledge of equivalence to understand and use equivalent ratios.

What to do

- On the board, write: $\frac{1}{2} =$.
- Ask learners to come to the board to write an equivalent fraction, for example: $\frac{1}{2} = \frac{2}{4}$.
- They continue to extend the chain of equivalent fractions for as long as possible.
- Introduce a new fraction, for example: $\frac{3}{4} =$.

Variation

 Introduce less common fractions, for example: $\frac{3}{11}$ and $\frac{9}{13}$.

Revise

Quadrilaterals

Learning objective

Code	Learning objective
6Gg.01	Identify, describe, classify and sketch quadrilaterals, including reference to angles, symmetrical properties, parallel sides and diagonals.

What to do [TWM.05]

- Display the **Geometry set tool**, showing a parallelogram and a trapezium.
- Remind learners that a four-sided shape is called a quadrilateral.
- Say: **Look at the two quadrilaterals. Tell your partner one property both shapes have in common and one difference.**

- Repeat for a different pair of quadrilaterals.
- Invite learners to share the properties they discuss.

Variation

1 Ask learners to make comparisons between the square, rectangle and T-shape only.

Drawing a circle

Learning objective

Code	Learning objective
6Gg.02	Know the parts of a circle: – centre – radius – diameter – circumference.

Resources

pencil (per learner); string (per learner); card (per learner); adhesive tape or a pin (per learner)

What to do

- Give the class the following instructions on how to draw a circle:
  - Tie one end of the string around a pencil.
  - Pin down the other end of the string where you want the centre of your circle to be (make sure the string rotates freely as opposed to enveloping around the pin).
  - Keeping the string taught and the pencil upright, use the string to outline the circle.
- Learners practise drawing circles. Demonstrate how to adjust the length of the string to draw circles of different sizes.

Variation

3 Ask learners to draw a circle with a width of 10 cm. Then challenge them to draw circles half the width and double the width.

Revise

Circles in a row 👥 ⚁

Learning objective

Code	Learning objective
6Gg.11	Construct circles of a specified radius or diameter.

Resources

pencil (per learner); string (per learner); card (per learner); adhesive tape or pin (per learner)

What to do

- Give the class the following instructions:
  ○ Draw a line across a piece of paper.
  ○ Draw a circle at one end of the line with its centre on the line. (Learners should use the method described in the Revise activity *Drawing a circle*.)
  ○ Draw a few circles with the same diameter along a line.
  ○ Keep your string the same length and draw another circle with the centre where the first circle crossed the line (see diagram).

- Learners repeat as many times as they wish.

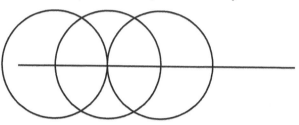

Variation

1 Learners practise drawing multiple circles across the page. They try to make some of them overlap.

Rotating objects 👥 ⚁

Learning objective

Code	Learning objective
6Gg.08	Identify rotational symmetry in familiar shapes, patterns or images with maximum order 4. Describe rotational symmetry as 'order *x*'.

Resources

various classroom objects: ruler, small circular hoop, equilateral triangular-shaped object (per group)

What to do

- Arrange the learners round a table of objects.
- Have a volunteer gradually horizontally rotate each object in turn. Ask learners to call out 'stop' when they see each object resume the same shape and position as before the rotation began.
- Elicit that the ruler looks the same after every half rotation, the triangle every third of a full rotation and a circle looks the same, no matter how much it is turned.

- Ask learners to find other objects and explore how often they look the same when rotated through a full circle.

Variation

2 Ask learners to find an object that looks the same every quarter rotation.

Revise

Guess the 3D shape

Learning objective

Code	Learning objective
6Gg.04	Identify, describe [and sketch] compound 3D shapes.

Resources

3D shapes: cube, cuboid, sphere, cone, cylinder, one type of pyramid (per class); bag (per class)

What to do [TWM.06]

- Choose a 3D shape and place it inside a bag without learners seeing it.
- Provide learners with clues about the faces of the shape, for example: **It has six faces, all of which are rectangles (or 4 rectangular faces and 2 square faces).** (cuboid)

- Ask learners to guess the name of the shape and praise the learner who is first to identify the shape.
- Repeat for different shapes.

Variation

 Include different types of pyramids and prisms. Provide clues about the number of edges and vertices.

Sketch the shape

Learning objective

Code	Learning objective
6Gg.04	Identify, describe and sketch compound 3D shapes.

Resources

3D shapes: cube, cuboid, sphere, cone, cylinder, one type of pyramid (per class); paper or mini whiteboard and pen (per learner)

What to do

- Name a 3D shape and ask learners to sketch it.
- Reveal the actual shape and ask learners to comment on the accuracy of their sketches.
- Repeat for different shapes.

Variation

1 Display the shape for learners to sketch.

Revise

Identify the net

Learning objective

Code	Learning objective
6Gg.06	Identify [and sketch] different nets for cubes, cuboids, prisms and pyramids.

Resources

paper or mini whiteboard and pen (per learner)
(for variation)

What to do

* Prepare the **Nets tool** as follows before showing it to learners: Display two shapes on screen, side-by-side and fast forward the videos to show the nets of the shapes.
* Show learners the two nets and ask: **Which 3D shapes are these the nets for?**
* Take answers and play the animations from the beginning to see if learners are correct.
* Repeat for another pair of 3D shapes.

Variation

2 Display the **Nets tool**. Select a 3D shape and pause the video at the beginning. Ask learners to sketch a net of the shape. Complete watching the video to see if their net is correct.

Drawing a dice

Learning objective

Code	Learning objective
6Gg.06	Identify and sketch different nets for cubes, cuboids, prisms and pyramids.

Resources

1–6 dice – preferably a large dice (per class); paper or mini whiteboard and pen (per learner) (for variation)

What to do

* Display the **Shape set tool** showing a cube.
* Hold up a dice. Show learners how the numbers on opposite sides of the dice add up to 7.
* Ask learners to say which faces of the cube on display would have the numbers 1 to 6 so that they obey the rules.
* Correct any misunderstandings.

Variation

3 Learners draw the net of a cube, labelling the faces correctly with numbers 1 to 6 as if it were to be made into a dice.

Geometry and Measure – Geometrical reasoning, shapes and measurements

Revise

Book angles

Learning objective

Code	Learning objective
6Gg.09	Classify, estimate, measure and draw angles.

Resources

book (per pair)

What to do

- Distribute books.
- Ask learners to open their books to show angles that are acute, right, obtuse and reflex.
- Play 'Simon says'. Say: **Simon says show me an acute/obtuse/right/straight/reflex angle.**
- Continue with instructions, but vary the language. For example, say: **Simon says show me an angle** that is not an obtuse angle/Show me a reflex angle that is less than 270°.

Variation

3 Increase the complexity of the language. Say, for example: **Simon says show me an angle that is about halfway between an obtuse and a full 360° angle/Show me an angle that is about halfway between 0° and a right angle.**

Estimating angles

Learning objective

Code	Learning objective
6Gg.09	Classify, estimate, measure and draw angles.

What to do

- Display the **Geometry set tool** showing the green triangle.
- Point to an angle. Say: **Estimate the size of this angle and record it.**
- Use the protractor to measure the angle. Remind learners of the procedure involved.
- Announce the size of the angle. Ask: **How close was your estimate? Award yourself one point for an estimate within 10° and two points for an estimate within 5°.**
- Repeat for other angles in the same shape and other shapes.
- The winner is the learner with the most points.

Variation

1 Learners say whether the angle is acute, obtuse or a right angle. Use the protractor to measure the angle. They score one point for each correct type of angle.

Revise

Angles on a straight line

Learning objective

Code	Learning objective
6Gg.10	Know that the sum of the angles in a triangle is 180°, and use this to calculate missing angles in a triangle.

What to do

- On the board, draw a straight line. The line is one arm of an angle. Mark a point in the centre of the line, the vertex of the angle. Draw a line from the vertex to form the angle.
- Remind learners that angles on a straight line add up to 180°.
- Draw angle markers.

- Label one angle, 53°. Ask: **What is the size of the other angle?** (127°)
- Repeat for different angles.

Variation

3 Draw more than two angles meeting at a point on a straight line. Label the size of all angles apart from one.

Guess the triangle

Learning objective

Code	Learning objective
6Gg.10	Know that the sum of the angles in a triangle is 180°, and use this to calculate missing angles in a triangle.

What to do [TWM.06]

- Display the **Geometry set tool**.
- Say: **I will describe a triangle by its properties. I want you to identify the type of triangle I am describing.**
  - **The triangle has one right angle.** (right-angled isosceles triangle)
  - **The triangle has three equal angles.** (equilateral triangle)
  - **The triangle has only two equal angles.** (isosceles triangle)
  - **The triangle has all unequal angles.** (scalene triangle)

- Each time, invite a learner to come to the board and identify the triangle that matches the description.

Variation

1 Include descriptions of side lengths alongside the angle clues.

Revise

Ordering units of time

Learning objective

Code	Learning objective
6Gt.01	Convert between time intervals expressed as a decimal and in mixed units.

Resources

mini whiteboard and pen (per learner)

What to do

- On the board, write: decades, seconds, days, centuries, weeks, minutes, months, years.
- Challenge learners to write the units of time in order, from least to greatest.
- Ask them to raise their mini whiteboards to confirm they have found the correct order.

Variation

3 Give each unit a value, for example: 307 seconds, 5 minutes, 49 hours, 2 days, and so on.

Train arrivals

Learning objective

Code	Learning objective
6Gt.01	Convert between time intervals expressed as a decimal and in mixed units.

What to do 📊

- Display the **Clock tool** displaying one 12-hour analogue clock and one 24-hour digital clock. Set the time on both clocks to 12:23.
- Say: **A train is due to arrive at a station at 12:23 but may be early or late. I will tell you how early or late it is and I want you to write the adjusted time of arrival.**
- Say: **The train is running 15 minutes late.** Ask: **What time is it expected to arrive?** (12:38)

- Say: **The train is running 10 minutes early.** Ask: **What time is it expected to arrive?** (12:13)
- Continue asking similar questions.
- For each time adjustment, ask a volunteer to come to the board to change both clocks to the new time.

Variation

3 Give early/late time adjustments in mixed hours and minutes.

Revise

Capacity conversions

Learning objective

Code	Learning objective
6Gg.05	Understand the difference between capacity and volume.

Resources

mini whiteboard and pen (per learner)

What to do

- On the board, write three decimal measurements in litres to two places, for example: 3·57 ℓ, 8·09 ℓ and 16·73 ℓ.
- Ask learners to convert and record each measurement as mixed units of litres and millilitres. For example: 7·07 ℓ is 7 ℓ 70 ml.
- Reveal the correct answers and discuss the calculations involved.

Variation

Learners convert mixed units given in litres and millilitres to two-place decimals, for example: 4 ℓ 400 ml (4·4 ℓ) or 6 ℓ 850 ml (6·85 ℓ).

Find the volume

Learning objective

Code	Learning objective
6Gg.05	Understand the difference between capacity and volume.

Resources

mini whiteboard and pen (per learner)

What to do

- Display the **Capacity tool** set to the 2000 ml measuring container and fill it to any numbered division, for example 1300 ml.
- Give learners ten seconds to record the volume of water in the container on their whiteboards
- Ask the learners to hold up their whiteboards and confirm the readings.
- Repeat for different volumes, for example 750 ml, 1350 ml.

- Then, on the board write: 1650 ml.
- Invite a learner to the board to fill the **Capacity tool** container with the correct volume of water.
- Repeat for different written volumes expressed as decimals or mixed units, for example, 1·9 ml, 1 ℓ 350 ml, 1·75 ml.

Variations

Repeat with containers of different sizes

Introduce volume given as a percentage or fraction, for example, 'Fill the container $\frac{3}{4}$ full. What is the volume?', 'Fill the container 30% full. What is the volume?'

Revise

Double the size, half the size

Learning objective

Code	Learning objective
6Gg.03	Use knowledge of area of rectangles to estimate and calculate the area of right-angled triangles.

Resources

paper or mini whiteboard and pen (optional)
(per learner)

What to do

- Remind learners that they can find the area of a rectangle by multiplying the length by the width.
- Ask: **What is the area of a rectangle 2 cm by 6 cm?** Expect learners to find the answer, 12 cm².
- Ask: **What is the area of a rectangle with sides double the length of a 2 cm by 6 cm rectangle?** (48 cm²)

- Ask: **What is the area of a rectangle with sides half the length of a 2 cm by 6 cm rectangle?** (3 cm²)
- Repeat for other dimensions, for example a rectangle 8 cm by 12 cm.

Variation

3 Include decimal lengths and widths (3, 4·5), (5·5, 8).

Field areas

Learning objective

Code	Learning objective
6Gg.03	Use knowledge of area of rectangles to estimate and calculate the area of right-angled triangles.

Resources

paper or mini whiteboard and pen (optional)
(per learner)

What to do

- On the board, draw a rectangle and label the sides 8 m and 20 m.
- Say: **The rectangle outlines a field. What is the area of the field?** (160 m²)
- Repeat for fields with other dimensions: 20 m and 40 m (800 m²); 15 m and 30 m (450 m²).

Variation

3 Ask learners to determine the missing dimension of a field given its area and one dimension.

Revise

Guess the net

Learning objective

Code	Learning objective
6Gg.07	Understand the relationship between area of 2D shapes and surface area of 3D shapes.

Resources

unfolded and flattened cereal box (per class)
(for variation)

What to do [TWM.06]

- Say: **I am thinking of the net for a shape. The net is a pattern of six rectangles made of three pairs that have different dimensions. What is the shape?** (a cuboid)

- Say: **I am thinking of the net for a shape. The net is a pattern of six squares all of which have identical dimensions. What is the shape?** (a cube)

- Repeat for other prisms, for example a triangular prism.

Variation

2 Show learners an unfolded and flattened cereal box and ask them to identify which shape it is the net for (a cuboid).

Revise

Kick the ball

Learning objective

Code	Learning objective
6Gp.01	Read and plot coordinates including integers, fractions and decimals, in all four quadrants (with the aid of a grid).

What to do

- Display the **Coordinates tool** set to one quadrant but with the data table not shown.
- Plot points at four different coordinate positions on the grid – these represent footballs.
- Say: **I am going to call out a set of coordinates. If there is a football in this location, then put your hand up.**
- Call out three coordinates, of which the last set identifies the position of a ball. Learners put up their hands.

- Repeat for different sets of coordinates, three of which identify the positions of the remaining three balls.

Variation

 Plot the four points at the vertices of a quadrilateral and ask learners to identify the shape.

Find the treasure

Learning objective

Code	Learning objective
6Gp.01	Read and plot coordinates including integers, fractions and decimals, in all four quadrants (with the aid of a grid).

What to do

- Divide the class into two teams, Team A and Team B.
- Display the **Coordinates tool** set to four quadrants.
- Secretly choose ten points on the grid to be the location of a treasure chest and write them down on a slip of paper. All the points should have x-coordinates and y-coordinates between 5 and –5.
- Say: **I have hidden ten treasure chests at points on the grid. The x- and y-coordinates are all between 5 and –5.** Ask: **Can you find these 'secret' locations?**
- Learners in each team take turns to call out a set of coordinates and plot them on the grid.

- Call out 'treasure' and reveal the slip of paper when a learner plots the point where a chest is located.
- The winner is the first team to find three treasure chests.

Variation

Work with coordinates in two quadrants only, for example I and II, or I and IV.

Geometry and Measure – Position and transformation

Geometry and Measure – Position and transformation

Revise

Around the shape

Learning objective

Code	Learning objective
6Gp.02	Use knowledge of 2D shapes and coordinates to plot points to form lines and shapes in all four quadrants.

What to do

- Display the **Coordinates tool** set to four quadrants.
- Plot a point in any quadrant on the grid.
- Divide the class into two teams.
- Name a shape, for example a square, and the length of its sides.
- The teams take turns to provide directions that will move the point to form the vertices of a square, for example: 'right four' or 'down four'.
- Repeat with a rectangle.

Variation

The team must provide all the instructions to complete the square before they plot the points.

Completing quadrilaterals

Learning objective

Code	Learning objective
6Gp.02	Use knowledge of 2D shapes and coordinates to plot points to form lines and shapes in all four quadrants.

What to do

- Display the **Coordinates tool** set to four quadrants.
- Plot these points on the grid: A (–3, 4), B (3, 4), C (5, 2), D (–3, –2), E (5, –4).
- Say: **The points A, B and D are three vertices of a quadrilateral.**
- Ask: **What are the coordinates of the fourth point that will make a square?** (3, –2)
- Say: **The points B, C and D are three vertices of a quadrilateral.**
- Ask: **What are the coordinates of the fourth point that will make a rectangle?** (–1, –4)

- Say: **The points B, C and E are three vertices of a quadrilateral.**
- Ask: **What are the coordinates of the fourth point that will make a parallelogram?** (3, –2)

Variation

Say: **The points (3, 4) and (5, 2) are two of the vertices of a square.** Ask: **What are the possible coordinates for the other two vertices?**

Revise

Translating shapes

Learning objective

Code	Learning objective
6Gp.03	Translate 2D shapes, identifying the corresponding points between the original and the translated image, on coordinate grids.

What to do

- Display the **Coordinates tool** set to one quadrant.
- Plot a triangle ABC with coordinates A (1, 1), B (2, 3), C (3, 1). Then plot an image of the triangle in another position.
- Ask learners to identify the translation in terms of up/down and left/right movements.
- Repeat for other shapes in different positions under different translations.

Variation

3 Ask learners to translate the shape two or three times.

Reflecting shapes

Learning objective

Code	Learning objective
6Gp.04	Reflect 2D shapes in a given mirror line (vertical, horizontal and diagonal), on square grids.

Resources

squared paper (per pair) ruler (per pair)

What to do

- Provide pairs of learners with squared paper and a ruler. Working together learners draw a 6 by 6 square and then draw a horizontal line and a vertical line so that the square is divided into 3 by 3 quadrants.
- Ask them to shade three or four cells in the top left quadrant to form a simple pattern. Learners reflect the pattern across the vertical mirror line, and then reflect this image across the horizontal mirror line. Finally, they reflect the second image across the vertical mirror line (see diagram).

- Choose pairs to share their reflection patterns with the class.

Variation

3 Learners use a 12 by 12 square to draw reflection patterns.

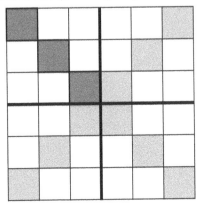

Geometry and Measure – Position and transformation

Geometry and Measure – Position and transformation

Revise

What's the shape?

Learning objective

Code	Learning objective
6Gp.02	Use knowledge of 2D shapes and coordinates to plot points to form lines and shapes in all four quadrants.

What to do

- Display the **Rotate and reflect tool**, with a horizontal mirror line.
- Place the blue square adjacent to the mirror line.
- Ask: **When the 3 by 3 square is reflected, what will the combined shape be?** (a rectangle) **What will its length and height be?** (3 by 6)
- Now set a vertical mirror line next to the 3 by 6 rectangle.
- Ask: **When the rectangle is reflected, what will the combined shape be?** (a square) **What will its length and height be?** (6 by 6)
- Repeat for other shapes, such as a right-angled isosceles triangle.

Variation

3 Show two mirror lines arranged at right angles. Ask learners to reflect shapes positioned adjacent to one mirror line across both lines. Ask: **What is the combined shape? What will its length and height be?**

Quarter turns

Learning objective

Code	Learning objective
6Gp.02	Use knowledge of 2D shapes and coordinates to plot points to form lines and shapes in all four quadrants.

What to do

- Ask learners to stand and point their arm in a named direction, for example, 'straight ahead towards the whiteboard' or 'directly to their right or left towards a window'. Remind learners of the meaning of the terms 'clockwise' and 'anticlockwise'.
- Ask them to make a quarter turn clockwise.
- Repeat for a second quarter turn clockwise, then a quarter turn anticlockwise.
- Mark where learners are pointing.
- Ask: **Can you reach the same point if you make a quarter turn anticlockwise?** (yes)
- Ask: **How many quarter turns clockwise will be the same as a quarter turn anticlockwise?** (three)

Variation

2 Ask learners to point to the corner of the room before making quarter turns. This will help learners establish that a quarter turn can be made from any starting point.

Revise

Cycling or scootering?

Learning objectives

Code	Learning objective
6Ss.01	Plan and conduct an investigation and make predictions for a set of related statistical questions, considering what data to collect (categorical, discrete and continuous data).
6Ss.02	Record, organise and represent categorical, discrete and continuous data. Choose and explain which representation to use in a given situation: - Venn and Carroll diagrams [- frequency diagrams for continuous data - tally charts and frequency tables - line graphs - bar charts - scatter graphs - waffle diagrams and pie charts - dot plots.]
6Ss.04	Interpret data, identifying patterns, within and between data sets, to answer statistical questions. Discuss conclusions, considering the sources of variation and check predictions.

What to do

- Display the **Venn diagram tool**. Explain to learners that they are going to complete a class survey that asks the statistical question: 'What do most learners in our class prefer, cycling or scootering, both or neither?'
- Choose learners to comment on what they think the results will show: **Which category will receive the most votes? The least votes? Why?**
- Go around the class and ask each learner to pick a category: cycling or scootering, both or neither?

- Record the response by placing a circle in the correct part of the Venn diagram. If the category 'neither' is chosen, the circle should be placed inside the rectangle (universal set) but outside the overlapping sets.
- When all the data is entered, ask learners to say what conclusions they can draw from the results and how well they did with their predictions.

Variation

Change the statistical question to: 'What do most learners in our class prefer, gymnastics or dance, both or neither?'

Finding information

Learning objectives

Code	Learning objective
6Ss.01	Plan and conduct an investigation and make predictions for a set of related statistical questions, considering what data to collect (categorical, discrete and continuous data).
6Ss.02	Record, organise and represent categorical, discrete and continuous data. Choose and explain which representation to use in a given situation: - Venn and Carroll diagrams [- frequency diagrams for continuous data - tally charts and frequency tables - line graphs - bar charts - scatter graphs - waffle diagrams and pie charts - dot plots.]
6Ss.04	Interpret data, identifying patterns, within and between data sets, to answer statistical questions. Discuss conclusions, considering the sources of variation and check predictions.

What to do

- Display the **Bar charter tool**. Explain to learners that they are going to complete a class survey that asks the statistical question: 'How do most learners in our class prefer to find out information about the world: "from books", "from websites", "from newspapers", "from talking to adults"?'
- Choose learners to comment on what they think the results will show: **Which category will receive the most votes? The fewest votes? Why?**

- Go around the class and record the data in a frequency table. Populate the **Bar charter tool** with the results.
- Choose learners to state two conclusions that they can draw from the data. Ask them to say how well they did with their predictions.

Variation

Change the statistical question to: 'What is your favourite school subject? Science, English, Maths, Art, History or PE?'

Statistics and Probability – Statistics

Statistics and Probability – Statistics

Revise

Predicting the weather

Learning objectives

Code	Learning objective
6Ss.01	Plan and conduct an investigation and make predictions for a set of related statistical questions, considering what data to collect (categorical, discrete and continuous data).
6Ss.02	Record, organise and represent categorical, discrete and continuous data. Choose and explain which representation to use in a given situation: - Venn and Carroll diagrams - tally charts and frequency tables - bar charts - waffle diagrams and pie charts [- frequency diagrams for continuous data - line graphs - scatter graphs - dot plots.]
6Ss.04	Interpret data, identifying patterns, within and between data sets, to answer statistical questions. Discuss conclusions, considering the sources of variation and check predictions.

What to do 📊

- Remind learners that a waffle diagram is like a 100 square that uses squares to represent percentages.
- Display the **Number square tool** set to 'Hide all' and 'Colour square'. Colour 67 squares blue, 24 squares red and 9 squares yellow.
- Explain that the waffle diagram shows the percentages of people who voted for what they thought the weather would be like tomorrow: blue indicating mild, cool weather; red indicating stormy weather; and yellow indicating warm, sunny weather.

- Ask learners questions that involve data comparison, for example: **What is the difference in percentage between people predicting sunny weather and those predicting stormy weather? If 200 people voted in the survey, how many people predicted that the weather would be mild?**

Variation

 Change the waffle diagram to show the percentages of people who like different PE activities, for example: games, gymnastics or dance.

Find the median

Learning objective

Code	Learning objective
6Ss.03	Understand that the mode, median, mean and range are ways to describe and summarise data sets. Find and interpret the mode (including bimodal data), median, mean and range, and consider their appropriateness for the context.

What to do

- Remind learners that to find the median of a data set, you put the numbers in ascending order and pick the middle number.
- On the board, write: 98 cm, 86 cm, 90 cm, 84 cm, 97 cm, 93 cm, 91 cm, 99 cm, 88 cm.
- Ask learners to find the median of the data set. (91)
- Repeat for a second data set: 495, 437, 581, 449, 452, 516, 478. (478)

Variation

 Ask learners to find the mode of the following data set: 13, 18, 13, 14, 13, 16, 14, 21, 13. (13)

Revise

Time before bed

Learning objectives

Code	Learning objective
6Ss.01	Plan and conduct an investigation and make predictions for a set of related statistical questions, considering what data to collect (categorical, discrete and continuous data).
6Ss.02	Record, organise and represent categorical, discrete and continuous data. Choose and explain which representation to use in a given situation: frequency diagrams for continuous data
6Ss.04	Interpret data, identifying patterns, within and between data sets, to answer statistical questions. Discuss conclusions, considering the sources of variation and check predictions.

Resources

paper (per pair); squared paper or Resource sheet 30: 2 cm squared paper (per pair); ruler (per pair); coloured pencil (per pair)

What to do

- Prior to the activity, write on the board the question: 'What is the most common amount of time learners have between coming home from school and going to bed?'
- Learners work in pairs. They construct a tally chart and collect the data from the class responding to the question: 'How long was your evening: the time between coming home from school and going to bed?' The results should be given as a decimal, for example, 4·5 ($4\frac{1}{2}$ hours), 4·75 ($4\frac{3}{4}$ hours), 5·25 ($5\frac{1}{4}$ hours).

- Learners construct a frequency table for the data deciding on suitable intervals for the horizontal (number of hours) axis, such as 3·5–4·5, 4·5–5·5, 5·5–6·5. Remind learners that numbers at the end of each interval go in the lower value group.
- They then use the data to construct a frequency diagram and answer the question on the board.

Variation

2 Change the statistical question to: 'How much time do learners spend working on homework each week?'

Revise

Greenhouse temperature

Learning objectives

Code	Learning objective
6Ss.01	Plan and conduct an investigation and make predictions for a set of related statistical questions, considering what data to collect (categorical, discrete and continuous data).
6Ss.02	Record, organise and represent categorical, discrete and continuous data. Choose and explain which representation to use in a given situation: line graph.
6Ss.04	Interpret data, identifying patterns, within and between data sets, to answer statistical questions. Discuss conclusions, considering the sources of variation and check predictions.

What to do

- Maisie wants to know the time of the day that the temperature in her greenhouse is at its highest. Every hour, she reads a thermometer in the greenhouse and records the measurement in a table.

Time	Temperature (°C)
12 p.m.	21
1 p.m.	22
2 p.m.	24
3 p.m.	28
4 p.m.	34
5 p.m.	36
6 p.m.	35
7 p.m.	32

- Display the **Line grapher tool**. Go to 'set up' and set the x-axis to text and the y-axis to 20 to 35 (increment: 1, sub-divide: 1). Go to the main screen and double click the x-axis labels to change the text to '12 p.m.', '1 p.m.', and so on. Ask learners to assist in plotting a line graph to show the change in temperature over time.

- Ask questions such as:
  - **At what times was the temperature below/above 33°C?**
  - **By how many degrees did the temperature rise between 1 p.m. and 4 p.m.?**
  - **When was the greenhouse at its hottest?**
  - **How does the graph help to answer Maisie's question?**
  - **What other conclusions can you draw from the graph?**

Variation

 Use the **Line grapher tool** to display a line graph of the temperatures in a holiday resort.

Revise

Is there a relationship?

Learning objectives

Code	Learning objective
6Ss.01	Plan and conduct an investigation and make predictions for a set of related statistical questions, considering what data to collect (categorical, discrete and continuous data).
6Ss.02	Record, organise and represent categorical, discrete and continuous data. Choose and explain which representation to use in a given situation: line graph.
6Ss.04	Interpret data, identifying patterns, within and between data sets, to answer statistical questions. Discuss conclusions, considering the sources of variation and check predictions.

Resources

paper (per learner); ruler (per learner)

What to do

- Investigate the statistical question: Is there a relationship between the side length of a square and its perimeter?
- Each learner draws four squares of sides 1 cm, 2 cm, 3 cm and 4 cm across the top of the paper and writes below each square its perimeter in centimetres.
- In the lower part of the paper, they use a ruler to draw the horizontal and vertical axes for a graph.
- They use the information in their diagrams to plot the points on the graph. They join the points with a straight line and continue the line to the last point that will fit their axes.
- They take turns to pose questions that require their partner to interpret points on the plotted line, for example: 'What is the perimeter of the square with sides of 6 cm/8·5 cm?'
- Ask: **How does the graph help to answer the statistical question that began the investigation? What conclusions can you draw from the data?**

Variation

Investigate the statistical question: 'Is there a relationship between the side length of an equilateral triangle and its perimeter?'

Siblings

Learning objectives

Code	Learning objective
6Ss.01	Plan and conduct an investigation and make predictions for a set of related statistical questions, considering what data to collect (categorical, discrete and continuous data).
6Ss.02	Record, organise and represent categorical, discrete and continuous data. Choose and explain which representation to use in a given situation: dot plots.
6Ss.04	Interpret data, identifying patterns, within and between data sets, to answer statistical questions. Discuss conclusions, considering the sources of variation and check predictions.

Resources

squared paper or Resource sheet 30: 2 cm squared paper (per learner); ruler (per learner)

What to do

- Explain to learners that Aisha has made the following prediction: 'I think most learners in my class will have three brothers and sisters.'
- Tell them that Aisha has carried out a class survey and recorded the data in a table:

Number of brothers and sisters	0	1	2	3	4	5
Frequency	6	9	12	8	4	1

- Ask them to construct a dot plot of the data and use it to investigate Aisha's prediction.
- What other conclusions can they draw from the data?

Variation

Learners investigate another prediction that Aisha makes: 'I think most learners in my class travel to school by car.'

Revise

Probability scale

Learning objective

Code	Learning objective
6Sp.01	Use the language associated with probability and proportion to describe and compare possible outcomes.

What to do

- Draw a probability scale on the board. Label the left end of the scale 'impossible', the middle point 'even chance' and the right end 'certain'. Ask learners to copy the scale.
- Mark the scale with two events, 'Event A' halfway between the left end and middle and 'Event B', halfway between the middle and the right end.
- [TWM.02/05/06] Ask: **What can you say about these two events?** Take responses. Remind learners that the further an event is to the left of the scale, the more unlikely it is to happen. The further it is to the right, the more likely it is to happen. Event A is therefore more likely to happen than Event B.

- Call out probability events and ask volunteer learners to mark them on the scale, for example:
  - Rolling a '7' on a 1–6 dice.
  - Spinning 'green' on a spinner that is half green and half blue.
  - Picking out a green counter in a bag of 10 green counters.
  - Picking out a blue counter in a bag of 3 blue counters and 7 red counters.
  - Picking out a red counter in a bag of 3 blue counters and 7 red counters.

Variation

Ask learners to suggest other probability events that could be marked on the scale.

Likely thumbs

Learning objectives

Code	Learning objective
6Sp.01	Use the language associated with probability and proportion to describe and compare possible outcomes.
6Sp.04	Conduct chance experiments or simulations, using small and large numbers of trials. Predict, analyse and describe the frequency of outcomes using the language of probability.

Resources

bag (per class); blue and red coloured counters (per class)

What to do

- Tell learners that they are going to think about events and how likely they are to happen. They hold their thumb up if an event is likely to happen, and thumb down if it is unlikely to happen.
- Show learners the bag and coloured counters.
- Place one blue counter and three red counters in the bag. Say: **I am going to pull out a counter.** Ask: **What are the chances it will be blue?** Remind learners that they need to express the probability as a proportion or a percentage. (1 in 4 or 25%)

- Empty the bag and then place three blue counters and three red counters in the bag. Say: **I am going to pull out a counter.** Ask: **What are the chances it will be red?** (even chance, 50%) Ask: **If there are six counters in the bag in total, why is the probability of picking a red counter 50%?** Take responses. Establish that 3 out of 6 is equivalent to a half ($\frac{3}{6} = \frac{1}{2} = 50\%$).
- Repeat for different red and clue counter combinations, for example, 8 counters: 6 red, 2 blue; 10 counters: 1 red, 9 blue.

Variation

Use a wider range of probabilities, for example, place four blue counters and six red counters in the bag. Ask: **What is the probability of pulling out a blue counter?** ($\frac{4}{10} = \frac{2}{5} = 40\%$)

Revise

Independent or mutually exclusive?

Learning objective

Code	Learning objective
6Sp.02	Identify when two events can happen at the same time and when they cannot, and know that the latter are called 'mutually exclusive'.

What to do

- Call out pairs of events and ask learners to say whether they are independent or mutually exclusive.
  - Flip a coin and get heads and get tails.
  - Roll a 1–6 dice and get a '5' and flip a coin and get 'heads'.
  - Roll a 1–6 dice and get a number that is odd and even.
  - Walk to the end of a street and turn left and right.

  - Pick a fruit each from two bowls of mixed fruits and get a banana and an orange.
  - Choose a '4' from a pack of 1–9 digit cards, replace it, and then choose a '5'.

Variation

 Choose learners to call out their own pair of probability events for the class to decide whether the events are independent or mutually exclusive.

Higher or lower? (2)

Learning objective

Code	Learning objective
6Sp.04	Conduct chance experiments or simulations, using small and large numbers of trials. Predict, analyse and describe the frequency of outcomes using the language of probability.

Resources

1–10 number cards from Resource sheet 2: 0–100 number cards (per pair)

What to do

- Working in pairs, learners shuffle the cards and place them face down in a pile.
- Player A turns over a card and player B decides whether the next card will be higher or lower, stating the probability of the outcome. For example, if the first card is a '3', then possible outcomes where the next card is higher will be '4', '5', '6', '7' '8', '9' or '10' (7 outcomes) and where the next card is the same or lower will be '1', '2' or '3' (3 outcomes). The probability of the next card being higher is therefore $\frac{7}{10}$.

- If player B is correct in their choice of 'higher' or 'lower' and the probability they gave (to be decided if correct by both players), they score a point and can continue their turn. If they are incorrect, then play passes to the other player.
- The winner of the game is the first person to score ten points.

Variation

1 Learners play with a pack of cards with digits 1 to 5 only.

Unit 1: Counting and sequences

Collins International Primary Maths
Recommended Teaching and
Learning Sequence: Term 1, Week 1

Learning objectives

Code	Learning objective
6Nc.01	Count on and count back in steps of constant size, including fractions and decimals, and extend beyond zero to include negative numbers.
6Nc.03	Use the relationship between repeated addition of a constant and multiplication to find and use a position-to-term rule.
6Nc.04	Use knowledge of square numbers to generate terms in a sequence, given its position.

Unit overview

In this unit, learners investigate counting on and back in decimal steps and fractions that use small denominators. They count in tenths and hundredths and examine how digits change when ones, tenths and hundredths boundaries are crossed. Counting sequences extends to beyond zero to include negative numbers.

Given a sequence of multiples, learners recognise the pattern and use their knowledge of a times table to find the term in any given position; for example, the 10th term in the sequence 4, 8, 12, 16... can be found with $4 \times 10 = 40$. Tables help learners see the connection between the term and its position in the sequence.

The work on square numbers in Stage 5 is extended, with learners deducing that a particular term in a square number sequence can be found by taking the square of the position.

Prerequisites for learning

Learners need to:
- understand the place value of whole numbers and decimals to two decimal places
- have experience with fractions represented on number lines
- understand that numbers extend beyond zero
- recognise, recreate and extend the patterns of square numbers, relating them to geometric sequences.

Vocabulary

counting on, counting back, ones boundary, tenths boundary, hundredths boundary, negative number, sequence, term, rule, position-to-term rule, square number

Common difficulties and remediation

Some learners find the concept of the hundredths place difficult to grasp. Use models, such as Base 10 equipment, in which the flat represents one whole, the long represents one tenth and the cube represents one hundredth, to illustrate the relationship between the numbers and help build understanding about the size of decimals.

Some learners have difficulty understanding and applying the rule for the nth term. Provide plenty of examples to establish this link. Help learners understand that unlike a term-to-term rule, when given the rule for the nth term in a sequence of multiples, we do not need to know the previous term.

Supporting language awareness

Ensure learners understand the difference between a term and a position, as this will assist them when they are formally introduced to algebra in the Lower Secondary curriculum. Use mapping diagrams and tables to illustrate this.

Encourage frequent use of mathematical language to help embed vocabulary. For example, when learners describe the next number in a sequence, encourage them to use the more mathematically correct phrase 'next term in the sequence'.

Promoting Thinking and Working Mathematically

Opportunities to develop all four pairs of TWM characteristics are provided throughout the unit; however, because of the sequential nature of arithmetic number patterns (counting on/back sequences) and special number patterns (square numbers) particular emphasis is given to Generalising (TWM.02). Learners identify familiar sequences (multiples or square numbers) and use the position-to-term rule to determine the value of a term for a given position.

In Lessons 3 and 4, learners are given opportunities to investigate whether statements about sequences are true or not and provide convincing (TWM.04) evidence to support their conclusions. When providing proof, they are asked to compare and evaluate representations or solutions to identify advantages and disadvantages (TWM.07).

Success criteria

Learners can:
- count on or back in decimals and fractions and understand how the digits change when crossing the ones, tenths and hundredths boundaries
- identify the position-to-term rule for a sequence and use it to find any given term in the sequence
- say the value of the term when given the position in the sequence of square numbers.

Number – Counting and sequences

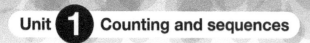

Number – Counting and sequences

Lesson 1: **Counting on and back in fractions and decimals**

Learning objective

Code	Learning objective
6Nc.01	Count on and count back in steps of constant size, including fractions and decimals[, and extend beyond zero to include negative numbers].

Resources

mini whiteboard and pen (per learner)

Revise

Use the activity *Decimals go round* or *Fractions on the line* from Unit 1: *Counting and sequences* in the Revise activities.

Teach [SB] ⬜ ⬛ [TWM.02]

- Discuss the Let's learn section in the Student's Book.
- Display the **Place value tool**, showing 3·9. Ask: **Which number is one tenth less/more than this number?** (3·8, 4) On the tool, illustrate that the tenths digit changes each time you add or subtract tenths and crossing the ones boundary increases or decreases the ones digit by 1.
- Show 7·49 on the **Place value tool**. Ask: **What is the decimal that is one tenth less/more than this number?** (7·48, 7·5) On the tool, illustrate that the hundredths digit changes each time you add or subtract hundredths, and crossing the tenths boundary increases or decreases the tenths digit by 1.
- Display the **Number line tool**, start: 4, end: 7, increment: 0·1, subdivide: 0·1. Lead a count forwards from 4·5 in steps of 0·2 as far as 4·9. Ask: **What is the next number?** (5·1)
- Position the pointer at 4·3 Ask: **If you count on in steps of 0·3 from 4·3, what will the fourth number be? Convince me.** (5·5)
- Repeat for different starting points and increments, for example: **Count back in steps of 0·04 from 5·79. [TWM.02] [T&T] Ask: What is the number in the counting sequence just before the tenths boundary is crossed?** (5·71) **How do the digits in the number change when the boundary is crossed?**
- Lead a count in thirds: $3\frac{1}{3}$, $3\frac{2}{3}$, 4, $4\frac{1}{3}$... up to 10. Ensure that learners understand what happens when counting across whole number boundaries, for example interpreting $4\frac{3}{3}$ as 5.
- Display **Slide 1**. Circle $6\frac{2}{3}$ on the thirds number line. **[T&T]** Say: **I start at $6\frac{2}{3}$ and count in two-thirds. Will $8\frac{1}{3}$ be part of this sequence? How do you know? Will 10? Convince me.**
- Use the number lines to count in other fractions, for example back in three-fifths from 6, asking similar questions to those above.
- On the board, write: 2·5. Say: **We are going to count on in 0·6s from 2·5.** Choose learners to come to the

board to write the next number in the sequence for the next five terms (2·5, 3·1, 3·7, and so on). Label the numbers in the sequence '1st term', '2nd term', '3rd term', and so on. Remind learners that each number in the sequence is called a term.
- 🖐 Say: **I start at $3\frac{1}{8}$ and count in three-eighths to 5. How many steps is that?** (5)
- Introduce the paired activity in the Student's Book. Then bring the class back together for pairs to share their methods.
- Discuss the Guided practice example in the Student's Book.

Practise [WB]

- Workbook

Title: Counting on and back in fractions and decimals

Pages: 6–7

- Refer to Activity 1 from the Additional practice activities.

Apply 👥 ⬜ [TWM.03/04]

- Display **Slide 2**. Learners discuss, calculate and record the statements that are **not** true.

Review ⬜

- Display the **Number line tool**, start: 2, end: 5, increment: 0·1, subdivide: 0·1. Explain that the scale measures temperature in degrees Celsius.
- Ask questions of the type: **If the temperature increases/decreases in five steps of 0·3 degrees, what will the temperature reading be?**

Assessment for learning

- The volume of water in a jug is 4·77 *l*. Every hour, 0·02 *l* of water evaporates from the jug. What is the volume of water after 4 hours? (4·69 *l*).

Same day intervention
Enrichment

- Develop TWM characteristics of Critiquing/Improving by asking learners for alternative strategies and to evaluate their own strategies and those suggested by others.

Lesson 2: **Counting on and back beyond zero**

Learning objective

Code	Learning objective
6Nc.01	Count on and count back in steps of constant size, including fractions and decimals, and extend beyond zero to include negative numbers.

Revise

Use the activity *Stepping back* from Unit 1: *Counting and sequences* in the Revise activities to remind learners of the relationship between positive and negative numbers.

Teach 🔲 🖵 [TWM.02/04]

- Discuss the Let's learn section in the Student's Book.
- Display **Slide 1**. Point to the first number line. Ask: **If you count back in steps of 4 from 10, what will the sixth number be?** Take responses. (−14)
- Point to the second number line. Say: **I want to count back from 1 in steps of 0·4. How does counting back on the whole number line help me to count back on the decimal number line?** Invite a learner to the board to explain. Elicit that the decimal count is the same as the whole number count, just ten times smaller.
- Say: **I want to move from 0·7 to −1·8. How many steps of 0·5 is that?** (5) **[TWM.04] [T&T]** Say: **I count on in steps of 0·6 from −1·6. Convince me that 0·8 will be part of the count.** Point to −0·3. Ask: **Is this number part of the count? How do you know? Is there a quick way to determine if an odd or even number is part of the count?** Take responses. Remind learners that adding or subtracting two even numbers always gives an even number and therefore, numbers with odd numbers of tenths will not be part of the count.
- Point to the number line in quarters on **Slide 1**. Ask: **If I count on from −1$\frac{1}{4}$ in steps of a quarter, what will the sixth number be?** Take responses. ($\frac{1}{4}$) Ask: **If I count back from $\frac{1}{4}$ in steps of a half, what will the fourth number be?** (−1$\frac{3}{4}$) **How do you know?**
- **[TWM.04]** Point to the number line in thirds on **Slide 1**. On the board, write the statement: 'If I count on from −3 in two-thirds, none of the numbers in the count will be whole numbers.' Ask: **True or false? How do you know?**
- **[TWM.04] [T&T]** Say: **I start at 2$\frac{1}{3}$ and count back in $\frac{2}{3}$s. Will I say −2$\frac{1}{3}$? Convince me.**
- 🔁 Say: **I start at −1·7 and count forwards in 0·4s. Will the fourth number be positive?** (no, −0·1)
- **[TWM.02]** Introduce the paired activity in the Student's Book, asking learners to work in pairs and discuss the best method for solving the problem. Then bring the class back together for pairs to share their thoughts with the whole class.

- Discuss the Guided practice example in the Student's Book.

Practise 🔲

- Workbook

Title: Counting on and back beyond zero

Pages: 8–9

- Refer to Activity 1 (Variation) from the Additional practice activities.

Apply 👥 🖵

- Display **Slide 2** and read the text to the class. Establish that learners understand the instructions and ask them to record their working. Ask: **Which lift ends up on the highest floor? How do you know?**

Review

- Display **Slide 2**. Divide the class into two teams.
- Ask questions of the type: **The lift rises from floor −1·1 in three stages of 0·5 floor units. On which floor is the lift now?** (0·4)
- Ask learners to raise a hand to respond. If correct, they score a point for their team.
- Play ten rounds. The team with most points at the end is the winner.

Assessment for learning

- The temperature falls 0·4°C every hour. Given a temperature of 1·5°C, what will the temperature be in four hours? (−0·1°C) Explain your solution.
- Toby counts back in $\frac{3}{4}$s from 1$\frac{1}{4}$. Will he say −2$\frac{1}{4}$? (no)
- A mole climbs up from a depth of 3·3 metres (−3·3 m) below ground in five steps of 0·3 metres. How far below the ground is the mole after its climb? (−1·8 m)

Same day intervention
Enrichment

- Ask learners to devise their own lift problems similar to the ones used on **Slide 2**. They swap papers with a partner and return them for marking.

Number – Counting and sequences

113

Number – Counting and sequences

Lesson 3: Finding the position-to-term rule

Learning objective

Code	Learning objective
6Nc.03	Use the relationship between repeated addition of a constant and multiplication to find and use a position-to-term rule.

Resources

paper or mini whiteboard and pen (per learner)

Revise

Use the activity *Name your terms* from Unit 1: *Counting and sequences* in the Revise activities to introduce learners to the vocabulary 'term' and 'position'.

Teach [SB] 🖥 [TWM.04]

- Discuss the Let's learn section in the Student's Book.
- Display **Slide 1**. Point to Table A. Remind learners of the difference between term and position. Ask: **What is the term in position 5?** (15) **Which position is the term with a value of 18?** (6) **[T&T]** Ask: **What is the position-to-term rule?** (multiply by 3) **How do you know? How would you use the rule to find the value of the term in the 9th position?** Take responses. Establish that you multiply the position by the multiple (9 × 3 = 27).
- Point to Table B on **Slide 1**. **[TWM.04]** Say: **I think the 7th term in the sequence will be 49. Convince me.** Take responses. Confirm that the position-to-term rule is 'multiply by 7' and the 7th term can be found with 7 × 7 = 49. Ask: **What would be the position of the term with the value 63? How do you know?** (9) **[T&T]** Ask: **What position is the term 91 for this sequence?** Take responses. Ask learners to share the strategies they used and to discuss the advantages and disadvantages of each method.
- Point to Table C on **Slide 1**. **[TWM.04]** Say: **Prove to me that this is a sequence of multiples.** Take responses. Establish that you can determine if a sequence is a set of multiples by confirming there is a common difference between consecutive terms (30 – 15 = 15; 45 – 30 = 15, and so on). Confirm the position-to-term rule is 'multiply by 15'. Say: **Calculate the value of the term in position 9.** (135)
- Introduce the paired activity in the Student's Book then bring the class back together for pairs to share their solutions.
- 📕 Say: **The terms in positions 4, 5, 6 and 7 of a sequence are 44, 55, 66, 77.** Ask: **What are the values of the terms in positions 2 and 9?** (22, 99)
- Discuss the Guided practice example in the Student's Book.

Practise [WB] [TWM.02/07]

- Workbook

Title: Finding the position-to-term rule

Pages: 10–11

- Refer to Activity 2 from the Additional practice activities.

Apply 👥 🖥 [TWM.01/02/04]

- Display **Slide 2** Ask: **How many panels of glass will there be in the next window in the sequence? How do you know?** (30)
- Discuss the different possible arrays for 30. (1 × 30, 2 × 15, 3 × 10, 6 × 5, and so on) Say: **Convince me that you have found all of the arrangements.**

Review

- On the board, write the following: 9 km, 18 km, 27 km, 36 km, 45 km. Explain that the measurements are sections of road that are constructed end-to-end. They are the first lengths of a sequence determined by a position-to-term rule. Ask: **What will the length of the seventh section be?** (63 km) **And the ninth section?** (81 km)
- Learners record their answers on paper or their whiteboards. Invite a learner to explain their solution.
- Ask learners to work in pairs. They draw a table with two rows labelled 'position' and 'term'. They enter a sequence of multiples in the table up to the 6th position. They swap papers and ask their partner to find the 9th and 11th terms in the sequence.

Assessment for learning

- Groups of hexagons are placed in a row. One hexagon is in the first position, two hexagons are in the second position, and so on. For all the hexagons of the group in the 8th position, how many sides will they have in total? (48)
- The terms in positions 4, 5, 6 and 7 of a sequence are 28, 35, 42, 49. What are the values of the terms in positions 3 and 9? (21, 63)

Same day intervention
Support

- Some learners may find differentiating between 'term' and 'position' challenging. Showing the ordinal numbers alongside the sequence can help learners become familiar with the idea.

Lesson 4: **Finding terms of a square number sequence**

Learning objective

Code	Learning objective
6Nc.04	Use knowledge of square numbers to generate terms in a sequence, given its position.

Resources

paper or mini whiteboard and pen (per pair); counters (per pair)

Revise

Use an activity from Unit 1: *Counting and sequences* in the Revise activities.

Teach [SB] 🖵 [TWM.04]

- Discuss the Let's learn section in the Student's Book.
- Display **Slide 1**. Invite learners to the board to complete the square number patterns and term values for positions 6, 7 and 8 (drawing each Pattern and writing each Term underneath the table). **[T&T]** Ask: **How do you use the position of a term in the sequence of square numbers to calculate its value?** Take responses. Confirm that the term is found by calculating the square of the position in the sequence.
- Ask learners to work in pairs. On the board, write: 81, 100, 121. **[T&T] [TWM.04]** Ask: **Prove that the numbers on the board are square numbers.** Provide paper or mini whiteboards, and counters. Establish that learners may request other resources around the class, if available. Give them time to answer the problem and then take responses. Discuss which of the model(s) were most effective in demonstrating that a number is square and how they could be improved visually.
- Ask: **Which position in the sequence of square numbers is each term? How do you know?** (9, 10, 11)
- On the board, write: 'Doubling the sequence position NEVER doubles the value of the term'. Working in pairs, ask learners to investigate the truth of this statement for square numbers. Invite learners to explain their conclusions. Ask: **What did your investigation reveal? How can you be certain that this never happens?**
- ⏃ Ask: **What is the value of the 20th term in the square number sequence?** (400)
- Ask: **How would you calculate the value of the 21st term in the square number sequence?** Take responses and discuss possible strategies, for example, compensation: 21 × 21 = 21 × 20 + 21 = 420 + 21 = 441.
- Introduce the paired activity in the Student's Book, asking learners to work in pairs to answer the problem. Bring the class back together for pairs to share their solutions. (91)

- Discuss the Guided practice example in the Student's Book.

Practise 📱 [TWM.01/08]

- Workbook

Title: Finding terms of a square number sequence

Pages: 12–13

- Refer to Activity 2 (Variation) from the Additional practice activities.

Apply 👥 🖵 [TWM.04]

- Display **Slide 2** and read the text to the class. Give learners time to complete the task and then ask them to discuss their ideas. Learners who finish early could investigate proof for the statement: 'Tripling the position number of a term in the square number sequence always increases the value of the term by nine times.'

Review

- Call out a position number in the square number sequence and ask learners to raise their hands with the corresponding square number.
- Call out a square number and ask learners to give the corresponding position number in the square number sequence.

Assessment for learning

- Explain the rule for determining the value of a square number from its position in the sequence of square numbers.
- How many square numbers have a value between 80 and 150? (4)
- What is the sum of the square numbers in positions 6, 7, 8? (149)

Same day intervention
Support

- Some learners may find differentiating between 'term' and 'position' challenging. Showing the ordinal numbers alongside the sequence can help learners become familiar with the idea.

Additional practice activities

Activity 1 ☷ ⚠2

Learning objective
• Count on and count back in steps of constant size, including fractions and decimals, and extend beyond zero to include negative numbers.

Resources
mini whiteboard and pen (per learner); counters (per pair)

What to do
• Learners draw a number line in intervals of 0·01 from 4·45 to 4·65.
• They take turns to place a counter on the number line.
• They choose a number that is the sum of 0·02s, 0·03s or 0·04s only, for example, 4·57 (4·45 + 0·03 + 0·03 + 0·03 + 0·03); or 4·54 (4·45 + 0·03 + 0·03 + 0·03).

• Their partner has to work out the sum, saying the number steps of 0·02, 0·03 or 0·04 used to reach the number on the number line.

Variations [TWM.01]

⚠2 Extend the number line beyond zero, for example, –0·1 to 0·1 with intervals of 0·01.
Learners take turns to choose a positive or negative number on the number line and ask their partner which number would be reached, given a number of steps on or back of 0·02, 0·03 or 0·04.

3 Learners ask their partner questions of the form: 'I count in steps of 0·03 from –0·11 to 0·07. How many steps is that?' (6)

Activity 2 ☷☷ ⚠2

Learning objective
• Use the relationship between repeated addition of a constant and multiplication to find and use a position-to-term rule.

Resources
mini whiteboard and pen (per learner)

What to do
• Arrange learners in groups of seven, with each group standing in a line.
• One learner, the 'detective' leaves the group and stands at a distance facing away from the group.
• Learners decide on a sequence of multiples, for example 7, 14, 21, 28, 45, 42 (multiples of 7). Each learner in the group is allocated a position number and term. The sequence should match the order of learners in the line.
• The 'detective' returns to the group. Learners take turns to say their position number and term, for example 'position 3, term 15' (multiples of 5). They

do this in random order and step out of line when making their announcement.
• Count how many announcements it takes for the detective to identify the position-to-term rule.
• The game is repeated with a new 'detective'. The winner of the game is the 'detective' who determines the rule with the fewest 'announcements'.

Variations

 The game is played with multiples of 5, 6, 7 or 8. Learners write out the sequences before the game and use this to help identify sequences during the game.

 Learners play the game as given but include special sequences, such as square or triangular numbers.

Unit 2: Addition and subtraction of whole numbers (A)

Collins International Primary Maths Recommended Teaching and Learning Sequence Term 1, Week 2

Learning objectives

Code	Learning objective
6Ni.01	Estimate, add [and subtract] integers.
6Nc.02	Recognise the use of letters to represent quantities that vary in addition and subtraction calculations.

Unit overview

In this unit, learners consolidate their knowledge of strategies for adding positive and negative integers. They now work with larger numbers and use number lines or scales to support calculations, for example thermometers and sea level scales.

Guided examples are provided, together with practical activities in which learners use an appropriate strategy to solve problems. In so doing, they learn that some strategies are more effective and efficient than others, depending on the calculation.

The unit concludes with two lessons that develop pre-algebra skills: finding the value of symbols used in addition or subtraction calculation to represent unknown values. Examples are included where the unknown quantity is in different positions, for example: $14 + a = 17$, $a + 3 = 17$, $17 = a + 3$, $17 = 14 + a$.

Prerequisites for learning

Learners need to:
- know how to support their mental calculations by the use of jottings to record intermediate steps, for example using a blank number line
- be able to select an appropriate mental method from a bank of strategies, including bridging, partitioning and compensation
- know how to write a calculation vertically, adding separate place values together and regrouping when required.

Vocabulary

positive, negative, temperature, degree, Celsius, unknown number, inverse operation

Common difficulties and remediation

Focus on addition problems with a negative augend and a positive addend. This order helps learners avoid interpreting movements along the number line as 'more negative' or 'less negative'. In so doing, they sometimes ignore the direction of 'positive' or 'negative' and move more or less based on the positive number line; or they focus on moving in the positive or negative direction, without paying attention to whether they should move 'more' or 'less' in that direction.

Supporting language awareness

At every stage, learners require mathematical vocabulary to access questions and problem-solving exercises. If appropriate, when a new key word is introduced, ask learners to write a definition in their books, drawing a box around it for emphasis.

Encourage learners to write the definition in their own words. Adapt language when it becomes a barrier to learning, for example using 'missing number' rather than 'unknown value'.

Promoting Thinking and Working Mathematically

Opportunities to develop all four pairs of TWM characteristics are provided throughout the unit.

In Lesson 1, they critique (TWM.07) when they discuss the effectiveness of strategies to find two numbers that have the sum of −12.

In Lesson 2, learners conjecture (TWM.03) when they prove that the original temperature before a rise of 32 degrees was −11°C given the new temperature of 21°C.

Success criteria

Learners can:
- work out the sum of a positive number and a negative number using a number line
- find the values of unknown quantities in a calculation by choosing the required inverse operation.

Unit 2 Addition and subtraction of whole numbers (A)

Number – Integers and powers

Lesson 1: **Adding positive and negative numbers (1)**

Learning objective

Code	Learning objective
6Ni.01	Estimate, add [and subtract] integers.

Resources

paper or mini whiteboard and pen (per learner)

Revise

Use the activity *Up above, down below* from Unit 2: *Addition and subtraction of whole numbers (A)* in the Revise activities.

Teach [SB] 🖥 📊 [TWM.05/07]

- Discuss the Let's learn section in the Student's Book.
- Display **Slide 1**. Read the first problem. **[T&T]** Ask: **How would you use the number line to calculate negative 13 add 9?** Take responses. Ask: **Would you expect the answer to be negative or positive? How do you know?** Invite a learner to the board to mark the number line to demonstrate the calculation. Expect them to circle –13 and count on 9 in ones (–4). Ask: **Is there a quicker way to count on 9 on the number line?** Take responses. Elicit that you can count on one and then the rest as four steps of 2 (+ 1 + 2 + 2 + 2 + 2).
- Display the **Number line tool**, scaled from –20 to 20. Ask: **How would you use the number line to calculate negative 17 add 20?** Take responses. Invite a learner to the board to mark the number line to demonstrate the calculation (3). Discuss different strategies for adding 20, for example four steps of 5.
- Return to **Slide 1**. **[TWM.05] [T&T]** Ask: **Which of the four problems, b to e, would you expect to give a positive answer? Give a rule to determine if the sum of a positive number and a negative number is positive.** Take responses. Establish that the sum will be a positive number if the magnitude of the positive addend is greater than that of the negative augend. Remind learners that magnitude is the distance the number is from zero. For example, both 15 and –15 have a magnitude of 15.
- With learners helping, work out the answers to problems **b** to **e**.
- **[TWM.07]** Say: **The sum of a positive and negative number is –12. What are the two numbers?** Ask learners to explain how they found number pairs with the correct total. Invite other learners to suggest a more efficient method and to explain how their strategy works.
- 📄 Say: **A submarine is 18 metres below the water surface. It rises 16 degrees. How far is it from the surface now?** (2 m)

- Introduce the paired activity in the Student's Book, asking learners to work out the heights reached by the fish.
- Discuss the Guided practice example in the Student's Book.

Practise [WB] [TWM.01]

- Workbook

Title: Adding positive and negative numbers (1)

Pages: 14–15

- Refer to Activity 1 from the Additional practice activities.

Apply 👥 🖥

- Display **Slide 2** and read the text to the class. Ask: **Which floor will Daisy and her family arrive at? How do you know?**

Review

- Ask learners to work in pairs. They take turns to say a 'target' number between –20 and 20. They race each other to write on paper or their whiteboard five addition calculations with a negative augend and a positive addend that give a sum equal to the 'target' number. Learners do not write the sum of each calculation but give the finished list of questions to their partner to answer. The winner is the learner who writes the most correct calculations in the shortest time.

Assessment for learning

- How would you use a number line to show that –17 + 33 is 16?
- On the board, write: –19 + 25, –14 + 13, –19 + 20, –17 + 15. Which of these calculations will give a negative sum? How do you know? (–14 + 13, –17 + 15)

Same day intervention

Enrichment

- Introduce addition problems where learners work with numbers with three-digit numbers, for example: –137 + 88, –276 + 425.

Unit **2** Addition and subtraction of whole numbers (A)

Lesson 2: **Adding positive and negative numbers (2)**

Learning objective

Code	Learning objective
6Ni.01	Estimate, add [and subtract] integers.

Resources

calculator (per learner)

Revise

Use the activity *Getting warmer* from Unit 2: *Addition and subtraction of whole numbers (A)* in the Revise activities.

Teach [SB] [II] [TWM.03]

- Discuss the Let's learn section in the Student's Book.
- Display the **Thermometer tool**, scaled from –20 to 30°C. Set the pointer to –14°C. Ask: **If the temperature rises from –14°C by 24 degrees, what is the new temperature?** Take responses. Ask: **Would you expect the answer to be negative or positive? How do you know?** Choose a learner to come to the board to demonstrate how the pointer can be moved in increments to count on 24. Discuss the different ways to count on 24, for example 12 jumps of 2. Say: **Write the temperature change as an addition calculation.** Ask learners to raise their whiteboards and confirm their calculations are correct (–14 + 24 = 10).
- **[TWM.03]** Say: **The new temperature is 21°C after a rise of 32 degrees. Convince me that the original temperature was –11°C.** Ask: **How did you work this out?**
- Say: **A scientist performs an experiment where the temperature of a material is raised from a very low temperature to a very high one, increasing by 317 degrees from –133°C. Before you work out the new temperature, estimate the answer.** Choose learners to explain the strategy they used to get an estimate, for example: 300 (rounding 317 to 300) – 100 (rounding –133 to –100) = 200. **[T&T]** Learners work in pairs to calculate the new temperature. Choose a pair to explain how they calculated the answer. Provide calculators and ask learners to confirm the answer (184°C).
- Ask similar questions based on large temperature changes but bear in mind that the temperature cannot fall below –273·15 (absolute zero). Each time, ask learners to estimate the answer and confirm with a calculator.
- ⏎ Say: **Lucy puts a thermometer in a beaker of water and places it in the freezer for a day. When she removes the beaker, the temperature reading is –13°C. She leaves the beaker by a radiator and returns to take the temperature. The reading tells Lucy that the water temperature has**

increased by 25 degrees. Ask: **What is the new temperature?** (12°C; –13 + 25 = 12)

- Introduce the paired activity in the Student's Book, asking learners to demonstrate an example of how they calculated the final temperature after two rises in temperature.
- Discuss the Guided practice example in the Student's Book.

Practise [WB]

- Workbook

Title: Adding positive and negative numbers (2)

Pages: 16–17

- Refer to Activity 1 (Variation) from the Additional practice activities.

Apply [II] [▢]

- Display **Slide 1** and read the text to the class. Choose learners to demonstrate how they worked out the temperature changes for one of the experiments.

Review [II]

- Display the **Thermometer tool**. Ask learners to work in threes. One learner says a negative temperature. A second learner says how much the temperature will rise in degrees. The third learner must work out the new temperature after the rise. The group confirms the answer and then they swap roles.

Assessment for learning

- How would you use a thermometer scale to show that a rise in temperature of 36 degrees from –17°C gives a temperature of 19°C?
- Liam's bank account balance is –$26. He deposits $53 into the account. How much does he have now? ($27)

Same day intervention
Enrichment

- Introduce addition problems where learners work with three-digit numbers, for example –137 + 88, –276 + 425.

Number – Integers and powers

Unit **2** Addition and subtraction of whole numbers (A)

Number – Integers and powers

Lesson 3: **Identifying values for symbols in addition calculations**

Learning objective

Code	Learning objective
6Nc.02	Recognise the use of letters to represent quantities that vary in addition and subtraction calculations.

Resources

paper or mini whiteboard and pen (per learner)

Revise

Use the activity *Missing numbers* from Unit 2: *Addition and subtraction of whole numbers (A)* in the Revise activities.

Teach [SB] 🖥 [TWM.05]

• Discuss the Let's learn section in the Student's Book.
• Display **Slide 1**. Ask learners to explain the first problem in their own words. Take responses. **[T&T]** Ask: **What calculation would give you the price of the pen? Write it down.** Choose learners to share their calculations. Ask: **How did you represent the unknown value, the price of the pen?** Remind learners that you can use symbols to represent unknown values in a calculation. Learners used a box symbol in Stage 5 but explain that they will now use letters. **[TWM.05]** Ask: **What advantage does using letters have over using a box?** Praise any learner who identifies that the range of letters available means that you can have more unknown values represented. A box symbol limits this to one unknown.
• Ask a learner to come to the board to write a number sentence for the first problem on **Slide 1** with the unknown represented by the letter 'a'. Expect: 12 + a = 17. Ask: **How would you find the value of a?** Take responses. Expect some learners to say that they know 12 + 5 = 17 and therefore, the price of the pen is $5. Remind them that they could use their knowledge of subtraction as the inverse operation of addition to work out the unknown. Write: 17 – 12 = a, therefore 'a' is 5.
• Repeat for the second problem on **Slide 1**. Ask learners to write a number sentence to represent the problem using the letter 'b' to represent the unknown. Expect: b + 18 = 35. Ask: **How would you find the value of b?** Remind them to use the inverse operation of subtraction. Choose a learner to explain their solution. Expect: b = 35 – 18 = 17. ($17)
• 👥 Say: **A wallet and a magazine cost $31. If the price of the magazine is $6, what is the price of the wallet? Write a number sentence to represent the problem and solve it.** ($31 – $6 = $25)
• Introduce the paired activity in the Student's Book, asking learners to find the price of the diary. Choose a learner to show the number sentence they wrote

to represent the problem and how they used it to find the unknown value ($24).
• Discuss the Guided practice example in the Student's Book.

Practise [WB]

• Workbook

Title: Identifying values for symbols in addition calculations

Pages: 18–19

• Refer to Activity 2 from the Additional practice activities.

Apply 👥 🖥

• Display **Slide 2** and read the text to the class. Give learners time to complete the task. Choose a pair to explain how they found the unknown values and the person who received the most money this month.

Review

• On the board, write: a + 24 = 53, 37 + b = 62 and 86 = 67 + c. Ask learners to find the value of each symbol (a: 29, b: 25, c: 19). Remind learners to estimate the answer before calculation.
• Ask learners to write three similar addition problems with unknown values. They swap papers with a partner to answer the questions and then return them for marking.

Assessment for learning

• Ryan travels 46 km in a car and then stops. He sets off again. By the time he stops for a second time, he has travelled a total of 82 km. How far did he travel on his second journey? (36 km)
• c + 77 = 133. What is the value of c? (56)

Same day intervention
Enrichment

• Provide more examples of number stories for learners to represent as number sentences. For example: 'A red string and a green string are tied together, end-to-end. The total length of the combined strings is 144 cm. The red string is 86 cm. What is the length of the green string?' (58 cm)

Lesson 4: Identifying values for symbols in subtraction calculations

Learning objective

Code	Learning objective
6Nc.02	Recognise the use of letters to represent quantities that vary in addition and subtraction calculations.

Resources

paper or mini whiteboard and pen (per learner)

Revise

Use the activity *Going shopping* from Unit 2: *Addition and subtraction of whole numbers (A)* in the Revise activities.

Teach [SB] 🖥 [TWM.03]

- Discuss the Let's learn section in the Student's Book.
- Display **Slide 1**. Ask learners to explain the first problem in their own words. Take responses. **[TWM.03] [T&T]** Ask: **What calculation would give you the original length of the ribbon? Write it down.** Choose learners to share their calculations. Ask: **How did you represent the unknown value, the original length?**
- Ask a learner to come to the board to write a number sentence with the unknown represented by the letter 'c'. Expect: c – 32 = 47. Ask: **How would you find the value of 'c'?** Take responses. Expect some learners to say that they know: 79 – 32 = 47 and therefore, the original length of the ribbon was 79 cm. **[TWM.03]** Ask: **If you calculated this mentally, how exactly did you do it?** Establish that you can use your knowledge of addition as the inverse operation of subtraction to work out the unknown length. Write: c = 47 + 32, therefore c is 79.
- Repeat for the second problem on **Slide 1**. Ask: **Can you use an inverse operation to solve this problem? How would you do it?** Choose a learner to come to the board to write a number sentence to represent the problem. Expect: 93 – c = 56 (or any letter). Ask: **What do you do next?** Take responses. Explain that if you know 93 subtract c is 56, then you also know that 56 add c is 93. Write: c + 56 = 93. Ask: **How do you find c?** Expect learners to use the inverse operation of subtraction: c = 93 – 56 = 37. Remind learners to estimate the answer before calculation.
- 🗣 Say: **A builder tips 63 kg of sand into a large container. Later, some sand is removed and used to make cement. 18 kg of sand remains in the container.** Ask: **How much was used to make cement?** (45 kg)
- Introduce the paired activity in the Student's Book, asking learners to find the price of the umbrella. Choose a learner to show the number sentence they wrote to represent the problem and how they used it to find the unknown value.
- Discuss the Guided practice example in the Student's Book.

Practise 📘

- Workbook

Title: Identifying values for symbols in subtraction calculations

Pages: 20–21

- Refer to Activity 2 (Variation) from the Additional practice activities.

Apply 👥 🖥 [TWM.04]

- Display **Slide 2** and read the text to the class. Give learners time to complete the task. Choose a pair to say how they found the missing numbers in the table.

Review

- On the board, write: a – 46 = 28, 52 – b = 27 and 16 = 82 – c. Ask learners to find the value of each symbol (a: 74, b 25, c: 66).
- Ask learners to write three similar subtraction problems with unknown values. They swap papers with a partner to answer the questions and then return them for marking.

Assessment for learning

- Zikra shortens a string to 39 cm. The string was originally 84 cm in length. How much did Zikra cut off? (45 cm)
- 92 – c = 37. What is the value of c? (55)

Same day intervention

Enrichment

- Provide more examples of number stories for learners to represent as number sentences. For example: 'A large water tank has a leak in it. 283 litres of water leaks out of the tank in one day. The tank now has 178 litres of water. How much did it have at the beginning of the day?' (461 litres)

Number – Integers and powers

121

Additional practice activities

Activity 1 👥 ▲2

Learning objective
• Estimate and add integers, where one integer is negative.

Resources
Resource sheet 1: 1–20 number cards; Resource sheet 2: 0–100 number cards (per pair); paper or mini whiteboard and pen (per learner)

What to do
• Learners shuffle the digit cards and place them face down in a pile.
• They take turns to pick two cards. They consider the first number negative and the second, positive.
• They find the sum of the two numbers, a negative augend and a positive addend. Learners may use a number line.

• In each round, they add the sum to their total score.
• The winner is the player with the higher score after six rounds.

Variations
2 Learners consider the augend as a negative temperature and the addend as a rise in temperature. They use a thermometer scale to calculate the new temperature each time.

3 Expand the number card range to 1–50.

Activity 2 👥 ▲2

Learning objective
• Recognise the use of letters to represent quantities that vary in addition and subtraction calculations.

Resources
two sheets of paper (per learner)

What to do
• Ask learners to work in pairs. They each draw a picture of a shop with six items on display priced between $10 and $100, without their partner seeing.
• Next, they draw three picture sentences on a different piece of paper. Each sentence should represent the addition of two different items from the shop where one item is an unknown value. For example, if a purse is $18 and a toothbrush is $8, they could draw:

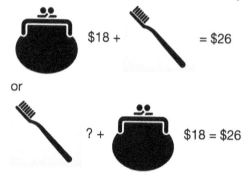

• Learners swap papers. They work out the missing values by rewriting the picture sentence as a number sentence where the unknown is represented by a letter.
• Once completed, they look at the picture of the shop to confirm the answers.

Variations
• Learners repeat the activity but this time, they draw and write subtraction questions that represent change calculations. For example:

1 The price of items in the shop is kept at $10 or less.

Unit 3: Addition and subtraction of whole numbers (B)

Collins International Primary Maths Recommended Teaching and Learning Sequence Term 1, Week 3

Learning objectives

Code	Learning objective
6Ni.01	Estimate[, add] and subtract integers.
6Nc.02	Recognise the use of letters to represent quantities that vary in addition and subtraction calculations.

Unit overview

In this unit, learners consolidate their knowledge of strategies for finding the difference between positive and negative integers, and the difference between two negative integers. They use number lines or scales to support calculations, for example thermometers and sea level scales.

Guided examples are provided, together with practical activities in which learners use an appropriate strategy to solve problems. In so doing, they learn that some strategies are more effective and efficient than others, depending on the calculation.

The unit concludes with two lessons that lay the foundations for algebra (which is introduced in Stage 7), including finding the value of variables used in addition or subtraction calculations and formulae. Examples include finding all the positive integer values for x and y given the equation $x + y = 10$, and using the perimeter formula for a square, $p = s + s + s + s$.

Prerequisites for learning

Learners need to:
- know how to work out the sum of a positive and negative number using a number line
- be able to find the values of unknown quantities in a calculation by choosing the required inverse operation
- know how to write and solve a one-step word problem as a simple algebraic equation.

Vocabulary

difference, negative, positive, temperature, degree, Celsius, unknown number, variable, equation, substitute, formula (formulae)

Common difficulties and remediation

The difference between two numbers is the same, whether we talk about 'the difference between 6 and 13' or '13 and 6'. In the real world, negative numbers come only in contexts of scales, like a temperature scale, sea level or basement levels of buildings. Guided examples should be given as simple contextual tasks involving negative numbers (for example, temperature rise/drop) where the problems are presented using words and not formal notation.

To help learners find the difference between positive and negative numbers, use a number line to demonstrate how to bridge across zero. For example, to find the difference between 10 and –5, show that this is 10 + 5. This works because, if you count from the positive number to 0 and then count from 0 to the negative number, you have the difference.

Supporting language awareness

At every stage, learners require mathematical vocabulary to access questions and problem-solving exercises. If appropriate, when a new key word is introduced, ask learners to write a definition in their books, drawing a box around it for emphasis.

Encourage learners to write the definition in their own words. Adapt language when it becomes a barrier to learning, for example using 'missing number' rather than 'unknown value'.

Promoting Thinking and Working Mathematically

Opportunities to develop all four pairs of TWM characteristics are provided throughout the unit.

In Lesson 1, learners improve (TWM.08) when they discuss the effectiveness of strategies to calculate the difference between a negative integer and a positive integer without the need to count on.

In Lesson 4, they have opportunities to characterise (TWM.05) when they discuss how to write and use a formula that contains a single variable.

Success criteria

Learners can:
- find the difference between a positive and negative integer using a number line
- find the difference between two negative integers using a number line
- understand what a variable is and how to use it
- solve an equation that contains two variables by substituting values for the unknowns and checking the answer
- substitute values into a simple formula.

Number – Integers and powers

Unit ③ Addition and subtraction of whole numbers (B)

Lesson 1: Subtracting positive and negative integers

Learning objective

Code	Learning objective
6Ni.01	Estimate[, add] and subtract integers.

Resources

calculator (per learner); paper or mini whiteboard and pen (per learner)

Number – Integers and powers

Revise

Use the activity *Hot or cold* or *Temperature stories* from Unit 3: *Addition and subtraction of whole numbers (B)* in the Revise activities.

Teach 🆂🅱 💻 📊 [TWM.08]

- Display **Slide 1**. Read the first problem. **[T&T]** Ask: **How would you use the number line to calculate the difference between negative 7 and 5?** Take responses. Invite a learner to the board to mark the number line to demonstrate the calculation. Expect them to circle –7 and count on in ones to 5 (a difference of 12). Say: **There are a lot of ones to count. Is there a way to split the count into two stages?** Take responses. Elicit that you can bridge to and from zero. Demonstrate a count from –7 to 0 (7) and a count from 0 to 5 (5). The sum 7 + 5 gives the difference, 12.
- Referring to questions **b–e** on **Slide 1**, test learners' estimation skills by asking: **Which of the four problems, b–e, would you expect to give an answer smaller than 20? More than 20? How do you know?** Take responses. **[TWM.08]** Ask: **What do you notice about the difference between the two numbers and the values of the bridging stages? Is there a quick way to calculate the difference between a negative and positive number?** Elicit that the difference between the two numbers is equal to the sum of the two bridging stages.
- With learners helping, work out the answers to problems **b** to **e**. Use the number lines to count on.
- Display the **Thermometer tool**, scaled from –20 to 30°C. Set the pointer to 14°C. Say: **The temperature begins at 14°C and falls to –12°C. What is the difference between the two temperatures?** Take responses. Choose a learner to come to the board to demonstrate how to work out the difference by bridging to and from zero: 12 (–12 to 0) + 14 (0 to 14) = 26 degrees.
- Say: **Sally drops a metal ball into the sea from a height of 38 metres above sea level. If it drops to a depth of –87 metres, how far has the ball fallen?** Remind learners to estimate the answer before calculation. Ask learners to check their answer with a calculator. (125 m)
- Discuss the Let's learn section in the Student's Book.
- 👥 Say: **A lift rises from basement level –7 to upper level 16. How many floors has the lift risen?** (23)

- Introduce the paired activity in the Student's Book, asking learners to find the difference between daily temperatures for Canada and the UK. Choose a pair of learners to explain the strategy they used to find the differences. Ask: **Did anyone calculate the differences instead of counting on? How did you do it?**
- Discuss the Guided practice example in the Student's Book.

Practise 🆆🅱 [TWM.01/07]

- Workbook

Title: Subtracting positive and negative integers

Pages: 22–23

- Refer to Activity 1 from the Additional practice activities.

Apply 👥 💻

- Display **Slide 2** and read the text to the class. Give learners time to work on the problems. Then ask them which strategy they used to find the distance dropped by the ball.

Review

- Ask learners to work in pairs. They take turns to say a 'target' number between 10 and 30. They race each other to write five pairs of positive and negative numbers that have a difference equal to the target number.
- Learners swap papers or whiteboards and check whether each pair is correct. The winner is the learner who writes the most correct calculations in the shortest time.

Assessment for learning

- How would you use a number line to show that the difference between –15 and 6 is 21?
- Give me a pair of numbers, one positive and one negative, that have a difference of 23.

Same day intervention
Enrichment

- Provide problems with three-digit numbers and ask learners to use any calculation strategy they prefer, for example: find the difference between –127 and 132 (259).

Lesson 2: **Subtracting two negative integers**

Learning objective

Code	Learning objective
6Ni.01	Estimate[, add] and subtract integers.

Resources

calculator (per learner)

Revise

Use the activity *Hot or cold* or *Temperature stories* from Unit 3: *Addition and subtraction of whole numbers (B)* in the Revise activities.

Teach 🆂🅱 ▢ ▥ [TWM.03]

- Display **Slide 1**. Remind learners of the terminology of subtraction. On the board, write: 15 – 7 = 8. Ask: **Which number is the subtrahend?** (7) **The minuend?** (15) **The difference?** (8)
- Read the first problem on **Slide 2**. [T&T] Ask: **How would you use the number line to calculate the difference between –7 and –2?** Take responses. Invite a learner to the board to mark the number line to demonstrate the calculation. Expect them to circle –7 and count on in ones to –2 (a difference of 5).
- Referring to questions **b–e** on **Slide 2**, test learners' estimation skills by asking: **Which of the four problems, b–e, would you expect to give an answer greater than 10? How do you know?** (**c** and **e**)
- With learners helping, work out the answers to problems **b–e**.
- Display the **Thermometer tool**, scaled from –20 to 30°C. Set the pointer to –19°C. [TWM.03] Ask: **If the temperature starts at –19°C and reaches –5°C, is that a rise or fall in temperature? How do you know?** Establish that numbers towards the top of the scale are larger and therefore, the temperature is higher (warmer) towards the right of the scale. A change from –19°C to –5°C is therefore a rise in temperature. Ask: **How much does the temperature rise from –19°C to –5°C? Use the temperature scale to show me.** Choose a learner to demonstrate the difference between the two temperatures (14 degrees).
- Ask: **What is the difference between –58 and –133?** Remind learners to estimate the answer before calculation. Ask learners to check their answer with a calculator. (75)
- ⤵ Ask: **A lift moves down from basement level –2 to basement level –13. How many floors is that?** (11)
- Discuss the Let's learn section in the Student's Book.
- Introduce the paired activity in the Student's Book, asking learners to find the amount spent by each

person. Choose a pair of learners to explain the strategy they used to find the differences. Ask: **Did anyone calculate the differences? How did you do it?** Ask learners to comment on the effectiveness of the strategies proposed.
- Discuss the Guided practice example in the Student's Book.

Practise 🆆🅱 [TWM.04]

- Workbook

Title: Subtracting two negative integers

Pages: 24–25

- Refer to Activity 1 (Variation) from the Additional practice activities.

Apply 👥 ▢

- Display **Slide 3** and read the text to the class. Give learners time to work out how many degrees the temperature rises each day. Choose a pair to demonstrate how they worked out the difference between the two temperatures.

Review ▥

- Display the **Thermometer tool**. Ask learners to work in threes. One learner says a negative temperature. A second learner states a second negative temperature. The third learner must work out the difference between the two temperatures. The group confirms the answer and then they swap roles.

Assessment for learning

- How would you use a thermometer scale to show that a change in temperature of –16°C to –3°C is a difference of 13 degrees?
- Tilly's bank account balance is –$5. She buys a gift and her account drops to –$23. What was the price of the gift? ($18)

Same day intervention

Enrichment

- Provide problems with three-digit numbers and ask learners to use any calculation strategy you prefer, for example: find the difference between –143 and –277 (134).

Lesson 3: **Identifying values of variables in calculations (1)**

Learning objective

Code	Learning objective
6Nc.02	Recognise the use of letters to represent quantities that vary in addition and subtraction calculations.

Resources

paper or mini whiteboard and pen (per learner); two plastic cups (per pair); counters – red and blue (per pair)

Revise

Use the activity *Unknown numbers* from Unit 3: *Addition and subtraction of whole numbers (B)* in the Revise activities.

Teach [SB] 🖵 [TWM.01/03]

- Write the equation $x + 5 = 8$ on the board and ask learners for the solution. Establish that there is only one value possible for the unknown value x, namely 3.
- Now write $x + y = 10$ on the board. Ask: **Can you find values for x and y?** Take responses and record possible solutions. Establish that there are many solutions, as the unknown values represented by the letters x and y are variables. Say: **You use variables to represent unknowns and quantities that vary. In other words, they may take on more than one value at any given time.** Explain that the best you can say for this equation is that x equals 10 minus y (or y equals 10 minus x) and you can deduce all the possible values for x and y. (If $x = 0$, $y = 10$; if $x = 2$, $y = 9$; and so on.)
- Write the equation: $b - a = 5$. Ask learners to find eight possible solutions for b and a. Choose learners to read out some of their solutions and record them on the board, for example: b: 7, a: 2; b: 13, a: 8. **[T&T]** Ask: **Why did I ask for just eight solutions and not every solution?** Take responses. Establish that there are an endless number of possible solutions!
- Display **Slide 1**. Read the text aloud. Ask: **Is there an equation you can write using variables that will allow you to find the answer?** Take responses. Invite a learner to the board to write an equation. For example, $r + b = 15$ where r is the number of red counters and b is the number of blue counters.
- Ask learners to work in pairs and distribute cups and counters. Groups place a blue counter in front of one cup and a red counter in front of the second cup to denote colours. They use the equation on the board to find all the possible solutions. Go round the class and ask pairs to say different solutions.
- Discuss the Let's learn section in the Student's Book.
- 📖 Say: **The sum of two variables s and t is 7. If s and t are whole numbers (positive integers), what are all the possible values of the two variables?** (0, 7; 1, 6; 2, 5; 3, 4; 4, 3; 5, 2; 6, 1; 7, 0)

- **[TWM.01/03]** Introduce the paired activity in the Student's Book, asking learners to find all the possible lengths of the two pieces of wood. Choose a pair to explain how they found all the possible solutions.
- Discuss the Guided practice example in the Student's Book.

Practise [WB] [TWM.01/02]

- Workbook

Title: Identifying values of variables in calculations (1)

Pages: 26–27

- Refer to Activity 2 from the Additional practice activities.

Apply 👥 🖵

- Display **Slide 2** and read the text to the class. Give learners time to complete the task. Move around the class and ask for different solutions for the values of A and B.

Review

- Explain that a robot has randomly picked five pieces of fruit from a mixed crate of apples and oranges.
- Ask learners to write a formula to show how many of each type of fruit there could be. ($a + o = 5$, where a = number of apples and o = number of oranges). Next, they draw up a grid to show all the possibilities.
- Choose a pair to share all the possible values. The rest of the class confirm the values are correct.

Assessment for learning

- p and q are two whole number variables. If $p + q = 6$, what are the possible values for the two variables?
- How many different whole number solutions are there to $c + d = 8$? How do you know?

Same day intervention
Enrichment

- Ask learners to enumerate possible values for a combination of three variables in a given range, for example: $x + y + z = 10$ where x, y, z are whole numbers less than 6.

Lesson 4: **Identifying values of variables in calculations (2)**

Learning objective

Code	Learning objective
6Nc.02	Recognise the use of letters to represent quantities that vary in addition and subtraction calculations.

Revise

Use the activity *Angles on a line* from Unit 3: *Addition and subtraction of whole numbers (B)* in the Revise activities.

Teach [SB] 🖵 [TWM.05]

- Display **Slide 1**. Point to the equations and explain that each is a formula for calculating the measurement of a square. Say that a formula is a fact or rule that uses variables. **[T&T]** Ask: **What do you think the first formula helps you calculate?** Take responses. Establish that this is the formula for calculating the perimeter of a square. **[TWM.05]** Ask: **What do you think '4 times s' means?** Establish that since all the sides of a square are the same length, the perimeter will equal s + s + s + s, and that this is equivalent to 4 lots of s, which you can write as: $4 \times s$. Underneath P = s + s + s + s write: $P = 4 \times s$
- Ask: **If the square has sides of 3 cm, what is its perimeter?** (12 cm) **How did you work it out?** (substitute 3 as the value of s in the formula $P = 4 \times s$). **What about a square with sides of 9 cm?** (36 cm)
- Point to the second formula on **Slide 1**. Ask: **What do you think this formula helps you calculate?** Take responses. Establish that this is the formula for calculating the area of a square. Remind learners that you use a superscript 2 to indicate that a number is squared.
- Ask: **If the square has sides of 4 cm, what is its area?** (12 cm^2) **How did you work it out?** (substitute 4 as the value of s in the formula $A = s^2$) **What about a square with sides of 11 cm?** (121 cm^2)
- On the board, write: g – 8 = h. Ask learners to write six possible solutions for g and h.
- Discuss the Let's learn section in the Student's Book.
- 🕮 Ask learners to calculate the perimeter and area of a square with sides of 7 cm, using formulae. (28 cm, 49 cm^2)
- Introduce the paired activity in the Student's Book, asking learners to write a formula to give the perimeter of a square.
- Discuss the Guided practice example in the Student's Book.

Practise [WB]

- Workbook

Title: Identifying values of variables in calculations (2)

Pages: 28–29

- Refer to Activity 2 (Variation) from the Additional practice activities.

Apply 👥 🖵

- Display **Slide 2** and read the text to the class. Give learners time to complete the task. Choose a pair to say how they found the combined area and combined perimeter of the three square buildings.

Review

- Ask pairs of learners to use the formula for calculating the area of a square to calculate the area of squares with the following sides: 3 cm, 8 cm, 12 cm. (9 cm^2, 64 cm^2, 144 cm^2)
- On the board write: p + 5 = q. Ask learners to write six possible solutions for p and q. Ask them to include at least two solutions where one of the variables is a negative number.

Assessment for learning

- A tray is filled with red and blue beads. If r is the number of red beads in the tray and b is the number of blue beads, write a formula that gives the total number of beads in the tray. If r is 48 and b is 37, what is the total number of beads? (85)
- Write a formula that will give the perimeter of a rectangle. Use it to find the perimeter of a rectangle 7 cm by 8 cm. (30 cm)

Same day intervention
Enrichment

- Ask learners to work with the formula for the area of a triangle. Introduce the formula $A = \frac{1}{2} \times b \times h$ where b is the base and h is the height. The algebraic expression of the formula $A = \frac{1}{2} bh$ is introduced in Stage 7.

Number – Integers and powers

Additional practice activities

Activity 1

Learning objective
- Estimate and subtract positive and negative integers.

Resources
Resource sheet 3: Spinners (1) (per pair); pencil and paper clip, for the spinner (per pair)

What to do
- Learners use the 2–9 spinner from Resource sheet 3.
- They take turns to spin twice. They take the first number as negative and the second as positive.

- Learners work out the difference between their pair of numbers.
- The player with the greater difference scores a point.
- The winner is the player with the higher score after nine rounds.

Variations
2 Learners treat both numbers spun as negative.

3 Learners double the numbers spun to make calculations more difficult.

Activity 2

Learning objective
- Recognise the use of letters to represent quantities that vary in addition and subtraction calculations.

Resources
mini whiteboard and pen (per learner)

What to do
- Learners write $x + y = z$ on their whiteboard.
- One learner decides on a value for z and a value for y.
- Both x and y should be positive and less than 10 but they could have the same value.
- The same learner completes the equations on the whiteboard and then their partner solves them.

- Learners swap roles and repeat, choosing different values for x and y.

Variation
2 Learners make up their own formula, for example $c = 6 \times d$. They give a range of numbers for d, for example 1 to 5, and ask their partner to find all the possible values for c.

Unit 4: Multiples, factors, divisibility, squares and cubes

Collins International Primary Maths
Recommended Teaching and
Learning Sequence Term 1, Week 4

Learning objectives

Code	Learning objective
6Ni.06	Understand common multiples and common factors.
6Ni.07	Use knowledge of factors and multiples to understand tests of divisibility by 3, 6 and 9.
6Ni.08	Use knowledge of multiplication and square numbers to recognise cube numbers (from 1 to 125).

Unit overview

In this unit, learners consolidate their knowledge of factors for multiples up to 10 × 10 and solve problems that require them to find common multiples of two numbers. They extend their understanding of divisibility rules to 3, 6 and 9.

In Lesson 4, learners revise their knowledge of square numbers and extend this to working with cube numbers. They recognise that if you multiply a number by itself and then by itself again, the result is a cube number. Learners also investigate the relationship between square and cube numbers, for example $5^3 = 5^2 \times 5$ and use the notation for squared (^{2}) and cubed (^{3}) numbers.

Prerequisites for learning

Learners need to:
- be able to recall the times tables up to 10 × 10
- know how to use their knowledge of times tables up to 10 × 10 to find multiples
- know how to use their knowledge of times tables up to 10 × 10 to find factors
- understand the term 'divisibility rule'
- be able to recall squares of all numbers up to 10 × 10.

Vocabulary

multiple, common multiple, factor, common factor, divisibility, square number, cube number

Common difficulties and remediation

Problems recalling the 6 times table may stem from inability with quick recall of the 2, 3 and 5 times tables. Problems recalling answers to the 7 times and 9 times tables may result from a lack of confidence with other tables. Return regularly to the other tables to develop confidence, playing step-counting games and solving missing-number problems, and reminding learners of the commutative property of multiplication.

Look out for learners who mistakenly assume that a square number is a number multiplied by 2 and a cubed number is a number multiplied by 3. This misconception might be the result of learners misinterpreting the superscript number notation (y^2, y^3) for the square and cube of a number. Address this misconception by modelling square numbers using arrays and grids, and modelling cube numbers by constructing 3D cubes from square layers of blocks.

Supporting language awareness

Learners will learn many terms relating to the concepts of multiplication and division. Some of these words, including 'factor', 'square' and 'cube', have more than one meaning and some are used imprecisely outside of mathematics. It is important that learners experience a variety of terms and that the terms are used accurately.

Promoting Thinking and Working Mathematically

Opportunities to develop all four pairs of TWM characteristics are provided throughout the unit.

In Lesson 1, learners generalise (TWM.02) when they explain why a number is not a common multiple of two numbers.

In Lesson 2, learners improve (TWM.08) when they suggest an alternative method to organise the listing of common factors of two numbers.

Opportunities to conjecture (TWM.03) are developed in Lesson 3 when learners reason why it is more efficient for divisibility tests for 3, 6 and 9 to be applied in a particular order.

Success criteria

Learners can:
- use their knowledge of times tables up to 10 × 10 to find factors
- find the common multiples of two numbers by comparing lists of multiples
- find the common factors of two numbers by comparing lists of factors
- apply divisibility tests for 3, 6 and 9 to numbers
- use knowledge of multiplication and square numbers to recognise cube numbers (from 1 to 125).

Lesson 1: **Common multiples**

Learning objective

Code	Learning objective
6Ni.06	Understand common multiples [and common factors].

Resources

paper or mini whiteboard and pen (per learner)

Revise

Use the activity *Mixed up multiples* from Unit 4: *Multiples, factors, divisibility, squares and cubes* in the Revise activities.

Teach 🆂🅱 🖥 [TWM.02]

* Display **Slide 1**. Discuss the text and the steps involved in finding common multiples. Remind learners that a multiple is the product of a number and any whole number.
* Write 6 and 8 on the board. Ask learners to list the multiples of each number, up to 100, on paper or their whiteboards and then to circle the multiples that both lists have in common. Choose learners to list the common multiples one at a time (24, 48, 72, 96).
* [T&T] Ask: **Is there a pattern to the multiples?** Elicit that the numbers increase by increments of 24. Ask: **Can you use this value (24) to work out the next common multiple of 6 and 8?** (120) **How did you calculate it?** (previous common multiple 96 + 24) [TWM.02] Ask: **Is 142 a common multiple of 6 and 8? If not, why?** (previous common multiple 120 + 24 is 144, not 142)
* Ask learners to help make a list of the multiples of 4 and 9. Ask them to identify common multiples up to 100. (36, 72)
* Discuss the Let's learn section in the Student's Book.
* 🗣 Say: **A lift rises from the ground floor and stops every two floors. A second lift rises from the ground floor and stops every five floors. What are the first two floors that both the lifts stop at?** (10, 20)
* Introduce the paired activity in the Student's Book, asking learners to work out how often Nathan will have falafels and meatballs on the same day. Hint that the solution can be found by using common multiples. Choose a learner to explain how they solved the problem (4 × 5 = 20; every 20 days).
* Discuss the Guided practice example in the Student's Book.

Practise 🆆🅱 [TWM.07]

* Workbook

Title: Common multiples

Pages: 30–31

* Refer to Activity 1 from the Additional practice activities.

Apply 👥 🖥

* Display **Slide 2** and read the text to the class. Learners find the first common multiple of 3 and 4. (12 days) Invite learners to share their answers and solutions.

Review

* Ask learners to make a list of the multiples of 6 and 9. Then ask them to identify common multiples up to 100. (18, 36, 54, 72, 90)

Assessment for learning

* Explain to a partner how you would find the common multiples for the numbers 10 and 15 up to 100.
* Is 95 a common multiple of 5 and 6? How do you know?

Same day intervention
Enrichment

* Ask learners to find common multiples of numbers greater than 10, for example 14 and 17, or 23 and 26.

Lesson 2: **Common factors**

Learning objective

Code	Learning objective
6Ni.06	Understand common multiples and common factors.

Resources

paper or mini whiteboard and pen (per learner)

Revise

Use the activity *Factor find* from Unit 4: *Multiples, factors, divisibility, squares and cubes* in the Revise activities.

Teach 📖 📊 🖥 [TWM.08]

- Remind learners that a factor is a number that divides exactly into another number. On the board, write: 16, 42. Say: **Write the factors of each number.** Invite learners to list the factors and explain how they found them.
- Display the **Number square tool**, set to 'Colour square'; 'Hide all'. Colour a grid of eight columns by four rows. Revise arrays. Say: **Multiplication can be represented as an array – a set of rows and columns. This array shows the number 32.** Discuss how the grid models the number fact $8 \times 4 = 32$ and the factor pair: 8, 4. Demonstrate the commutative number facts that are revealed when an array is turned through a quarter turn. Show that different multiples of 4 and 8 can be modelled by adding or removing rows or columns in the array. Ask learners to use the array method to find the factors of 18. For example, colour a grid of six columns by three rows. Say: **This array shows the number 18.** (factors: 3×6) **[T&T]** Ask: **What other arrays are there for 18?** (arrays: 1×18, 2×9; factors: 1, 2, 3, 6, 9)
- Display **Slide 1**. Discuss the text and the steps involved in finding common factors. **[TWM.08]** Ask learners, working in pairs, to design an alternative method to organise finding common factors. Choose pairs to demonstrate their methods and invite comments from the class regarding the strengths and weaknesses of the system proposed. Ask: **How would you improve this method?**
- On the board, write: 24 and 56. Ask learners to list the factors of each number on paper or their whiteboards and then circle the factors that both lists have in common. Choose learners to list the common factors one at a time. (1, 2, 4, 8)
- 🕮 Ask: **What are all the common factors of 28 and 56?** (1, 2, 4, 7, 14, 28)

- Discuss the Let's learn section in the Student's Book.
- Introduce the paired activity in the Student's Book, asking learners to work in pairs to find the greatest number of piles of stickers possible. Choose learners to explain how they solved the problem (6 piles).
- Discuss the Guided practice example in the Student's Book.

Practise 📝

- Workbook

Title: Common factors

Pages: 32–33

- Refer to Activity 1 (Variation) from the Additional practice activities.

Apply 👥 🖥

- Display **Slide 2** and read the text to the class. If necessary, provide a clue for solving the problem. Learners will need to use what they know about common factors. (18 plants in each row)

Review

- Ask learners to make a list of the factors of 32 and 80. Then ask them to identify all of the common factors (1, 2, 4, 8, 16).

Assessment for learning

- Explain to a partner how you would find the common factors for the numbers 18 and 42 (1, 2, 3, 6).
- Is 8 a common multiple of 28 and 56? How do you know?

Same day intervention
Enrichment

- Ask learners to find common factors of pairs of three-digit numbers, for example 126 and 144 (1, 2, 3, 6, 9, 18).

Number – Integers and powers

Lesson 3: **Tests of divisibility by 3, 6 and 9**

Learning objective

Code	Learning objective
6Ni.07	Use knowledge of factors and multiples to understand tests of divisibility by 3, 6 and 9.

Resources

mini whiteboard and pen (per pair)

Revise

Use the activity *Three in a row* from Unit 4: *Multiples, factors, divisibility, squares and cubes* in the Revise activities.

Teach SB 🖵 [TWM.03/08]

- Display **Slide 1**. Discuss the divisibility rules for 3, 6 and 9.
- On the board, write: 459. **[T&T]** Ask: **Is this number divisible by 3? How do you know?** Take responses. Confirm that 459 is divisible by 3, recording: 4 + 5 + 9 = 18 and 1 + 8 = 9. Explain that by finding the sum of the digits, which is 18 and the sum of 1 and 8, you get 9, so it is divisible by 3. Ask: **Is this number also divisible by 6? What tests do you need to apply?** Take responses. Confirm that you need to test that 459 is divisible by both 2 and 3. Ask: **Which test should you apply first?** Agree the test for divisibility by 2. Ask: **Why?** Establish that it is the easiest test to apply since you only need to confirm whether the number is even (no). Agree that 459 is not divisible by 6. Ask: **Is this number also divisible by 9? How do you know?** Take responses. Confirm that 459 is divisible by 9 as the sum of its digits ends up with 9 (already confirmed by the divisibility test for 3).
- **[TWM.03]** Ask: **When testing a number for divisibility by 3, 6 or 9, which order should you apply the tests? Why?** Take responses. Elicit that you should first test for 9 since a positive answer will also tell you that the number is divisible by 3. Explain that this is because all multiples of 9 are multiples of 3. Ask: **Can you assume that if a number is divisible by 3, it is also divisible by 9?** (no, because not all multiples of 3 are multiples of 9) Also, since the divisibility test for 6 can use the information gathered from the divisibility test for 3, it should follow after that test.
- On the board, draw a table with columns headed 'Divisible by 9?', 'Divisible by 3?' and 'Divisible by 6?' and rows for the numbers 288, 483 and 825. Ask learners to work in pairs to complete the table. Confirm the answers as a class. (9: 288 only; 3: 288, 483, 825; 6: 288)
- 🄟 Ask: **Is 567 divisible by 9? How do you know?** (yes, 5 + 6 + 7 = 18, 1 + 8 = 9)
- Discuss the Let's learn section in the Student's Book.
- **[TWM.08]** Introduce the paired activity in the Student's Book, asking learners to find five

three-digit numbers that are divisible by 3, 6 and 9. Ask: **How did you work out which numbers would be divisible by all three divisors? Did anyone use multiplication to find the numbers?** Choose a pair of learners who used multiplication to explain their solution. Learners should have found multiples of 162 (multiples of 3 × 6 × 9).
- Discuss the Guided practice example in the Student's Book.

Practise WB [TWM.01]

- Workbook

Title: Tests of divisibility by 3, 6 and 9

Pages: 34–35

- Refer to Activity 2 from the Additional practice activities.

Apply 👥 🖵

- Display **Slide 2** and read the text to the class. Ask learners to find the crate of drinks that can be stacked into arrangements of 3, 6 or 9. Choose a pair of learners to explain how they found the solution (648 drinks).

Review

- Ask learners to work in pairs to write down on their whiteboard three three-digit numbers divisible by 3, three three-digit numbers divisible by 6 and three three-digit numbers divisible by 9. They organise and label the numbers in groups: 'Divisible by 3', 'Divisible by 6', 'Divisible by 9'. Next, they replace one digit in each number with a line. They swap whiteboards with another pair and try to find the missing digit. Learners then return their whiteboards for marking.

Assessment for learning

- Is 954 divisible by 6? How do you know? (yes)
- Give me a three-digit number that is divisible by 3 but not 6.
- Give me a three-digit number that is divisible by 6 but not 9.

Same day intervention

Enrichment

- Ask learners to test four- and five-digit numbers for divisibility by 3, 6 or 9, for example: 17 835 (divisible by 3 only).

Lesson 4: **Cube numbers**

Learning objective

Code	Learning objective
6Ni.08	Use knowledge of multiplication and square numbers to recognise cube numbers (from 1 to 125).

Resources

interlocking cubes (per group); paper or mini whiteboard and pen (per pair)

Revise

Use the activity *Multiplication race* from Unit 4: *Multiples, factors, divisibility, squares and cubes* in the Revise activities.

Teach [SB] 🖵 [TWM.03]

- Display **Slide 1**. Discuss the text and the diagrams. Ensure that learners understand what a cube number is, and how it is made. Ask: **What multiplication would give you the next cube number after 5?** ($6 \times 6 \times 6$) **How would you write this?** Choose a learner to come to the board to write the multiplication for the cube of 6 ($6 \times 6 \times 6$). Ensure that learners know to use the superscript 3 to show that a number is cubed. Write 6^3 on the board alongside $6 \times 6 \times 6$.

- Organise learners into groups and provide interlocking cubes. Ask them to model the first four cube numbers (1, 8, 27, 64). Ask: **Why are these numbers called cube numbers?** Take responses. Establish that they get their name from the fact that an equivalent number of smaller cubes can be arranged to make the 3D shape of a larger cube.

- Say: **When you measure 3D shapes, you use cubic measures, such as cubic centimetres and cubic metres.** Write both abbreviations, cm^3 and m^3, on the board and compare to area measures of cm^2 and m^2. Also point out the superscript 3, as with cube numbers. Point to **Slide 1** and show learners the measures on the cubes. Take one example and say: **This cube measures $3 \times 3 \times 3$. If each of the small cubes is 1 cm^3, then the cube measures 27 cm^3.** If appropriate, discuss the link between cube numbers and volume.

- [T&T] [TWM.03] Say: **Toby's teacher gave him a clue for finding cube numbers. She said that he should use a square number to find a cube number. What do you think she meant?** Take responses. Elicit that if you know the square of a number, then you can use it to find the cube of that number, for example: $4^3 = 4^2 \times 4 = 16 \times 4 = 64$.

- 🖐 Ask: **What is the cube of 5?** (125)

- Discuss the Let's learn section in the Student's Book.

- Introduce the paired activity in the Student's Book, asking pairs to find a square number that is also a cube number (for example, 1 and 64).

- Discuss the Guided practice example in the Student's Book.

Practise [WB]

- Workbook

Title: Cube numbers

Pages: 36–37

- Refer to Activity 2 (Variation) from the Additional practice activities.

Apply 👥 🖵

- Display **Slide 2** and read the text to the class. Choose learners to explain how they found the total number of smaller cubes in each larger cube.

Review

- Pairs of learners race each other to list numbers 1 to 5 in a row on paper or their whiteboard. Above this, they write the corresponding sequence of square numbers. In a third row, they write the sequence of corresponding cube numbers (1, 1, 1; 2, 4, 8; 3, 9, 27; 4, 16, 64; 5, 25, 125). The first pair to put the sequence in order raises their hands.

Assessment for learning

- What is the cube of 4? How do I write this as a number sentence using the cubed symbol? ($4^3 = 64$)
- 125 is the cube of which number? (5)

Same day intervention
Enrichment

- Ask learners to solve the following problem: 'A large cube is constructed from smaller, equal-sized cubes. How many smaller cubes will be needed to construct a larger cube with a side length of 8 smaller cubes?' (512)

Number – Integers and powers

Additional practice activities

Activity 1

> **Learning objective**
> • Understand common multiples and common factors.

Resources

Resource sheet 4: Gameboard (per learner); counters (per group); two 1–6 dice or Resource sheet 5: 1–6 spinner (per learner); pencil and paper clip, for the spinner (per learner); Resource sheet 6: 10–100 number cards (per learner) (for variation)

What to do

- Arrange the learners into groups of three or four and provide each learner with a gameboard and two dice or a 1–6 spinner, pencil and paper clip.
- Each learner rolls both dice or spins their spinner twice.
- Each learner then finds the common multiples of the two numbers they generated up to 60.

- They cover these numbers on their gameboard with counters.
- The winner is the player who has the most numbers covered on their gameboard after five rounds.

Variation

Each learner takes two cards from a pile of 10–100 number cards.

They find the common factors for their pair of numbers.

Each player scores points equivalent to the number of common factors found.

The winner is the player with the most points after five rounds.

Activity 2

> **Learning objectives**
> • Use knowledge of factors and multiples to understand tests of divisibility by 3, 6 and 9.
> • Use knowledge of multiplication and square numbers to recognise cube numbers (from 1 to 125).

What to do

- On the board, write: 5 × 75 =, 4 × 108 =, 6 × 24 =, 8 × 86 =, 7 × 96 =, 4 × 142 =, 6 × 239 =, 8 × 644 =, 7 × 688 =.
- Ask learners to work in pairs to find the product of each multiplication and say whether it is divisible by 3, 6 or 9.

Variation [TWM.04]

Learners investigate the statement: 'All cube numbers are either multiples of 9 or within one of a multiple of 9.' Is this statement correct? How do you know?

Unit 5: Whole number calculations

Collins International Primary Maths
Recommended Teaching and
Learning Sequence Term 1, Week 5

Learning objectives

Code	Learning objective
6Ni.02	Use knowledge of laws of arithmetic and order of operations to simplify calculations.
6Ni.03	Understand that brackets can be used to alter the order of operations.

Unit overview

This unit extends learners' knowledge and understanding of how numbers behave and leads learners to making generalisations. Learners are reminded that the commutative and associative properties of addition and multiplication, and the distributive property of multiplication over addition, lead to any-order principles for addition and multiplication. For example, the terms in the expression $25 \times 3 \times 4 + 5$ can be rearranged to make the calculation easier to solve: $25 \times 4 \times 3 + 5 = 100 \times 3 + 5 = 305$. When calculating, learners draw on these strategies and apply them.

In Lesson 3 and 4, learners are introduced to the use of brackets in expressions. They investigate how putting brackets in different places can alter the answer to a calculation and explore the order of operations using brackets, for example, $4 + 5 \times 2 = 14$ and $(4 + 5) \times 2 = 18$.

Prerequisites for learning

Learners need to:
- be able to name, describe and apply the commutative property of addition and multiplication
- be able to name, describe and apply the associative property of addition and multiplication
- be able to name, describe and apply the distributive property of multiplication over addition.

Vocabulary

commutative property, associative property, distributive property, multiplier, multiplicand, augend, addend, term, brackets, order of operations

Common difficulties and remediation

Provide scaffolded support for learners who find the properties of number difficult to grasp. For the commutative property, provide tiles or counters and ask learners to model multiplications; for example, constructing two arrays to model 8×2 and 2×8. Learners compare the product of each factor pair and identify that products are identical.

For the associative property, ask learners to construct visual models with counters or tiles, for example modelling $3 \times 2 \times 2$ as either (two 3 by 2 arrays or three 2 by 2 arrays). Again, learners find that the

products are the same and begin to generalise that changing the groupings does not change the product.

For the distributive property, explain that we use this a lot when multiplying mentally. For example, if asked to find the product of 4×37 mentally, many of us would decompose 37 into 30 and 7. We would then say that $(4 \times 30) + (4 \times 7) = 120 + 28 = 148$.

Ask learners to model two arrays of 4×37 counters. They split the second array into 4×30 and 4×7 and find that the products are the same.

Supporting language awareness

Learners should first describe the behaviour of numbers in their own words, for example: 'I can multiply by breaking the bigger number apart, multiplying the numbers separately, and then adding the results together to get the answer.' With further investigation, and using similar examples, learners make generalisations and recognise that this property is true for any (real) numbers.

Promoting Thinking and Working Mathematically

Opportunities to develop all four pairs of TWM characteristics are provided throughout the unit.

In Lesson 1, learners classify (TWM.06) when they identify the property of number that can be used to simplify a calculation.

In Lesson 3, learners generalise (TWM.02) when they identify that the steps involved in solving a calculation have failed to follow the correct order of operations.

Opportunities to critique (TWM.07) are developed in Lesson 4 when learners rewrite a calculation so that it follows the order of operations.

Success criteria

Learners can:
- understand that a generalisation of a behaviour of number can be expressed as a written rule or property
- recognise how commutative, associative and distributive properties can be applied when simplifying calculations
- apply the correct order of operations to evaluate calculations including the use of brackets.

Number – Integers and powers

Lesson 1: Simplifying calculations (1)

Number – Integers and powers

Learning objective

Code	Learning objective
6Ni.02	Use knowledge of laws of arithmetic and order of operations to simplify calculations.

Resources

paper or mini whiteboard and pen (per learner)

Revise

Use the activity *Reordering additions* from Unit 5: *Whole number calculations* in the Revise activities.

Teach [SB] 🖥 [TWM.06]

- Display **Slide 1**. Discuss the three properties of number that can be used to help simplify calculations. Take learners through the examples under each property heading. Ask learners to work in pairs. **[T&T] [TWM.06]** Name each property in turn and ask pairs of learners to write on paper or their whiteboard an example of a calculation that can be made easier by applying the property. For example, reordering the terms in the addition $8 + 27 + 32 =$ to $32 + 8 + 27 =$ because $32 + 8$ makes 40, an easier number to add.

- Point to calculations **a–d** on **Slide 1**. **[T&T]** Ask learners to discuss each calculation and the strategy they would use to simplify and solve it. Choose pairs to explain their solution, encouraging them to use the language of number properties as they do so. Agree the following simplifications: **a** $16 + 7 + 14 = 16 + 14 + 7 = 30 + 7 = 37$ (associative); **b** $4 \times 7 \times 25 = 4 \times 25 \times 7 = 100 \times 7 = 700$ (associative); **c** $25 \times 3 \times 4 + 5 = 25 \times 4 \times 3 + 5 = 100 \times 3 + 5 = 300 + 5 = 305$ (associative); **d**

$$33 \times 7 \begin{cases} 30 \times 7 = 210 \\ 3 \times 7 = \underline{21} + \\ 221 \end{cases}$$

$7 \times 33 + 9 = 221 + 9 = 230$ (distributive)

- Ask learners to simplify and solve calculations **e–h**. Choose pairs to present their solutions and confirm the answers as a class (**e** 120; **f** 630; **g** 425; **h** 234).

- ▶ On the board, write $7 \times 27 + 26 =$. Say: **Explain to your partner how you would solve this problem.** (215)

- Discuss the Let's learn section in the Student's Book.

- Introduce the paired activity in the Student's Book. Ask learners to discuss the best strategy for solving each calculation using the language of number properties.

- Discuss the Guided practice example in the Student's Book.

Practise [WB]

- Workbook

Title: Simplifying calculations (1)

Pages: 38–39

- Refer to Activity 1 from the Additional practice activities.

Apply 🖥 👥

- Display **Slide 2** and read the text to the class. Learners write and solve a calculation to find the total number of eggs.

Review

- Write the following calculations on the board and challenge learners to answer them as quickly as possible: $38 + 34 + 22 + 46 =$, $7 \times 54 + 3 =$, $6 \times 12 \times 50 =$. (140, 381, 3600)

- Choose a calculation and ask learners to explain the strategy they used to solve it.

Assessment for learning

- What is $3 \times 54 + 13$? (175)
- A calculation is simplified to $6 \times 70 + 6 \times 8 + 36$. What was the calculation? ($6 \times 78 + 36$)

Same day intervention

Support

- Some learners interpret the order of operations incorrectly, thinking that multiplication and division do not have equal priority, and that one comes before the other. Explain that the two operations have equal priority and must be performed in order, from left to right. The same is true for addition and subtraction.

Lesson 2: **Simplifying calculations (2)**

Learning objective

Code	Learning objective
6Ni.02	Use knowledge of laws of arithmetic and order of operations to simplify calculations.

Resources

paper or mini whiteboard and pen (per learner); Resource sheet 7: 3–99 number cards (per pair)

Revise

Use the activity *Reordering multiplications* from Unit 5: *Whole number calculations* in the Revise activities.

Teach SB 🖵 [TWM.02]

- Display **Slide 1**. Discuss how the properties of number can be used to simplify a problem. Take learners through the example provided.
- **[T&T] [TWM.02]** Working in pairs, ask learners to think of a similar problem that could be simplified using the distributive property. Choose a pair of learners to explain the problem and how it could be simplified.
- Read problem A on **Slide 1**. Ask learners to write a calculation for the problem on paper or their whiteboard and then simplify it. Choose a pair of learners to explain how they solved it. Expect them to use the associative property to reorder the terms in the multiplication. Record:
  $200 \times 71 = 2 \times 100 \times 71 = 2 \times 71 \times 100 = 142 \times 100 = 14\,200$ cm. Ask: **What is the length in metres?** (142 m) Ask learners to think of a similar problem that could be simplified using the associative property. Choose a pair of learners to explain the problem and how it could be simplified.
- Read problem B on **Slide 1**. Ask learners to write a calculation for the problem. Choose a pair of learners to describe the calculation involved $(6 \times 9 \times 5 + 2^2)$. Ask: **How would you simplify this calculation?** Take responses. Expect them to use the associative property to reorder the terms in the multiplication. Record the reordering and simplification of the problem in stages: $6 \times 9 \times 5 + 2^2 = 5 \times 6 \times 9 + 4 = 30 \times 9 + 4 = 270 + 4 = 274$ cm^2. Ask learners to think of a similar problem that could be simplified using the associative property. Choose a pair of learners to explain the problem and how it could be simplified.
- 🗣 Say: **There are 35 bookshelves in a library. Each bookshelf has 7 shelves. There are 60 books on each shelf. How many books is that in total?** (14 700)
- Discuss the Let's learn section in the Student's Book.
- Introduce the paired activity in the Student's Book, asking learners to work in pairs to find the total amount received by the supermarket cashier. Choose a pair of learners to explain how they solved the addition problem.
- Discuss the Guided practice example in the Student's Book.

Practise WB

- Workbook

Title: Simplifying calculations (2)

Pages: 40–41

- Refer to Activity 1 (Variation) from the Additional practice activities.

Apply 👥 🖵

- Display **Slide 2** and read the text to the class. Give learners time to answer the problem and then choose pairs to explain how they found the total area of the field.

Review

- Distribute number cards to pairs of learners and ask them to sort the cards face down into three piles as follows: pile A – numbers 3 to 9; pile B – multiples of 10; pile C –numbers 11 to 99 (excluding multiples of 10 which are in pile B).
- Learners take one card from each pile A, B and C and one extra card from pile C to make the calculation: card A × card B × card C + card C.
- Learners race their partner to simplify and solve the calculation using properties of number. The player with the greater result scores a point. After five rounds, the winner is the player with the higher score.

Assessment for learning

- On the board write: $73 + 74 + 37 + 16$. Ask: **What is the answer? Which property of number did you use to solve this?** (200, associative)
- Tell me why 73×30 can be written as $70 \times 30 + 3 \times 30$.

Same day intervention
Support

- Guide learners to construct representations of problems, such as arrays and area models. These will help them make connections between the concrete and pictorial representations and the abstract mental procedures.

Number – Integers and powers

137

Unit **5** **Whole number calculations**

Lesson 3: **Using brackets (1)**

Learning objective

Code	Learning objective
6Ni.03	Understand that brackets can be used to alter the order of operations.

Resources

paper or mini whiteboard and pen (per learner); paper, for the Workbook (per learner)

Revise

Use the activity *Ordering operations* from Unit 5: *Whole number calculations* in the Revise activities.

Teach [SB] 🖵 [TWM.01]

- Display **Slide 1**. Remind learners that when a calculation involves a mix of operations, you need to apply a set of rules to tell you the order in which to complete them. On the board, write: $32 ÷ 8 + 4 × 3 =$. **[T&T]** Ask: **In which order do you complete each operation?** Take responses. Explain that division and multiplication operations are completed first, followed by the addition. Working in pairs, ask learners to solve the calculation. Choose a pair of learners to explain the steps involved. Record: $32 ÷ 8 + 4 × 3 = 4 + 12 = 16$.

- On the board, write: $4 × (7 + 2) =$. Referring to the Order of Operations diagram on **Slide 1**, explain that calculations in brackets must be completed first, before any other operation. They show which parts of the calculation go together and are important because if you work out an expression in the wrong order, it is quite possible that you will get at an incorrect answer. Ask learners to solve the calculation. Record: $= 4 × 9 = 36$.

- On the board, write: $15 – 9 ÷ 3 =$. **[TWM.01]** Say: **Leah calculates the answer to $15 – 9 ÷ 3$ as 2. She asks her friend Florence to calculate the answer. Florence says that it is 12. What has gone wrong? What does this tell you about the importance of the order of operations and why brackets are important?** Take responses. Expect learners to explain that the order of operations was not used by Leah who completed the subtraction first, rather than the division. However, Leah may have intended for the subtraction to be completed first and if so, she should have used brackets to indicate this. Record on the board: $(15 – 9) ÷ 3 = 2$.

- On the board, write: $(6 × 8) + 13 =$ and $60 ÷ (42 – 27 =)$. Ask learners to simplify and solve the calculations. Choose a pair of learners to explain the steps involved (61, 4).

- Say: **Prove that the answer to $9 × (8 + 4)$ is 108.**

- Discuss the Let's learn section in the Student's Book.

- Introduce the paired activity in the Student's Book, asking learners to say the order in which they solved the operations for each calculation.

- Discuss the Guided practice example in the Student's Book.

Practise [WB]

- Workbook

Title: Using brackets (1)

Pages: 42–43

- Refer to Activity 2 from the Additional practice activities.

Apply 👥 🖵

- Display **Slide 2** and read the text to the class. Choose a pair of learners to read out the calculation they wrote and the operation they included in brackets. Ask: **Why is it important that this operation was placed in brackets?**

Review

- On the board, write: 7, 28 and 57. Ask learners to work in pairs to write three calculations that give these answers. Each calculation must include two operations with one operation inside brackets. Each calculation should include one multiplication or division operation *and* one addition or subtraction operation.

- Choose learners to share the calculations they have written.

Assessment for learning

- I calculate the answer to $18 + 5 × 3$ as 69. My friend calculates the answer as 33. Explain why we have two different answers and how the calculation could have been written differently to avoid this confusion.

- Explain to a partner how you would solve $60 ÷ (15 – 3)$. (5)

Same day intervention
Support

- Remind learners that the order of operations is multiplication or division before addition or subtraction. If part of the calculation includes brackets, then this should be completed first before the other operations.

Number – Integers and powers

Lesson 4: **Using brackets (2)**

Learning objective

Code	Learning objective
6Ni.03	Understand that brackets can be used to alter the order of operations.

Resources

paper or mini whiteboard and pen (per learner)

Revise

Use the activity *Spinning operations* from Unit 5: *Whole number calculations* in the Revise activities.

Teach 🖥 [TWM.07]

- Discuss the following problem: 33 learners in a class are each given 5 red stars and 3 blue stars for an art project. **[T&T]** Ask: **What calculation would you write that gives you the total number of stars received by the learners?** Give learners time to answer the question. On the board, write $33 \times 5 + 3 =$. **[TWM.07]** Say: **I have written a calculation to solve. Can anyone see a problem with it?** Take responses. Establish that following the order of operations, the problem will be completed with the multiplication step first followed by the addition. Record: $33 \times 5 + 3 = 165 + 3 = 168$. The correct calculation would find the total number of stars given to each child multiplied by the total number of children. Ask: **What would this calculation look like?** Take responses. Praise learners who point out that the addition in the calculation needs to be written inside a set of brackets to ensure it is completed first. Record: $33 \times (5 + 3) = 33 \times 8 = 30 \times 8 + 3 \times 8 = 264$.

- 🗣 Say: **I place some counters on a table and arrange them into 12 groups of 21. I then remove 6 counters from each group. How many counters are on the table now?** $(12 \times (21 - 6) = 180)$

- Discuss the Let's learn section in the Student's Book.

- Introduce the paired activity in the Student's Book, asking pairs to find the total number of parents who attend the school picnic.

- Discuss the Guided practice example in the Student's Book.

Practise 📘 [TWM.04]

- Workbook

Title: Using brackets (2)

Pages: 44–45

- Refer to Activity 2 (Variation) from the Additional practice activities.

Apply 👥 🖥

- Display **Slide 1** and read the text to the class. Choose learners to explain how they found the total amount of water remaining in all 7 tanks.

Review

- Explain a problem to learners: **Ben fills each of 6 small plant pots with 235 g of soil. Unfortunately, he spills 65 g of soil out of each pot. What is the total amount of soil left in all 6 pots?**

- Learners write a calculation to represent the problem and solve it. Ensure that they write the subtraction operation in brackets, representing the amount of soil that remains in one pot.

Assessment for learning

- Tell me a word problem that is represented by the calculation $5 \times (32 + 24)$.

- Write and solve the calculation that will tell me how many bananas are in 17 bunches of 14 bananas where 8 bananas have been removed from each bunch $(17 \times (14 - 8))$.

Same day intervention
Support

- Look out for learners who once told they need to insert brackets in a calculation (to indicate an operation that should be completed first), place them at the very beginning of the expression. Explain that brackets can be placed around any operation in a calculation, whether it is first, middle or last.

Number – Integers and powers

Additional practice activities

Activity 1 :: △2

> **Learning objective**
> • Use knowledge of laws of arithmetic and order of operations to simplify calculations.

Resources

paper (per pair)

What to do

• Learners work in pairs to write three calculations, each of which can be simplified using a property of number: commutative, associative or distributive.

• Learners swap papers with another pair and solve the problems. They write their working and name the property of number used in the solution.

• Learners return their papers for marking.

Variation

△2 Learners write three word problems, each of which can be simplified using a property of number: commutative, associative or distributive.

Activity 2 :: △2

> **Learning objective**
> • Understand that brackets can be used to alter the order of operations.

Resources

paper (per pair)

What to do

• Learners work in pairs to write three calculations, each of which has two operations, a combination of a multiplication or division, and an addition or subtraction.

• One of the operations is written in a set of brackets to show that it must be calculated first.

• Learners swap papers with another pair and solve the problems. They return their papers for marking.

Variations

1 The teacher can use a highlighter pen to highlight the bracketed section of the calculation to help learners remember its priority.

△2 Learners write three word problems, each of which has two operations, a combination of a multiplication or division, and an addition or subtraction.

Unit 6: Multiplication of whole numbers

Collins International Primary Maths
Recommended Teaching and
Learning Sequence Term 2, Week 1

Learning objective

Code	Learning objective
6Ni.04	Estimate and multiply whole numbers up to 10 000 by 1-digit or 2-digit whole numbers.

Unit overview

In this unit, learners consolidate their knowledge of written methods of multiplication that were introduced in Stages 3 to 5 (partitioning, and the expanded written method) and extend these to include the formal written method. They develop understanding of how the algorithm uses the distributive property and place value to help solve a multiplication problem.

This unit focuses on multiplication of numbers up to 10 000 by one- or two-digit numbers, including estimating and checking the answer to a calculation. Learners use this knowledge to answer simple problems.

Learners are encouraged to look carefully at the numbers in the calculation and decide on the most efficient, effective and appropriate method, including using mental strategies such as partitioning.

Prerequisites for learning

Learners need to:
- be able to make a reasonable estimate for the answer
- partition a multi-digit number in order to multiply by a one-digit number
- multiply a multi-digit number by a one-digit number, using a range of strategies
- apply the correct order of operations to evaluate calculations.

Vocabulary

partitioning, grid method, expanded written method, formal written method (short multiplication), formal written method (long multiplication)

Common difficulties and remediation

Some learners find regrouping in multi-digit multiplication difficult to master and require additional experiences with the expanded written forms. Once they are confidently multiplying, using partial products, they are ready to reduce the amount of recorded information and appreciate how regrouping can help.

It is important to identify learners who make errors when performing written calculations. The errors may be procedural, indicating that the learner has not yet fully grasped how to use the algorithm. Several interventions should be considered. Give learners multiplications with deliberate errors. Ask them to identify the mistakes, then talk through what is wrong and how errors should be corrected. Encourage learners to estimate answers and use approximations. Assess their progress by careful questioning, for example: **What answer do you expect to get? How did you reach that estimate? Do you expect your answer to be less than or greater than your estimate? Why?**

Supporting language awareness

If appropriate, when a new key word is introduced, ask learners to write a definition in their books, drawing a box around it for emphasis. Encourage learners to write the definition in their own words. It is also important to adapt language when it becomes a barrier to learning; for example, using simpler terms to describe the formal written method: 'Multiplying each place value of a number and then finding the sum.'

Promoting Thinking and Working Mathematically

Opportunities to develop all four pairs of TWM characteristics are provided throughout the unit.

In Lesson 1, learners specialise (TWM.01) when they determine a ThHTO × O calculation given a specific set of digits and an answer range.

In Lesson 2, they critique (TWM.07) when they compare the steps involved in the expanded written method and formal written method of multiplication.

Learners improve (TWM.08) in Lesson 3 when they discuss the most efficient methods to solve a ThHTO × TO multiplication calculation.

Success criteria

Learners can:
- estimate the answer to a multiplication of a whole number up to 10 000 by a one-digit or two-digit whole number
- multiply whole numbers up to 10 000 by one-digit numbers using the formal written method (short multiplication)
- multiply whole numbers up to 10 000 by two-digit numbers using the formal written method (long multiplication).

Lesson 1: **Multiplying by 1-digit numbers (1)**

Learning objective

Code	Learning objective
6Ni.04	Estimate and multiply whole numbers up to 10 000 by 1-digit [or 2-digit] whole numbers.

Resources

paper or mini whiteboard and pen (per learner); paper for the Workbook (per learner)

Revise

Use the activity *Complete the table (1)* from Unit 6: *Multiplication of whole numbers* in the Revise activities.

Teach [SB] 🖵 [TWM.01/04]

- On the board write: $63 \times 7 =$. Ask learners to estimate the product (420). Say: **Solve the problem mentally. What is the product?** (441) Ask: **Which strategy did you use?** Take responses. Expect learners to mention partitioning. Choose a learner to come to the board and explain how they solved the problem mentally, writing the mental steps they took: $63 \times 7 = (60 \times 7) + (3 \times 7) = 420 + 21 = 441$. **[TWM.04]** Ask: **Is your answer reasonable? How do you know?** (yes, it is close to the estimate) On the board, write: $56 \times 6 =$ and $83 \times 8 =$ and ask learners to mentally calculate the products (336, 664).
- Display **Slide 1**. Point to the first problem. **[T&T]** Ask: **How would you estimate the answer?** Review an estimation method: round the three-digit number to the nearest 100 and multiply (600×4). Ask: **What is your estimate?** (2400)
- Remind learners of the expanded written method of multiplication that involves calculating and recording the products of each place value multiplication. Take them through the recording involved in the first problem $568 \times 4 =$. Refer to the guided example but write out the calculation on the board, step by step. First, multiply the ones of the three-digit number by the one-digit number ($O \times O$). Ask: **What is 8 multiplied by 4?** (32) Write 32 in the first row below the top line, making sure the digits are lined up with the tens and ones columns. Next multiply $T \times O$. Point to the '6' in 568. Ask: **What does the '6' digit represent?** (60) Ask: **What is 60 multiplied by 4?** (240) Write 240 in the second line. Then multiply $H \times O$. Ask: **What is 500 multiplied by 4?** (2000) Write 2000 in the third line. Remind learners that the answer is found by adding the products of each place value multiplication. Ask: **What is the answer?** (2272: 2000 + 240 + 32) Compare the answer with the estimate. (2272 is close to 2400 and so the answer is likely to be correct)
- Ask learners to copy and complete problems **b** and **c** (2322, 6632).
- Display **Slide 2**. Take learners through the guided example of a ThHTO × O multiplication. Explain that the only difference compared to a HTO × O multiplication is that there is an extra row for the partial product of the Th × O multiplication. Ask

learners to copy and complete problems **b** and **c** (22 685, 62 608).

- **[TWM.01]** Say: **Give me a ThHTO × O problem where all the digits are 2s or 3s and the answer is between 6900 and 7000.** Take responses and ask learners to confirm the answer is in the given range ($2323 \times 3 = 6969$).
- 🖥 Say: **I solve the calculation 3453×4 to give the answer 13 712. Is that correct?** (no, 13 812)
- Discuss the Let's learn section in the Student's Book.
- Introduce the paired activity in the Student's Book, asking learners to calculate the total distance flown by the aircraft.
- Discuss the Guided practice example in the Student's Book.

Practice 👥

- Workbook

Title: Multiplying by 1-digit numbers (1)

Pages: 46–47

- Refer to Activity 1 from the Additional practice activities.

Apply 👥 🖵

- Display **Slide 3**. Remind learners to estimate the answer before calculating. Choose a pair to explain how they found the answer.

Review

- On the board, write and solve different ThHTO × O calculations using the expanded written method. In each example, make one mistake. Ask learners to identify the mistake in each calculation and then correct it.

Assessment for learning

- How would you estimate the answer to 3926×8?
- Which is greater, 4743×6 or 3554×8? (28 458/28 432)

Same day intervention
Support

- Some learners may prefer to continue using the grid method for calculation until they are confident with the expanded written method. Others will continue to practise and refine skills of formal multiplication methods.

Number – Integers and powers

Lesson 2: **Multiplying by 1-digit numbers (2)**

Learning objective

Code	Learning objective
6Ni.04	Estimate and multiply whole numbers up to 10 000 by 1-digit [or 2-digit] whole numbers.

Resources

paper or mini whiteboard and pen (per leaner)

Revise

Use the activity *Expanded written method (1)* from Unit 6: *Multiplication of whole numbers* in the Revise activities.

Teach [SB] 🖥 [TWM.07]

- Display **Slide 1**. Point to the first calculation 23 × 7 =. Ask: **What is a good estimate for this calculation?** (140) Explain that the examples introduce a new method called the 'formal written method' also known as 'short multiplication'. Ask learners to comment on the similarities and differences.

- Go through the calculation, comparing the two methods. Say: **In the expanded written method, you multiply the ones of the two-digit number by the one-digit number.** Ask: **What is 3 multiplied by 7?** (21) Remind learners that you write the product in the first row below the top line, making sure the digits are lined up with the tens and ones columns. **[T&T] [TWM.07]** Ask: **How does this compare with the first calculation in the formal method?** Explain that in the formal method, you regroup the product (21) as two 10s and one 1. You write '1' (one 1) in the answer line and carry the two 10s to the tens column by writing a small '2' below the answer line. You must remember to add the '2' to the product of the multiplication of the tens in the next step.

- Return to the expanded written method and discuss the multiplication of the tens (20 × 7 = 140). Remind learners that the answer is found by adding the partial products 21 + 140 = 161. Compare this to the next step in the formal method where you multiply 2 (two 10s) by 7 to give 14 and then add the 2 that was carried over to give 16 (16 tens). You write '16' in the hundreds and tens columns of the answer line to give the answer 161.

- On the board, write: 68 × 4 = and 77 × 6 = . Ask learners to work in pairs. For the first calculation (68 × 4), one learner uses the expanded method and the other learner uses the compact method. They compare methods and confirm they both have the same answer (272). For the second calculation (77 × 6 =), they switch roles.

- Return to **Slide 1** and discuss the three-digit by one-digit multiplication. Ask: **What is a good estimate for 468 × 6?** (3000)

- Compare and contrast the two methods. Ask questions to develop understanding, for example: **Why is there a small '4' below the answer line in the tens column?** Say: **The 4 tens carried over**

must be added to the result of multiplying the 6 tens (60) by 6 (36 tens + 4 tens = 40 tens). You regroup 40 tens as 4 hundreds + 0 tens, record a 'zero' in the tens column of the answer line and carry the 4 hundreds to the hundreds column by writing a small '4' under the answer line. Explain that you complete the calculation by multiplying the 4 hundreds by 6 to give 24 hundred and then add the 4 hundreds carried over to make 28 hundreds (2 thousands and 8 hundreds). You write '28' below the thousands and hundreds columns of the answer line to give the answer, 2808.

- On the board, write: 5273 × 4 =. Take learners through the steps to multiply a four-digit number by a one-digit number using both methods (21 092). Compare and contrast each step.

- 👥 Say: **Libby solves the calculation 4373 × 4 using the formal method. In her calculation, she has written a small digit '2' below the answer line in the hundreds column. Why has she done this?** (the product of 7 tens × 4 + 1 carried over = 29 tens; 29 tens regrouped as 2 hundreds + 9 tens; 2 hundreds carried over to the hundreds column)

- Discuss the Let's learn, paired activity and Guided practice section in the Student's Book.

Practice 👥

- Workbook

Title: Multiplying by 1-digit numbers (2)

Pages: 48–49

- Refer to Activity 1 (Variation) from the Additional practice activities.

Apply 👥 🖥

- Display **Slide 2**. Remind learners to estimate.

Review

- On the board write: 76 × 8 = ; 674 × 7 = ; 8235 × 6 = and ask learners to use the formal written method. Have them confirm using the expanded written method.

Assessment for learning

- How you would solve 758 × 6?

Same day intervention

Support

- Return to informal methods.

Number – Integers and powers

Lesson 3: **Multiplying by 2-digit numbers (1)**

Learning objective

Code	Learning objective
6Ni.04	Estimate and multiply whole numbers up to 10 000 by [1-digit or] 2-digit whole numbers.

Resources

paper or mini whiteboard and pen (per leaner)

Revise

Use the activity *Complete the table (2)* from Unit 6: *Multiplication of whole numbers* in the Revise activities.

Teach [SB] 🖵 [TWM.08]

- On the board write: $3124 \times 60 =$. **[T&T] [TWM.08]** Ask: **What is the most efficient strategy to solve this problem?** Take responses. Learners might suggest partitioning, the grid method or column multiplication. Explain that this would involve a lot of steps. Ask: **Is there is a quicker method?** Provide a hint by asking: **How could you multiply 3124 by a one-digit number and use this to find the answer?** Praise any learner who suggests multiplying by 6 and then multiplying the product by 10. Divide the class into three groups. Ask each group to find the product of $3124 \times 6 =$ using a different method: partitioning, the grid method and the expanded written method. Choose learners from each group to discuss their calculation and confirm the answer (18 744). Ask: **How do you find the product of 3124×60?** (multiply the answer to 3124×6 by 10) **What is the answer?** (187 440)
- On the board, write: $6356 \times 70 =$. Switch the method used by each of the three groups and ask them to find the product (444 920). Switch the groups' roles a third time and ask them to calculate $8654 \times 80 =$ (692 320).
- Display **Slide 1**. Remind learners that they can use the grid method to multiply by two-digit numbers. Take learners through the three guided examples. Remind them that both numbers must be partitioned and the component place values split across the columns and rows. Invite a learner to the slide to complete this stage. Also, remind learners that they get the answer by finding the sum of the partial products.
- Ask learners to work individually to solve the three remaining problems on **Slide 1**. They copy and complete the grids on whiteboards or paper. Invite learners to share their answers and confirm with the class.
- ⏸ Say: **A publishing company transports books to shops in crates of 5783. How many books will there be in 40 crates? How many in 43 crates?** (231 320, 248 669)

- Discuss the Let's learn section in the Student's Book.
- Introduce the paired activity in the Student's Book, asking learners to calculate the total amount of water that flows over a waterfall for different time periods.
- Discuss the Guided practice example in the Student's Book.

Practice 👥

- Workbook

Title: Multiplying by 2-digit numbers (1)

Pages: 50–51

- Refer to Activity 2 from the Additional practice activities.

Apply 👥 🖵

- Display **Slide 2**. Remind learners to estimate the answer before calculating. Choose a pair to explain how they found the answer.

Review

- On the board, write and solve different HTO × TO and ThHTO × TO calculations, using the expanded written method. *In each example, make one mistake. Ask learners to identify the mistake in each calculation and then correct it.*

Assessment for learning

- Asif multiplies 4378 by 40 using the grid method, splitting 4378 into 4000, 300, 70 and 8, and 40 into 40 and 0. What is wrong with Asif's strategy? Is there a better method? Why? (a better strategy is to multiply 4378 by 4 using any preferred method and then to multiply the product by 10)
- Calculate 5867×60 using the grid method. What is the answer? (352 020)

Same day intervention
Support

- Some learners may continue to use the grid method for calculation until they are confident with the expanded written method. Others will continue to practise and refine skills of formal multiplication methods.

Lesson 4: **Multiplying by 2-digit numbers (2)**

Learning objective

Code	Learning objective
6Ni.04	Estimate and multiply whole numbers up to 10 000 by [1-digit or] 2-digit whole numbers.

Resources

paper or mini whiteboard and pen (per learner); squared paper, for the Workbook (per learner)

Revise

Use the activity *Expanded written method (2)* from Unit 6: *Multiplication of whole numbers* in the Revise activities.

Teach 📖 🖥

- Display **Slide 1**. Introduce the formal written method for multiplying by a two-digit number. This method is called 'long multiplication'. Take learners through the TO × TO calculation, comparing the steps of the two methods.
- Working with the same problem, explain the steps in the formal method (long multiplication). Follow the completed example but write out the entire calculation alongside as you move through the steps. Explain that when multiplying by a two-digit number, you need to have two rows in the work area for recording partial products. Say: **First, you multiply the ones digit in the multiplicand by the ones digit in the multiplier.** Ask: **What is 8 × 6?** (48) Say: **48 is 4 tens and 8 ones**. Write the 8 ones in the ones column and carry the 4 tens. Explain that you write a small '4' in the top of the tens column to show that it is a carried digit. Say: **Next, you multiply the 3 tens in the multiplicand by the 6 ones in the multiplier.** Ask: **What is 3 × 6?** (18) **Plus the 4 carried over?** (22) Write '22' in the hundreds and tens columns below the line. Explain that you now multiply the digits in the multiplicand by the tens digit of the multiplier. Also, explain that you begin a second row of answers. Remind learners that the digit 5 in the multiplier has a value of 50. Ask: **What is 8 × 50?** (400) Write '00' in the ones and tens columns of the second row and carry the 4 hundreds by writing a small '4' in the hundreds column. Explain that in long multiplication, you always insert a zero as a placeholder in the ones column of the second row as you are multiplying by a set of tens (see Support below). Now multiply the 3 tens (30) in the multiplicand by the 5 tens in the multiplier. Ask: **What is 30 × 50?** (1500) **[T&T] How many hundreds is that?** (15) **Plus 4 hundreds?** (19) Write '19' in the thousands and tens columns. Remind learners that you add the two partial products to get the answer: 228 + 1900 = 2128.
- On the board, write: 63 × 26 = and 84 × 53 =. Ask learners to work in pairs. For the first calculation

(63 × 26 =), one learner uses the expanded method and the other learner uses the formal method. They compare methods and confirm they both have the same answer (1638). For the second calculation (84 × 53 =), they switch roles (4452).

- Repeat the steps discussed when demonstrating the TO × TO calculation for the HTO × TO example given on **Slide 1**.
- 🗣 Say: **In a warehouse, there are 586 tins of paint on each of 37 shelves. How many tins of paint are there altogether?** (21 682)
- Work through Let's learn, the paired activity and Guided practice in the Student's Book.

Practice 👥

- Workbook

Title: Multiplying by 2-digit numbers (2)

Pages: 52–53

- Refer to Activity 2 (Variation) from the Additional practice activities.

Apply 👥 🖥

- Display **Slide 2**. Remind learners to estimate first.

Review

- On the board write: 83 × 77 =; 567 × 44 =; 6838 × 66 = and ask learners to use the formal written method to calculate the answers. Learners confirm answers using the expanded written method.

Assessment for learning

- The mass of a chair is 3628 g. What is the mass of 33 chairs? (119 724 g)

Same day intervention
Support

- A common error in long multiplication is for a learner to forget to insert a placeholder zero in the second row of the working area when multiplying by the tens digit. To remedy this, place value needs to be reinforced, so that when learners multiply by the tens digit, they recognise that they are actually multiplying by a set of tens.

Additional practice activities

Activity 1

Learning objective
- Estimate and multiply whole numbers up to 10 000 by 1-digit whole numbers.

Resources
3–9 spinner made from Resource sheet 3: Spinner (1) (per pair); pencil and paper clip, for the spinner (per pair); paper or mini whiteboard and pen (per leaner)

What to do
- Each learner writes five different four-digit numbers.
- They take turns to choose one of their numbers, spin the spinner and find the product of the two numbers, using the expanded written method.
- They continue until all of the calculations are complete.

- Learners compare the products of the multiplications in the order they were completed. The player with the greater product scores one point.
- The player with more points at the end is the winner.

Variations
1 Provide place value charts and counters to help learners construct concrete models for multiplication.

2 Learners calculate using the formal written method rather than the expanded method.

Activity 2

Learning objective
- Estimate and multiply whole numbers up to 10 000 by 2-digit whole numbers.

Resources
2–9 spinner made from Resource sheet 3: Spinner (1) (per learner); pencil and paper clip, for the spinner (per learner); paper or mini whiteboard and pen (per leaner)

What to do
- Each learner spins the spinner six times and records the digits.
- They use the digits to create a ThHTO × TO calculation.

- Learners use any written method they prefer to solve the calculation.
- The player with the greater number scores a point.
- Continue play for five rounds. The winner is the player with the higher score.

Variation
2 Learners calculate using the formal written method.

Number – Integers and powers

Unit 7: Division of whole numbers (A)

Collins International Primary Maths
Recommended Teaching and
Learning Sequence Term 2, Week 2

Learning objective

Code	Learning objective
6Ni.05	Estimate and divide whole numbers up to 1000 by 1-digit [or 2-digit] whole numbers.

Unit overview

This unit refreshes and consolidates the work completed in Stage 5 on dividing two-digit numbers by one-digit whole numbers. Learners are reminded of the 'chunking' method of division, at first as a concrete approach through the manipulation of place value counters, where they repeatedly subtract individual groups of the divisor. Once learners are secure with these approaches, they use written methods of division, the expanded written method and short division, to divide two- and three-digit numbers by one-digit numbers. Answers with remainders are expressed as a whole number or a fraction of the divisor.

Throughout the unit, learners are encouraged to look carefully at the numbers in the calculation and decide on the most efficient, effective and appropriate method, including mental strategies.

Prerequisites for learning

Learners need to:
- know how to make a reasonable estimate for the answer to a calculation
- be able to mentally partition two- and three-digit numbers into hundreds, tens and ones
- know how to divide a two-digit number by a one-digit number, using a range of strategies
- be able to solve a division calculation that has a remainder.

Vocabulary

dividend, divisor, place value, chunking method, expanded written method, short division

Common difficulties and remediation

It is important to identify learners who make errors when performing written calculations. The errors may be procedural, indicating that the learner has not yet fully grasped how to use the algorithm. Several interventions should be considered. Give learners concrete, pictorial and abstract examples that have errors. Ask them to identify the mistakes, then talk through what is wrong and how errors should be corrected. Encourage learners to estimate answers and use approximations by careful questioning, for example: **What answer do you expect to get? How did you reach that estimate? Do you expect your answer to be less than or greater than your estimate? Why?**

When working with division and multiplication, learners often think of the operations as discrete, not appreciating the relationship between the two. It is important for learners to think of division as determining a missing factor and associate this work with writing division number sentences. Some learners struggle when trying to solve division calculations, particularly those that are best solved using grouping strategies. Give these learners frequent opportunities to identify and extract the mathematical components of word problems. They should use concrete materials and pictorial representations, as well as formulating the right question to ask, for example: 'How many groups of six can I make from 42?'

Supporting language awareness

It is important to adapt language when it becomes a barrier to learning; for example, using simpler terms to describe 'chunking', a concrete/pictorial method of division: 'Subtracting "chunks" (multiples) of the number we are dividing by and counting the number of chunks.'

Promoting Thinking and Working Mathematically

In Lesson 1, learners generalise (TWM.02) when they determine that regrouping allows division to proceed in a concrete modelling of a TO ÷ O division calculation.

In Lesson 2, learners conjecture (TWM.03) when they identify that the small digits recorded in a short division calculation indicate a remainder carried forward.

Learners characterise (TWM.05) in Lesson 3 when they explain that division can only proceed in a HTO ÷ O where the hundreds digit of the dividend is less than the ones digit of the divisor if the hundreds and tens of the divisor are regrouped as a set of tens.

Success criteria

Learners can:
- divide a two- or three-digit number by a one-digit number using the 'chunking' method
- divide a two- or three-digit number by a one-digit number using the expanded written method
- divide a two- or three-digit number by a one-digit number using a formal written method (short division)
- represent a remainder of a division calculation as a whole number or a fraction of the divisor.

Number – Integers and powers

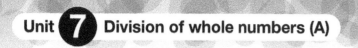

Unit **7** Division of whole numbers (A)

Lesson 1: Dividing 2-digit numbers by 1-digit numbers (1)

Learning objective

Code	Learning objective
6Ni.04	Estimate and divide whole numbers up to 1000 by 1-digit [or 2-digit] whole numbers.

Resources

paper or mini whiteboard and pen (per learner); place value counters or Resource sheet 8: Place value counters (1s and 10s) or other suitable classroom resources (per learner)

Revise

Use the activity *Division wheel (1)* from Unit 7: *Division of whole numbers (A)* in the Revise activities.

Teach 🆂🅱 ▥ 🖥 [TWM.02]

- On the board, write 85 ÷ 3 =. Ask learners to estimate the answer (between 20 (60 ÷ 3) and 30 (90 ÷ 3) but closer to 30). Display the **Place value counters tool**. Invite a learner to the board to arrange counters to represent 85. Explain that since you know 10 times the divisor (3) is 30, you should look for how many 30s there are in the tens column. You divide in one 'chunk' because it is quicker and more efficient. Ask a learner to circle sets of 30 (3 × 10). Ask: **How many 30s is that?** (2) **[T&T] [TWM.02]** Ask: **How do you proceed with division, given that you now have two '10' counters remaining in the tens column and five 'ones' in the ones column?** Take responses. Agree that you need to regroup the two tens and the five ones as 25. Invite a learner to exchange the two tens for twenty ones and circle groups of 3. Ask: **How many groups of 3 is that?** (8) Ask: **Is there a remainder?** (yes, 1) **What is the answer?** (28 r 1) Ask: **Is there another way you can write the answer?** Elicit that you can think of the remainder as a fraction of the divisor. Ask: **What is the remainder as a fraction?** ($\frac{1}{3}$, answer: $28\frac{1}{3}$)
- Display **Slide 1**. Discuss the two main steps involved in using the concrete 'chunking' method to calculate 97 ÷ 4 =. Remind learners that you partition the dividend into tens and ones, and circle and record groups of the divisor.
- Ask learners to draw a TO place value chart on paper or their whiteboards. Provide place value counters or Resource sheet 8: Place value counters or remind learners that you can use any object or symbol to represent place value, for example red counters for tens and yellow counters for ones. Invite learners to suggest their own representations, selecting from resources available in the classroom.
- Write: 87 ÷ 7 = and ask learners to work in pairs, using their chosen representations to model the division. Invite pairs to demonstrate the modelling and give the answer. (12 r 3 or $12\frac{3}{7}$)
- 🗣 Say: **Convince me that 92 ÷ 8 is 11 r 4.**

- Discuss the Let's learn section in the Student's Book.
- Introduce the paired activity in the Student's Book, asking learners to work out the length of each cut piece of material.
- Discuss the Guided practice example in the Student's Book.

Practice 🆆🅱

- Workbook

Title: Dividing 2-digit numbers by 1-digit numbers (1)

Pages: 54–55

- Refer to Activity 1 from the Additional practice activities.

Apply 👥 🖥 [TWM.08]

- Display **Slide 2**. Read the text aloud and ensure learners understand the problem. Learners should identify that they need to divide by 7 to find the total number of weeks and any remaining days. Choose pairs to explain how they answered each question.

Review

- Ask learners to solve the following calculations: 93 ÷ 8 =, 88 ÷ 3 =, 94 ÷ 6 =, 91 ÷ 8 = . Have them express the remainder in the answer as both a whole number and as a fraction (11 r 5, $11\frac{5}{8}$; 29 r 1, $29\frac{1}{3}$; 15 r 4, $15\frac{2}{3}$; 11 r 3, $11\frac{3}{8}$).

Assessment for learning

- Explain how you would use place value counters to calculate 89 ÷ 7. (12 r 5)
- Give me a two-digit divided by a one-digit calculation that leaves a remainder of 7.

Same day intervention
Support

- Learners who do not have recall of the multiplication facts will find division difficult. Ensure learners can also relate multiplication and division through finding the answer to missing number problems in multiplication number sentences, for example, 5 × ☐ = 35.

Number – Integers and powers

Lesson 2: **Dividing 2-digit numbers by 1-digit numbers (2)**

Learning objective

Code	Learning objective
6Ni.05	Estimate and divide whole numbers up to 1000 by 1-digit [or 2-digit] whole numbers.

Resources

paper or mini whiteboard and pen (per learner)

Revise

Use the activity *Division strategies (1)* from Unit 7: *Division of whole numbers (A)* in the Revise activities.

Teach 🔲 ⬜ [TWM.03]

- On the board, write: 93 ÷ 4 =. Ask learners to estimate the answer (between 20 (80 ÷ 4) and 30 (120 ÷ 4) but closer to 20). Invite learners to suggest strategies for solving the calculation including mental methods.
- Display **Slide 1**. Take learners through the expanded written method for division. Ask: **Is it possible to make 10 groups of 4?** (yes) **What about 20 groups?** (yes) **30 groups?** (no) Point to the '2' in the tens position of the answer line to represent 20 (20 groups of 4). Subtract the 20 groups of 4 (80) from 93, leaving 13. Ask: **How many fours are there in 13?** (3) Point to the '3' in the ones position of the answer line. Subtract the 3 groups of 4 (12) from 13, leaving 1. The answer is 23 r 1.
- On the board, write: 96 ÷ 4 =, 97 ÷ 7 =. Ask learners to solve the two calculations using the expanded written method. Choose learners to share their solutions, explaining the steps involved (24, 13 r 6).
- Display **Slide 2**. Introduce the formal written method of short division. Explain that in short division, the number of complete times that the divisor divides into each digit is written above the digit, and the remainder carried forward (as a multiple of 10) to the next number.
- Working with the calculations, explain the steps in the short division method comparing and contrasting it to the expanded written method. Follow the completed example but write out the entire calculation alongside as you move through the steps. Explain that, to set out the calculations correctly, they must place the divisor outside the long division bar. The dividend is written inside the long division bar. The quotient is recorded on top of the division bar.
- Begin the division. Ask: **How many 3s go into 8?** (2) Write '2' on the answer line in the tens column. **[T&T] [TWM.03]** Ask: **Why do we record the '2' in the tens column when we asked the question how many 3s go into 8?** Take responses. Explain that since the 8 digit in 86 has a value of 80 we are really asking how many 3s go into 80 (20). This leaves

a remainder of 20. Write a small '2' to the left of the next digit in the dividend (6) to make 26. Ask: **How many 3s go into 26?** (8) Write '8' in the answer line in the ones column. This leaves a remainder of 2. This is written on the answer line to give the quotient 28 r 2.
- On the board, write 97 ÷ 5 = and 95 ÷ 6 =. Ask learners to work in pairs. For the first calculation (97 ÷ 5 =), one learner uses the expanded method and the other learner uses the compact method. They compare methods and confirm they have the same answer (19 r 2). For the second calculation (95 ÷ 6 =), learners switch roles (15 r 5).
- Discuss the Let's learn section in the Student's Book.
- Introduce the paired activity in the Student's Book.
- Discuss the Guided practice example.

Practice 🔳

- Workbook

Title: Dividing 2-digit numbers by 1-digit numbers (2)

Pages: 56–57

- Refer to Activity 1 (Variation) from the Additional practice activities.

Apply 👥 ⬜

- Display **Slide 3**. Invite pairs to discuss the method they used to find the maximum number of tickets possible and the amount left over.

Review

- On the board, write: 95 ÷ 8 =; 86 ÷ 7 =. Have learners work out the solutions using short division, writing the remainders as fractions.

Assessment for learning

- Show me why the answer to the division 95 ÷ 7 is $13\frac{4}{7}$.
- Find the quotient of 82 ÷ 3. Express the remainder in two ways, as a whole number and as fraction.

Same day intervention

Support

- If learners make consistent errors with the formal written method, they should return to informal methods, such as the expanded written method.

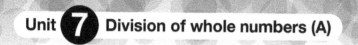

Unit **7** Division of whole numbers (A)

Lesson 3: **Dividing 3-digit numbers by 1-digit numbers (1)**

Number – Integers and powers

Learning objective

Code	Learning objective
6Ni.05	Estimate and divide whole numbers up to 1000 by 1-digit [or 2-digit] whole numbers.

Resources

paper or mini whiteboard and pen (per learner); counters or Resource sheet 8: Place value counters (1s, 10s and 100s) or other suitable classroom resources (per learner); paper, for the Workbook (per learner)

Revise

Use the activity *Spot the error* from Unit 7: *Division of whole numbers (A)* in the Revise activities.

Teach [SB] [📊] [🖥] [TWM.01/04/05]

- On the board, write 364 ÷ 8 =. Ask learners to estimate the answer (between 40 (320 ÷ 8) and 50 (400 ÷ 8)).
- Display the **Place value counters tool**. Invite a learner to the board to arrange counters to represent 364. **[TWM.05] [T&T]** Ask: **How do you begin division when you have 3 in the hundreds place of the dividend and must divide by 8?** Take responses. Establish that you need to exchange the hundreds for tens. Ask: **Three hundred – how many tens is that?** (30) The learner exchanges 3 hundreds for 30 tens and circles sets of 80 (8 × 10). Ask: **How many 80s is that?** (4) **[T&T] [TWM.01]** Ask: **How do you proceed with division given that you now have four tens counters remaining in the tens column and four ones in the ones column?** Take responses. Agree that you need to regroup the four tens and the four ones as 44. The learner completes the exchange. Ask: **How many groups of 8 is that?** (5) **Is there a remainder?** (yes, 4) Ask: **What is the answer?** (45 r 4) Involve learners in using the **Place value counters tool** to solve other division calculations, for example 246 ÷ 5 = (49 r 1).
- Display **Slide 1**. Discuss the two main steps involved in using the concrete 'chunking' method to calculate 343 ÷ 6. Remind learners that they partition the dividend into hundreds, tens and ones and circle and record groups of the divisor.
- Ask learners to draw a place value chart on paper or whiteboards. Provide counters or Resource sheet 8: Place value counters or remind learners that they can use any object or symbol to represent place value, for example green counters to represent hundreds, red counters for tens and yellow counters for ones. Invite learners to suggest their own representations, selecting from resources available in the classroom. On the board, write: 259 ÷ 8 = and ask learners to work in pairs to model the division, using their chosen representations. Invite pairs to demonstrate the modelling and give the answer (32 r 3).

- **[TWM.04]** [💬] Say: **Convince me that 235 ÷ 7 is 33 r 4.**
- Discuss the Let's learn section in the Student's Book.
- Introduce the paired activity in the Student's Book, asking learners to find the number of bricks Sam lays per day. Choose a pair of learners to explain how they solved the problem.
- Discuss the Guided practice example in the Student's Book.

Practice [WB] [TWM.04]

- Workbook

Title: Dividing 3-digit numbers by 1-digit numbers (1)

Pages: 58–59

- Refer to Activity 2 from the Additional practice activities.

Apply [👥] [🖥]

- Display **Slide 2**. Give learners time to work out the numbers of octopuses. Choose pairs of learners to explain their calculations.

Review

- Ask learners to work in pairs to solve a problem: **Mr Smith works in a shop. He packs apples into bags of 7. He has 326 apples to pack and all the apples must be placed in bags. How many bags does he need?** (326 ÷ 7 = 46 r 4, therefore 47 bags)

Assessment for learning

- I divide $468 into 9 equal shares by grouping. Explain how I do this ($52).
- 367 kg of sand is divided into 6 equal piles of equal mass. What is the mass of each pile? Give your answer as a mixed number ($61\frac{1}{6}$ kg).

Same day intervention
Support

- Learners who do not have recall of the multiplication facts will find division difficult. Ensure learners can also relate multiplication and division through finding the answer to missing number problems in multiplication number sentences, for example 90 × ☐ = 450.

Lesson 4: **Dividing 3-digit numbers by 1-digit numbers (2)**

Learning objective

Code	Learning objective
6Ni.05	Estimate and divide whole numbers up to 1000 by 1-digit [or 2-digit] whole numbers.

Resources

paper or mini whiteboard and pen (per learner)

Revise

Use the activity *Division strategies (2)* from Unit 7: *Division of whole numbers (A)* in the Revise activities.

Teach 🔲 💻 [TWM.03]

- Display **Slide 1**. Model how to answer a HTO ÷ O calculation using the expanded written method. Ask: **Is it possible to make 40 groups of 4?** (yes) **What about 50 groups?** (yes) **60 groups?** (no) Point to the '5' in the tens position of the answer line to represent 50 (50 groups of 4). Subtract the 50 groups of 4 (200) from 235, leaving 35. Ask: **How many fours are there in 35?** (8) Point to the '8' in the ones position of the answer line. Subtract the 8 groups of 4 (32) from 35, leaving 3. The answer is 58 r 3.

- On the board, write: 267 ÷ 3 =, 374 ÷ 8 =. Ask learners to solve the two calculations using the expanded written method. Choose learners to share their solutions, explaining the steps involved (89, 46 r 6).

- Display **Slide 2**. Model how to answer a HTO ÷ O calculations using the formal written method of short division. Explain the steps, comparing and contrasting it to the expanded written method. Follow the completed example but write out the entire calculation alongside as you move through the steps. Begin the division. Ask: **How many 6s go into 4?** (0) Write '0' on the answer line in the hundreds column. This leaves a remainder of 4 (400). Write a small '4' to the right of the next digit in the dividend (4) to make 44. Ask: **How many 6s go into 44?** (7) Write '7' on the answer line in the tens column. **[T&T]** **[TWM.03]** Ask: **What does the '7' represent?** (7 tens, 70) **Why do we record the '7' representing a value of 70 when we asked the question how many 6s go into 44?** Take responses. Explain that we are really asking how many 6s go into 440 (44 tens). This leaves a remainder of 2 (20). Write a small '2' to the left of the next digit in the dividend (7) to make 27. Ask: **How many 6s go into 27?** (4) Write '4' on the answer line in the ones column. This leaves a remainder of 3 which is written on the answer line to give the quotient 74 r 3.

- On the board, write 343 ÷ 4 = and 467 ÷ 8 =. Ask learners to work in pairs. For the first calculation (343 ÷ 4 =), one learner uses the expanded method and the other learner uses the compact method. They compare methods and confirm they have the same answer (85 r 3). For the second calculation (467 ÷ 8 =) learners switch roles (58 r 3).

- 📢 Say: **Explain to a partner how you would use short division to divide 223 by 4.** (55 r 3)

- Discuss the Let's learn section in the Student's Book.

- Introduce the paired activity in the Student's Book, asking learners to work out the distance run by each runner. Choose a pair of learners to explain how they solved the problem.

- Discuss the Guided practice example in the Student's Book.

Practice 🔲

- Workbook

Title: Dividing 3-digit numbers by 1-digit numbers (2)

Pages: 60–61

- Refer to Activity 2 (Variation) from the Additional practice activities.

Apply 👥 💻

- Display **Slide 3**. Invite pairs to discuss the method they used to find the answer. Ask: **Did you have to change the quotient in any way?** Establish that the answer needs to be rounded up to seat the remainder who do not form a group of six.

Review

- On the board, write: 261 ÷ 4 =, 396 ÷ 6 =, 635 ÷ 8 =. Have learners work out the solutions using short division, writing any remainders as a fraction ($65\frac{1}{4}$, 66, $79\frac{3}{8}$).

Assessment for learning

- Show me why the answer to the division 436 ÷ 9 is $48\frac{4}{9}$.

- Find the quotient of 765 ÷ 8. Express the remainder in two ways, as a whole number and as fraction (95 r 5 or $95\frac{5}{8}$).

Same day intervention

Support

- If learners make consistent errors with the formal written method of short division, they should return to informal methods such as the expanded written method until they have a greater understanding of the processes involved.

Additional practice activities

Activity 1

Learning objective
• Estimate and divide whole numbers up to 1000 by 1-digit whole numbers.

Resources
5–9 spinner from Resource sheet 9: Spinner (2) (per pair); paper clip and pencil, for the spinner; place value counters (1s and 10s) or Resource sheet 8: Place value counters (per pair); paper or mini whiteboard and pen (per learner)

What to do
• Each learner spins the spinner three times and arranges the three digits to make a TO ÷ O calculation.
• They model and solve the calculation using place value counters.
• The value of the remainder (if any) is taken as the score.

• Learners play the game for five rounds, adding the remainders to their score.
• The winner is the player with the higher score.

Variations

1 Provide place value charts and counters to help learners construct concrete models for multiplication.

2 Learners solve each TO ÷ O calculation using the expanded written method or short division.

Activity 2

Learning objective
• Estimate and divide whole numbers up to 1000 by 1-digit whole numbers.

Resources
3–6 spinner from Resource sheet 9: Spinner (2) (per pair); paper clip and pencil, for the spinner; place value counters (1s, 10s and 100s) or Resource sheet 8: Place value counters (per pair); paper or mini whiteboard and pen (per learner)

What to do
• Each learner spins the spinner four times and arranges the three digits to make an HTO ÷ O calculation.
• They model and solve the calculation using place value counters.

• The value of the remainder (if any) is taken as the score.
• Learners play the game for five rounds adding the remainders to their score.
• The winner is the player with the higher score.

Variation

2 Learners solve each HTO ÷ O calculation using the expanded written method or short division.

Unit 8: Division of whole numbers (B)

Collins International Primary Maths Recommended Teaching and Learning Sequence Term 2, Week 3

Learning objective

Code	Learning objective
6Ni.05	Estimate and divide whole numbers up to 1000 by [1-digit or] 2-digit whole numbers.

Unit overview

This unit extends the work completed in Unit 7 of dividing two- and three-digit numbers by two-digit whole numbers. Learners use the expanded written method to solve TO ÷ TO problems. This is then followed by an introduction to the compact form of the expanded written method. Learners are reminded to estimate the quotient first and use their estimate to check the answer to their calculation. Lessons 3 and 4 introduce HTO ÷ TO calculations and learners use the expanded method and the formal written method (long division) to solve related problems.

Throughout the unit, learners are encouraged to look carefully at the numbers in the calculation and decide on the most efficient, effective and appropriate method, including mental strategies.

Prerequisites for learning

Learners need to:
- know how to make a reasonable estimate for the answer to a calculation
- be able to mentally partition two- and three-digit numbers into hundreds, tens and ones
- know how to divide a three-digit number by a one-digit number, using a range of strategies
- be able to solve a division problem that has a remainder.

Vocabulary

dividend, divisor, quotient, expanded written method, compact form of the expanded written method, trial and improvement method, long division

Common difficulties and remediation

Some learners may find the long division algorithm challenging because of its many steps. Look out for learners who ignore place values and focus on individual digits rather than the whole numeral; for example, when dividing 896 by 28 they get a quotient of 302 and fail to realise their error. If the learner were to see the digits 8 and 9 as 89 tens rather than 89 hundreds, they would have realised that 28 can go into 89 three times, as opposed to 300 times.

Look out for learners who confuse some of the steps of the algorithm, for example, failing to bring down a digit. Another common error is for a learner not to notice that a difference is greater than or equal to the divisor and creates an extra place value in the quotient. To avoid these misconceptions, ask questions to elicit thinking: **What digits from the dividend have you already used? Which digits have yet to be used? Do the steps allow for a digit greater than 9 to be recorded in the quotient? If the difference is greater than the dividend, what does this tell you? Should the difference always be smaller than the divisor? Why?**

It is important that learners are reminded of the value of each digit in the quotient throughout the steps in the algorithm. For example, when dividing 896 by 28, the first digit written above the division box is 3, but this digit actually represents 30. Encourage learners to think of 30 as an estimate of the quotient. Multiplying back and subtracting in the next step will determine what remains to be divided, either by a whole number multiple of the divisor or the remainder.

Supporting language awareness

It is important to adapt language when it becomes a barrier to learning; for example, explaining that the 'long' in 'long division' implies the many steps involved in finding the quotient rather than being a method used exclusively for 'long numbers', i.e. a divisor or dividend (or both) having many digits.

Promoting Thinking and Working Mathematically

In Lessons 1 and 2, learners critique and improve (TWM.07/08) when they discuss the strengths and weaknesses of different rounding strategies and whether there is a better grouping strategy than 'trial and improvement'.

In Lesson 3, learners convince (TWM.04) when they discuss possible errors that can be made and how to spot them when using the expanded written method to divide.

Success criteria

Learners can:
- divide a two- or three-digit number by a two-digit number using the expanded written method
- divide a two-digit number using a compact form of the expanded written method
- divide a three-digit number by a two-digit number using the long division method.

Number – Integers and powers

Lesson 1: **Dividing 2-digit numbers by 2-digit numbers (1)**

Number – Integers and powers

Learning objective

Code	Learning objective
6Ni.05	Estimate and divide whole numbers up to 1000 by [1-digit or] 2-digit whole numbers.

Resources

paper or mini whiteboard and pen (per learner) paper, for the Workbook (per learner)

Revise

Use the activity *Division wheel (2)* from Unit 8: *Division of whole numbers (B)* in the Revise activities.

Teach 📖 🖥 [TWM.07]

- Write on the board: 72 ÷ 24 =. Ask: **What is your estimate for the quotient?** Take responses. Choose learners to discuss the calculation involved. Expect answers and strategies to vary. Discuss a rounding strategy: 72 ÷ 24 → 75 ÷ 25 = 3. **[T&T] [TWM.07]** Ask: **What do you think are the advantages of this strategy: rounding to the nearest '5'? Are there any disadvantages?** (advantage: more accurate than rounding to 10; disadvantage: unless you choose the numbers for their compatibility, dividing by multiples of 5 is more challenging than dividing by multiples of 10) Take responses. Explain that the rounding strategy you choose should 'fit' the numbers. For example, since you know 25 is a factor of 75, you round to the nearest '5' rather than '10'. Explain that this is an example of a compatible number strategy: you round to compatible or 'friendly' numbers that can be easily calculated mentally.
- Display **Slide 1**. Take learners through the four steps of the expanded written method for calculating 72 ÷ 24 =. Ask: **What are the differences between how you use this method for a TO ÷ TO calculation, compared to a TO ÷ O calculation?** Take responses. Explain that the layout is similar to the method used previously for a TO ÷ O division but for a two-digit divisor, you use repeated subtraction of only one group of the divisor until you find the quotient.
- Provide one more guided example where there is no remainder: 92 ÷ 23 =. Ask learners to estimate first (100 ÷ 25 = 4) and then take them through the four steps of the expanded method. Discuss how one group of the divisor (23) is subtracted from 92 with the difference carried forward. Four steps of multiplying back and subtracting gives a difference of 0 and therefore, a quotient of 4.
- Ask learners to use this form of the expanded method to solve the following calculations: 72 ÷ 18 = (4) and 98 ÷ 14 = (7).
- 🗣 Say: **Explain to a partner how you would use the expanded written method to divide 96 by 16.** (6)

- Discuss the Let's learn section in the Student's Book.
- Introduce the paired activity in the Student's Book.
- Discuss the Guided practice example in the Student's Book.

Practice 📓

- Workbook

Title: Dividing 2-digit numbers by 2-digit numbers (1)

Pages: 62–63

- Refer to Activity 1 from the Additional practice activities.

Apply 👥 🖥

- Display **Slide 2**. Invite pairs to discuss how they found the length of each section of the route.

Review

- Write various calculations on the board and ask learners to estimate the answer, for example: 96 ÷ 33 =; 90 ÷ 15 =; 78 ÷ 13 =. Invite learners to display their estimates. Ask: **How did you work out your estimate?** Discuss the rounding strategies used.
- Ask learners to calculate the answer to each of the division problems. Confirm answers.

Assessment for learning

- How many groups of 26 are there in 78? (3)
- How would you use the expanded written method to show that 54 ÷ 18 = 3?

Same day intervention

Support

- If, at any time, learners make an error in the steps involved in the expanded written method, return to the previous step and ask questions to develop understanding of the process: **Did you find all the groups of the divisor that can divide into the dividend? Did you remember to multiply back and subtract? Check the calculation of the difference– did you remember to regroup to allow subtraction to proceed?**

Lesson 2: **Dividing 2-digit numbers by 2-digit numbers (2)**

Learning objective

Code	Learning objective
6Ni.05	Estimate and divide whole numbers up to 1000 by [1-digit or] 2-digit whole numbers.

Resources

paper or mini whiteboard and pen (per learner)

Revise

Use the activity *Two-digit divisors* from Unit 8: *Division of whole numbers (B)* in the Revise activities.

Teach 📖 💻 [TWM.08]

• Display **Slide 1**. Take learners through the compact form of the expanded written method. Explain that the compact form reduces the number of steps involved. You only need to determine the number of groups of the divisor in the dividend and record the number in the answer line. To check this, you multiply back and subtract. A difference of zero tells you that the divisor divides evenly into the dividend, with no remainder.

• Set up the compact form layout to solve the calculation: 85 ÷ 17 =. Say: **Give me an estimate for this problem.** (4 or 5) **Describe the rounding strategy you used.** Ask: **How many groups of 17 go into 85?** (5) **How do you know?** Take responses. Discuss a trial and improvement strategy: 3 × 17 = 51 (no), 4 × 17 = 68 (no), 5 × 17 = 85 (yes). **[T&T] [TWM.08]** Ask: **Is there a more efficient method than trial and improvement? Describe this method.** Take responses and discuss the strengths and weaknesses of the suggestions. Ask: **What is the ones digit of the divisor?** (7) **What is the ones digit of the dividend?** (5) **How many groups of 7 would give a ones digit of 5?** (5; 7 × 5 = 35) **Does 5 times 17 make 85?** (yes) Explain that focusing on the ones digit of the divisor and the dividend will give a clue to the digits in the quotient.

• Ask learners to use the compact form of the expanded method to solve the following calculations: 96 ÷ 24 = (4) and 91 ÷ 13 = (7).

• 🗣 Say: **Explain to a partner how you would use the compact form of the expanded written method to divide 78 by 13.** (6)

• Discuss the Let's learn section in the Student's Book.

• Introduce the paired activity in the Student's Book, asking learners to find the number of full buckets of water required to remove all the water from a paddling pool.

• Discuss the Guided practice example in the Student's Book.

Practice 📘

• Workbook

Title: Dividing 2-digit numbers by 2-digit numbers (2)

Pages: 64–65

• Refer to Activity 1 (Variation) from the Additional practice activities.

Apply 👥 💻

• Display **Slide 2**. Invite pairs to discuss how they found the number of portions of mixture. (6).

Review

• Reinforce the compact form of the expanded written method of division taught in the lesson. Write various calculations on the board and ask learners to estimate the answer for example, 78 ÷ 39 =; 93 ÷ 31 =. Invite learners to display their answers. Ask: **How did you work out your estimate?** Discuss the rounding strategies used.

• Ask learners to calculate the answer to each division calculation. Confirm answers and choose a learner to explain how they solved one of the problems.

Assessment for learning

• How would you use the compact form of the expanded written method to show that 90 ÷ 15 = 6?

• Give me two TO ÷ TO divisions with a quotient of 5.

Same day intervention

Support

• If learners make consistent errors with the compact form of the expanded written method, they should return to the regular form of the expanded method until they have a greater understanding of the processes involved.

Number – Integers and powers

Lesson 3: **Dividing 3-digit numbers by 2-digit numbers (1)**

Learning objective

Code	Learning objective
6Ni.05	Estimate and divide whole numbers up to 1000 by [1-digit or] 2-digit whole numbers.

Resources

paper or mini whiteboard and pen (per learner)

Revise

Use the activity *Find the error (1)* from Unit 8: *Division of whole numbers (B)* in the Revise activities.

Teach [SB] 🖥 [TWM.04]

- Write on the board: 624 ÷ 16 =. Explain that learners are now going to use the expanded written method to divide three-digit dividends by two-digit divisors. First, make an estimate for this division. Ask: **What is 30 × 16?** (480) **40 × 16?** (640) **Is 624 closer to 480 or 640?** (640) Say: **So the estimate is 40.**

- Display **Slide 1**. Copy the layout for the expanded method to calculate 624 ÷ 16 =. Ask and record: **Can you make 30 groups of 16?** (yes, 16 × 30 = 480) Write 3 (three tens) on the answer line. Say: **Subtract the 30 groups of 16 from 624, which leaves 144.** Complete the subtraction. **[T&T]** Ask: **How many times will 16 go into 144?** (9) Write 9 (nine ones) on the answer line. Say: **Subtract 144 from 144.** Complete the subtraction. Ask: **How many are left?** (0) Say: **The answer is 39.** Ask: **How does this compare to our estimate?**

- Repeat for 552 ÷ 24 =. Work with learners to record each stage of the calculation, asking questions similar to those above to help identify the largest possible tens and ones multiples of the divisor (23). Say: **A learner began the expanded method for this problem. He recorded '1' in the tens column of answer line and calculated 552 – 240 in the working area. [TWM.04]** Ask: **What error has he made? Why do you think this happened?** Take responses. Confirm that the learner thought that 552 could only make 10 groups of 24, not 20. Agree that the calculation 552 – 240 gives 312, which makes more than 10 groups of 24. Recording a digit greater than 9 in the ones column of the answer line should alert the learner that he has made an error.

- 🗣 Say: **Explain to a partner how you would use the expanded written method to divide 532 by 28.** (19)

- Discuss the Let's learn section in the Student's Book.

- Introduce the paired activity in the Student's Book, asking learners to find the number of paint tins that should be placed on each shelf.

- Discuss the Guided practice example in the Student's Book.

Practice [WB]

- Workbook

Title: Dividing 3-digit numbers by 2-digit numbers (1)

Pages: 66–67

- Refer to Activity 2 from the Additional practice activities.

Apply 👥 🖥

- Display **Slide 2**. Invite pairs to discuss how they found the number of bottles with a combined sugar content of the value given.

Review

- On the board, write: 667 ÷ 23 =, 559 ÷ 13 =, 806 ÷ 26 =, 896 ÷ 32 =. Ask learners to order the calculations by their value, from the smallest quotient to the greatest. (896 ÷ 32 = 28, 667 ÷ 23 = 29, 806 ÷ 26 = 31, 559 ÷ 13 = 43)

Assessment for learning

- How many groups of 37 are there in 629? (17)
- How would you use the expanded written method to show that 784 ÷ 28 = 28?

Same day intervention
Support

- If, at any time, learners make an error in the steps involved in the expanded written method, return to the previous step and ask questions to develop understanding of the process: **Did you find all the multiples of (10 times) the value of the divisor that can divide into the dividend? Did you remember to multiply back and subtract? Check the calculation of the difference – did you remember to regroup to allow subtraction to proceed?**

Lesson 4: **Dividing 3-digit numbers by 2-digit numbers (2)**

Learning objective

Code	Learning objective
6Ni.05	Estimate and divide whole numbers up to 1000 by [1-digit or] 2-digit whole numbers.

Resources

paper or mini whiteboard and pen (per learner)

Revise

Use the activity *Find the error (2)* from Unit 8: *Division of whole numbers (B)* in the Revise activities.

Teach [SB] 🖥 [TWM.03]

- On the board, write: 528 ÷ 16 =. With help from learners, complete the division on the board using the expanded written method (33).
- Display **Slide 1**. Introduce the formal written method of long division. Explain that this method reduces the amount of information you need to record. Follow the layout on **Slide 1** and write the steps alongside the expanded method on the board. Say: **First, we look at the hundreds in the dividend.** Ask: **Does 16 divide into 5?** (no) Regrouping the 5 hundreds and 2 tens of the dividend as 52 tens we ask: **Does 16 divide into 52?** (yes, 3 × 16 = 48) Record the answer 3 in the tens column above the calculation. Write 48 under the 52. Say: **Subtract 48 tens from 52 tens. What is the answer?** (4 tens). Write '4' in the tens column under the 48. Say: **Now there are 4 tens.**
- [T&T] [TWM.03] Point to the arrow on the diagram and say: **The '8' of the dividend is brought down alongside the '4'. Why do you think this is done?** Take responses. Establish that 16 does not divide into 4 so you convert the 4 tens into ones and add the existing ones from the calculation. You indicate this by drawing an arrow from the 8 down alongside the 4 tens and write '8' in the ones column next to the 4 tens. Say: **Now there are 48 ones.** Ask: **Does 16 divide into 48?** (yes, 3 × 16 = 48) Record the answer 3 in the ones column on the answer line. Say: **The answer is 33.**
- Demonstrate use of the long division method to solve two more HTO ÷ TO problems: 544 ÷ 17 = (32); 992 ÷ 32 = (31).
- On the board, write: 874 ÷ 38 =, 928 ÷ 29 =. Ask learners to work in pairs. For the first calculation (874 ÷ 38 =), one learner uses the expanded method and the other learner uses the long division method. They compare methods and confirm they have the same answer (23). For the second calculation (928 ÷ 29 =) the learners switch roles (32). Remind them to estimate first before beginning the calculation.
- 👥 Say: **Explain to a partner how you would use long division to divide 731 by 17.** (43)

- Discuss the Let's learn section in the Student's Book.
- Introduce the paired activity in the Student's Book, asking learners to find the total number of tables required to seat all the people at the company meeting.
- Discuss the Guided practice example in the Student's Book.

Practice [WB]

- Workbook

Title: Dividing 3-digit numbers by 2-digit numbers (2)

Pages: 68–69

- Refer to Activity 2 (Variation) from the Additional practice activities.

Apply 👥 🖥

- Display **Slide 2**. Invite pairs to discuss how they found the number of groups of 38 cows.

Review

- On the board, write 814 ÷ 22 = 29; 954 ÷ 18 = 31; 899 ÷ 29 = 53, 928 ÷ 32 = 37. Explain that the quotients have got mixed up. Ask learners to use the long division method to work out the correct quotient for each problem. (37; 53; 31, 29)

Assessment for learning

- Sophie has completed a long division problem and has drawn an arrow in one of the steps. What might she have used the arrow for?
- How would you use the long division method to show that 888 ÷ 37 = 24?

Same day intervention

Support

- Provide learners with direct instruction on the use of the long division algorithm for division. When teaching the algorithm, emphasis should be placed on the first step in the cycle of repeated steps when a quotient is estimated. Deciding which multiple of the divisor to subtract at each stage can be a process of trial and improvement.

Number – Integers and powers

Additional practice activities

Activity 1

Learning objective
• Estimate and divide whole numbers up to 1000 by 2-digit whole numbers.

Resources
number cards: 12, 13, 19, 23, 7, 6, 5, 4, 84, 95, 92, 78 from Resource sheet 10: Division number cards (1) (per learner); ÷ and = cards from Resource sheet 11: Operator cards (per learner); paper or mini whiteboard and pen (per learner) (for variation)

What to do
• Each learner shuffles their set of cards.
• They race each other to arrange their cards into four division calculations of the form 'TO ÷ TO = quotient'.
• The winner is the player who correctly completes all four calculations in the shortest time.

Variation
Learners race each other to write five TO ÷ TO division calculations that give quotients of 2, 3, 4, 5 and 6.

Activity 2

Learning objective
• Estimate and divide whole numbers up to 1000 by 2-digit whole numbers.

Resources
number cards: 14, 16, 18, 22, 23, 25, 28, 31, 434, 448, 450, 506 from Resource sheet 12: Division number cards (2) (per learner); ÷ and = cards from Resource sheet 11: Operator cards (per learner); paper or mini whiteboard and pen (per learner) (for variation)

What to do
• Each learner shuffles their set of cards.
• They race each other to arrange their cards into four division calculations of the form 'HTO ÷ TO = quotient'.
• The winner is the player who correctly completes all five calculations in the shortest time.

Variation [TWM.07]
Each learner chooses one of the following calculations to solve:

$972 \div 12 =$ $805 \div 23 =$ $868 \div 28 =$ $972 \div 36 =$

Learners must not choose the same calculation. They complete the division calculation using the long division method but make a single error in their computations. They then swap papers to try to determine the mistake their partner has made and correct it.

Number – Integers and powers

Unit 9: Place value and ordering decimals

Collins International Primary Maths
Recommended Teaching and
Learning Sequence Term 3, Week 1

Learning objectives

Code	Learning objective
6Nc.01	Count on and count back in steps of constant size, including [fractions and] decimals, and extend beyond zero to include negative numbers.
6Np.01	Understand and explain the value of each digit in decimals (tenths, hundredths and thousandths).
6Np.03	Compose, decompose and regroup numbers, including decimals (tenths, hundredths and thousandths).
6Nf.08	Understand the relative size of quantities to compare and order numbers with one or two decimal places, [proper fractions with different denominators and percentages,] using the symbols =, > and <.

Unit overview

Learners extend their knowledge of decimals to include thousandths. They record numbers with three decimal places on a place value chart and explain the value of the digit in the thousandths position. They count back in thousandths and examine how digits change when ones, tenths and hundredths boundaries are crossed. Counting sequences extend beyond zero to include negative numbers.

Learners practise writing numbers with up to three decimal places in expanded form and compose and decompose in different ways.

They use number lines and place value grids to compare numbers with up to two decimal places. They extend this to ordering sets of numbers, adopting a systematic approach to find the greater decimal by comparing digits in different place value positions, from left to right, starting with the largest place.

Prerequisites for learning

Learners need to:
- be able to write tenths and hundredths as fractions and decimals, such as $\frac{7}{10}$ (0·7) and $\frac{3}{100}$ (0·03)
- count on and back in tenths and hundredths
- be able to write a decimal in expanded form
- know how to compose, decompose and regroup tenths and hundredths in different ways
- be able to compare the tenths and hundredths using a number line or place value grid.

Vocabulary

decimal, tenth, hundredth, thousandth, compose, decompose, place value, regroup, compare, order, trailing zero

Common difficulties and remediation

Look out for learners who incorrectly apply whole number thinking when comparing and ordering decimals. With whole numbers 6 < 71 < 584 but it is incorrect to order decimals by a strategy based on 'longer is bigger'. 0·71 is greater than 0·6 but 0·584 is not greater than 0·71. Similarly, methods of calculation used with whole numbers must be changed to accommodate decimal fractions. For example, 7 + 28 = 35 but 0·7 + 0·28 ≠ 0·35.

For learners to develop a secure understanding of decimals and how to use them, they must recognise that place values are nested in other place values, for example, 5·43 has 54 tenths (54·3 tenths to be exact) and 543 hundredths; 0·873 has 873 thousandths, 87 hundredths (87·3 hundredths to be exact) and 8 tenths (8·73 tenths to be exact).

Supporting language awareness

Encourage learners to write the definition in their own words, for example: 'Composing means coming together to make something, for example, 9·382 is composed of nine ones, three tenths, eight hundredths and two thousandths.' 'Decomposing means to separate into smaller parts, for example, 9·382 can be decomposed into nine ones, three tenths, eight hundredths and two thousandths.'

Promoting Thinking and Working Mathematically

In Lesson 1, learners generalise (TWM.02) when they identify the difference in value of the same digit occurring in different decimal place positions.

In Lesson 2, learners characterise (TWM.01) when they give examples of the composition of a three-place decimal.

Success criteria

Learners can:
- explain the value of the third decimal place in numbers such as 2·458, 0·405 or 3·003
- demonstrate their understanding of decimal place value by writing a number with three decimal places in expanded form
- compose, decompose and regroup tenths, hundredths and thousandths in different ways
- compare tenths and hundredths in a number by placing them on a number line or place value grid.

Number – Place value, ordering and rounding

Unit (9) Place value and ordering decimals

Lesson 1: **Decimal place value**

Learning objectives

Code	Learning objective
6Nc.01	Count on and count back in steps of constant size, including [fractions and] decimals, and extend beyond zero to include negative numbers.
6Np.01	Understand and explain the value of each digit in decimals (tenths, hundredths and thousandths).

Resources

mini whiteboard and pen (per learner)

Revise

Use the activity *Decimal-go-round* from Unit 9: *Place value and ordering decimals* in the Revise activities.

Teach [SB] 🖥 📊 [TWM.02/06]

- On the board, write 0·7 and 0·07. **[TWM.02] [T&T]** Ask: **What is the value of the digit 7 in each number?** Take responses. Remind learners that the first and second decimal places are the number of tenths and hundredths respectively. Ask: **What is the relationship between the two values?**

- On the board, write 0·777. Underline the 7 in the thousandths position. Ask: **What is the value of the underlined 7 digit?** Take responses.

- Display **Slide 1**. Read and discuss the text. Explain that the third digit to the right of the decimal point is the thousandths digit. It is also known as the 'third decimal place'. On the board, write 0·453 and 8·509 and underline the third decimal place in each number. Ask learners to identify the value of each underlined digit. Ask: **How would you record 3 thousandths as a decimal with no tenths or hundredths?** Choose a learner to come to the board to write the number. Expect: 0·003. Have learners write on their mini whiteboards the decimal number that is 9 thousandths.

- Display the **Place value tool**. Arrange the cards to display 0·001. Increase the number of thousandths, one at a time, and ask learners to comment on how the decimal digits change. For each number in the count up to 0·014, choose learners to come to the board to write the equivalent fraction: $\frac{1}{1000}$, $\frac{2}{1000}$, $\frac{3}{1000}$ ... $\frac{13}{1000}$, $\frac{14}{1000}$.

- **[TWM.06]** Stop at 0·014 and ask: **What does the digit 1 mean in the number 0·014? What does the digit 4 represent?** Ask: **This number has only four thousandths – am I right?** Take responses. Establish that 0·014 is also 14 thousandths, as one hundredth is equivalent to 10 thousandths (10 thousandths + four thousandths = 14 thousandths). Agree that 0·014 can also be read or interpreted in more than one way: 'zero point zero one four', '1·4 hundredths' or '14 thousandths'.

- Arrange the class in a circle. Start a forwards count from a three-place decimal in steps of 0·001, for example 3·751. Ask learners to say what the next number in the sequence is (3·752). Ensure that learners understand what happens to the digits

when the count crosses the hundredths boundary, from 3·759 to 3·76. Repeat for 6·393 and focus on the count crossing the tenths boundary, from 6·399 to 6·4. At any point, announce a change in the count, for example count back in steps of 0·002 or count forwards in steps of 0·003. Include examples where learners are asked to count on or back from a negative number, including counting through zero. For example, count on in steps of 0·001 from −3·464. **[TWM.02]** Ask: **How will the digits in the second and third decimal places change as you count on from a negative number?** (thousandths digit will decrease from 9 down to 0, then tenths digit will decrease by one and thousandths digit will reset to 9)

- 🖩 Ask: **What are the values of the digits in the number 2·786?**

- Discuss the Let's learn section in the Student's Book.

- Introduce the paired activity in the Student's Book, then discuss the Guided practice example.

Practise [WB] [TWM.01]

- Workbook

Title: Decimal place value

Pages: 70–71

- Refer to Activity 1 from the Additional practice activities.

Apply 👥 🖥 [TWM.02]

- Display **Slide 2**. Ask learners to explain how they worked out the first negative number.

Review

- Working in pairs, one learner says a three-place decimal. The other learner says the number backwards and then the value of digits in the number, for example, 4·732 reversed is 2·374: 2 ones, 3 tenths, 7 hundredths and 4 thousandths. They then swap roles.

Assessment for learning

- Which number has 5 hundredths, 4 ones and 2 thousandths? (4·052)

Same day intervention
Enrichment

Provide questions where the steps are larger, for example: **Count back in steps of 0·008 from 5·231.**

Lesson 2: Composing and decomposing decimals

Learning objective

Code	Learning objective
6Np.03	Compose, decompose and regroup numbers, including decimals (tenths, hundredths and thousandths).

Resources

paper or mini whiteboard and pen (per learner)

Revise

Use the activity *Compose and decompose* from Unit 9: *Place value and ordering decimals* in the Revise activities.

Teach 🄢 [TWM.01]

- Discuss the Let's learn section in the Student's Book.
- On the board, draw a place value grid with ones, tenths, hundredths and thousandths. Write 6·289 on the grid. Ask: **How would you write 6·289 as the sum of its separate place values – ones, tenths, hundredths and thousandths?**
- Take responses. Remind learners that this is called 'decomposition'. Decomposing a number means to separate it into smaller parts, such as splitting a number by place value. On the board, write: $6·289 = 6 + 0·2 + 0·08 + 0·009$.
- Repeat for 3·571 and 54·503. Point to the zero in the second decimal place in 54·503. Ask: **What is the value of this digit?** (0 hundredths) **[T&T] Why do you need it in the number if the value is zero?** Take responses. Elicit that zero hundredths is a placeholder to 'hold' the hundredths position. On the board, write: $54·503 = 50 + 4 + 0·5 + 0 + 0·003$.
- **[T&T] [TWM.01]** Say: **You are now going to practise composing three-place decimals.** Ask: **Who can remember what 'composing' a number means? Give me an example of composition.** Take responses. Remind learners that composing a number means making a number from component parts with different place values. 78·186 is composed of 7 tens, 8 ones, 1 tenth, 8 hundredths and 6 thousandths or 70, 8, 0·1, 0·08 and 0·006. Write: 2·699, 82·808 and ask learners to say what each of these numbers is composed of.
- **[TWM.01]** Ask: **What is the largest five-digit number with three decimal places that can be composed from the digits 6, 2, 7, 4 and 1?** Remind learners that three decimal places represent the number of tenths, hundredths and thousandths. State the place values of the composing parts. (76·421)
- ℗ Ask: **What number is composed from 0·3, 60, 0·005, 8 and 0·09?** (68·395)
- Introduce the paired activity in the Student's Book, asking learners to solve the problem.

- Discuss the Guided practice example in the Student's Book.

Practise 🄦🄱

- Workbook

Title: Composing and decomposing decimals

Pages: 72–73

- Refer to Activity 1 (Variation) from the Additional practice activities.

Apply 👥 🖥 [TWM.01]

- Display **Slide 1**. Ask learners to record as many different decompositions of the jug's capacity as they can. Choose learners to share their decompositions and confirm they are correct.

Review 📊

- Display the **Place value tool**. Without learners seeing, create some three-place decimals, such as 9·277. Learners write the numbers in the decomposed form. ($9·277 = 9 + 0·2 + 0·07 + 0·007$)
- Then, show some three-place decimals expanded as separate place values. Learners write them composed as one number, using correct decimal notation.

Assessment for learning

- Convince me that 18·403 is equivalent to $10 + 8 + 0·4 + 0·003$.
- Which number can be composed from 2 tenths, 6 tens, 8 thousandths, 7 ones and 5 hundredths? (67·258)

Same day intervention

Enrichment

- Working in pairs, each learner writes and completes two decompositions and then erases some of the information. For example, $28·211 = 20 + 8 + 0·2 + 0·01 + 0·001$ could become: $2\_·2\_1 = \_\_\_ + 8 + \_\_\_ + 0·01 + \_\_\_$. Learners swap papers to fill in the missing digits and then return them for marking.

Lesson 3: **Regrouping decimals**

Number – Place value, ordering and rounding

Learning objective

Code	Learning objective
6Np.03	Compose, decompose and regroup numbers, including decimals (tenths, hundredths and thousandths).

Resources

paper or mini whiteboard and pen (per learner)

Revise

Use the activity *Regrouping race* from Unit 9: *Place value and ordering decimals* in the Revise activities.

Teach [SB] 🖵 [TWM.04]

- Discuss the Let's learn section in the Student's Book.
- On the board, write: 9·824. **[T&T]** Ask: **How would you use place value to split this number into its component parts?** Expect: 9·824 = 9 + 0·8 + 0·02 + 0·004. Ask: **You can split numbers by their place value but are there any other ways?** Prompt learners by asking them to express the number as thousandths only. Ask: **How many thousandths make 9·824?** (9824) **How did you work this out?** Elicit that to convert a number to thousandths, you multiply by 1000. Ask: **Are any other decompositions possible?** Accept 9 ones and 824 thousandths, 98 tenths and 24 thousandths, 982 hundredths and 4 thousandths, and so on.
- Write on the board: 8·414, 76·297. Ask learners to work in pairs to find four different ways to decompose each number. Take responses and confirm as a class.
- Ask: **What is the decimal number with 28 tenths and 4 hundredths?** (2·84)
- Display **Slide 1**. Take learners through the guided examples of how to regroup and decompose negative numbers. On the board, write: –33. Ask: **How would you regroup this number?** Have learners draw a number line to show their working. They may choose to simply draw an empty number line and mark the groupings: from –33 to 0 and make jumps of 0 to –10 (–10), –10 to –20 (–10), –20 to –30 (–10) and –30 to –33 (–3). Ask: **What are the new groupings?** (–10, –10, –10, –3)
- On the board, write –6·235. Ask: **How would you decompose this number?** Take responses and confirm: –6 + –0·2 + –0·03 + –0·005.
- Write: –6·2 + 0·035. **[TWM.04]** Ask: **Is this a regrouping of –6·235? Explain why.** Take responses. Confirm that it is not, because the sign of 0·035 should be negative.
- 🖵 Ask: **How would you regroup –48?** (–10, –10, –10, –10, –8)

- Introduce the paired activity in the Student's Book, asking learners to recompose each set of regroupings.
- Discuss the Guided practice example in the Student's Book.

Practise [WB]

- Workbook

Title: Regrouping decimals

Pages: 74–75

- Refer to Activity 2 from the Additional practice activities.

Apply 👥 🖵

- Display **Slide 2**. Ask learners to regroup both temperatures. Choose learners to explain how they regrouped each number.

Review

- Learners work in pairs. They each secretly write a five-digit decimal number with three decimal places and then take turns to describe their number in a decomposed form. For example, 53·914 could be described as 5 tens, 39 tenths and 14 thousandths. Learners try to guess the composed number.
- Pairs then work out the sum of their numbers, using a regrouping strategy.

Assessment for learning

- Convince me that 7·281 decomposed is 72 tenths and 81 thousandths.
- Explain to a partner how you would solve 22·168 + 44·371 using a regrouping strategy.

Same day intervention
Enrichment

- Ask learners to solve subtraction problems where regrouping is necessary. For example: 3·287 – 1·533, 26·834 – 13·562. Discuss how 3·287 – 1·533 can be decomposed and regrouped as 3·2 – 1·5 + 0·087 – 0·033 = 1·7 + 0·054 = 1·754.

Lesson 4: **Comparing and ordering decimals**

Learning objective

Code	Learning objective
6Nf.08	Understand the relative size of quantities to compare and order numbers with one or two decimal places, [proper fractions with different denominators and percentages,] using the symbols =, > and <.

Resources

paper or mini whiteboard and pen (per learner)

Revise

Use the activity *Comparing hundredths* from Unit 9: *Place value and ordering decimals* in the Revise activities.

Teach [SB] 🖵 [TWM.04]

- On the board, write: 4·76 and 4·67. **[T&T] [TWM.04]** Ask: **How would you prove that 4·76 is greater than 4·67?** Take responses. Explain that a number line can be useful for comparing numbers, but another strategy is a place value grid.
- Draw a place value grid on the board with columns for tens, ones, tenths and hundredths. Write 4·76 and 4·67 in the grid, aligning the place values in the columns. Explain that to find the greater number, you compare the digits in the different place values from left to right, until you find one digit that is of greater value than the other. 4·76 has more tenths than 4·67, so 4·76 is greater. Write: 4·76 > 4·67.
- Display **Slide 1**. Discuss the steps for comparing decimals.
- Ask learners to compare 9·26 and 9·27. (9·26 < 9·27)
- Display **Slide 2** and discuss how the steps for comparing two decimals can be applied to ordering a set of decimal numbers.
- Write: 2·84, 2·78, 2·82, 2·87, 2·76. Ask learners, working in pairs, to draw a place value grid for ones, tenths and hundredths and use it to order the numbers, from least to greatest. Choose a pair to explain the method they used. Establish the order and record: 2·76, 2·78, 2·82, 2·84, 2·87.
- Draw a place value chart and write in rows: 7·98, 8, 7·9, 8·1, 7·99. Ask: **How do you order numbers when they have different numbers of decimal places?** Take responses. Establish that to make it easier to compare, make sure all the decimals have the same number of decimal places by adding trailing zeros to the end, if required. Explain that 'trailing zeros' are zeros inserted after the final decimal digit that do not change the value of a number. To compare the five numbers, insert a zero in the hundredths place for 7·9 and 8·1, and a zero in each of the tenths and hundredths places for 8. Ask learners to order the numbers. Confirm and record: 7·90, 7·98, 7·99, 8·00, 8·10.
- 🗫 Ask: **Which is greater, 6·43 or 6·34? How do you know?** (6·43 > 6·34 as 6 ones and 4 tenths is greater than 6 ones and 3 tenths)
- Discuss the Let's learn section in the Student's Book.

- Introduce the paired activity in the Student's Book, asking learners to order the times and answer the questions about the order.
- Discuss the Guided practice example in the Student's Book.

Practise [WB]

- Workbook

Title: Comparing and ordering decimals

Pages: 76–77

- Refer to Activity 2 (Variation) from the Additional practice activities.

Apply 👥 🖵

- Display **Slide 3** and read the text to the class. Choose a pair of learners to explain how they ordered the heights.

Review

- On the board, write several number pairs: 0·77, 0·76; 4·45, 4·54; 66·12, 66·1. Ask learners to compare the numbers in each pair, writing number statements and using the correct symbol of comparison, < or >. They share answers and explain solutions.

Assessment for learning

- Here are two decimals: 22·22 and 22·17. Explain how you would find the greater number.
- Here are five measurements in kilograms: 82·18, 82, 82·17, 82·1, 82·09. Explain how you would order them, from lightest to heaviest.

Same day intervention
Support

- Learners may develop several misconceptions or difficulties about comparing and ordering decimal numbers. Some learners believe the longer the decimal, the greater the number. They treat the decimal part the same as they would a whole number, for example, thinking that 27·45 > 27·5 because 45 > 5. Remediation exercises can be useful, including reading a decimal number and writing it in the expanded form, so learners are reminded of the importance of the position of the digits.

Number – Place value, ordering and rounding

163

Additional practice activities

Activity 1

Learning objective
• Understand and explain the value of each digit in decimals (tenths, hundredths and thousandths).

Resources
• Resource sheet 13: 0–9 Digit cards (per learner); plastic counter (per learner)

What to do
• Learners each have a set of digit cards. To start the activity, they take the 0 out of the set and place it next to a small plastic counter so that it reads 0·__ __ __ (the counter acting as the decimal point).
• The concept behind the activity is that in each round, learners use their digit cards to fill in the three decimal places. For example, if they have the digits 2, 4 and 8 they might make the number 0·482.
• Learner A decides whether the aim of the first round is to make the smallest or largest number. They tell Learner B this information.
• Both players then choose three of their digit cards to place to the right of the 0 to form a decimal number.
• On the count of three, they reveal their cards and the smaller/larger number wins (depending on the original aim).

• After each round, the three chosen cards are put to one side and play continues with the remaining cards each learner has.
• Each learner can use their opponent's cards to inform the next round's aim. For example, if they have 6, 3, 9 and 8 remaining and their opponent has the digits 5, 2, 4 and 3 remaining, Learner B might choose to have the aim of 'larger number wins', knowing that any decimal with 6, 8 or 9 in the tenths place will win.

Variations [TWM.05]

1 The game is played with two decimal places only.

2 Learners must prove their number is larger or smaller by saying: 'My number comprises [x tenths], [y hundredths] and [z thousandths] and is larger/smaller than your number because it has more/fewer tenths/hundredths/thousandths.'

Activity 2

Learning objective
• Compose, decompose and regroup numbers, including decimals (tenths, hundredths and thousandths).

Resources
paper or mini whiteboard and pen (per learner)

What to do
• Learners write down five three-place decimal numbers without their partner seeing them.
• They take turns to describe one of their numbers, using three clues. The first clue should be the most difficult; the last the easiest.
• For example, if the number is 5·382, the clues could be: (1) My number has 53 tenths and 82 thousandths; (2) My number has 5 ones and 382 thousandths; (3) My number has five ones, three tenths, eight hundredths and two thousandths.
• Each learner has around ten seconds to guess the number before their partner moves on to the next clue.
• Players score three points for guessing correctly given the first clue; two points for the second; and one for the third.
• Play continues with the remaining numbers.

• The winner is the player with more points.

Variations

2 Learners work in pairs to write a list of eight supermarket food items with prices in dollars and multiples of 10 cents for example, $4.30.

The prices should all be different and listed in a random order. Depending on the ability of pairs of learners, it may be appropriate to tell learners a price range, for example, between $1 and $10 (Challenge 1), or between $1 and $5 (Challenge 2), or between $10 and $12 (Challenge 3).

Learners swap papers with another pair and write the prices in order, from least to greatest.

They swap papers for marking.

1 Learners play with two decimal numbers each time. They provide clues such as 'My second number has 47 more tenths than my first number', 'My first number has 3 more ones than my second number', 'My first number has 16 tenths and my second number has 5 ones'.

Unit 10: Place value, ordering and rounding decimals

Collins International Primary Maths
Recommended Teaching and
Learning Sequence Term 3, Week 2

Learning objectives

Code	Learning objective
6Np.02	Use knowledge of place value to multiply and divide whole numbers and decimals by 10, 100 and 1000.
6Np.04	Round numbers with 2 decimal places to the nearest tenth or whole number.

Unit overview

In this unit, learners investigate multiplication and division of whole numbers and decimals by 10, 100 or 1000 and note the effect on the digits, moving them left or right a number of place positions.

In Lessons 3 and 4, learners practise rounding two-place decimals to the nearest tenth, and to the nearest whole number.

Prerequisites for learning

Learners need to:

• understand the place value of decimals as tenths, hundredths and thousandths
• know how to multiply a number by 10 by moving digits one place value to the left and multiply by 100 by moving digits two place values to the left
• know how to divide a number by 10 by moving digits one place value to the right and divide by 100 by moving digits two place values to the right
• be able to round a number by using a number line or by finding the rounding digit
• know that for rounding to the nearest whole number, the rounding digit is the ones place.

Vocabulary

multiply, place value, divide, round, rounding digit

Common difficulties and remediation

Some learners find rounding difficult, particularly the concept of using a 'rounding digit'. Encourage them to use a blank (empty) number line as a practical visual aid, allowing them to see the choice of numbers to which they could round the given number.

Supporting language awareness

At every stage, learners require mathematical vocabulary to access questions and problem-solving exercises. If appropriate, when a new key word is introduced, ask learners to write a definition in their books, drawing a box around it for emphasis. Encourage learners to write the definition in their own words, for example: 'When rounding to the nearest whole number, you are finding the closest set of ones to your number.' Sketching a diagram alongside will help embed the definition.

Promoting Thinking and Working Mathematically

Opportunities to develop all four pairs of TWM characteristics are provided throughout the unit.

In Lesson 1, learners generalise (TWM.02) when they determine the multiplier (a power of ten) in a calculation when given the multiplicand and product.

In Lesson 2, learners conjecture (TWM.03) when they ask questions of the form: 'Give me a division problem where the divisor is [power of ten] and the quotient is a [x]-place decimal.'

Learners classify (TWM.06) in Lesson 3 when they decide how to group numbers according to how they round.

Success criteria

Learners can:

• multiply a number by 10, 100 or 1000 by moving digits one, two or three place values to the left
• divide a number by 10, 100 or 1000 by moving digits one, two or three place values to the right
• know that for rounding to the nearest tenth, the rounding digit is the first decimal place
• know that for rounding to the nearest whole number, the rounding digit is the ones digit.

Number – Place value, ordering and rounding

Lesson 1: **Multiplying whole numbers and decimals by 10, 100 and 1000**

Learning objective

Code	Learning objective
6Np.02	Use knowledge of place value to multiply [and divide] whole numbers and decimals by 10, 100 and 1000.

Resources

paper or mini whiteboard and pen (per pair)

Revise

Use the activity *10, 100 or 1000 times larger* from Unit 10: *Place value, ordering and rounding decimals* in the Revise activities.

Teach 🔲 💻 📊 [TWM.01/02/04]

- Display **Slide 1**. Ask: **How would you use the chart to show 67 multiplied by 10?** Ask a volunteer to demonstrate movement of the digits. Confirm that the digits move one place to the left and the number becomes 10 times larger. (670) Repeat for 67 multiplied by 100. Ask: **How do the digits move across the place value columns when you multiply by 100?** (two places to the left, 6700) Repeat for 67 × 1000 =. (three places to the left, 67 000)
- Display the **Function machine tool**, set up with 485 and '× 10'. Ask: **What will the output be?** (4850) Repeat for '× 100' (48 500).
- Display **Slide 2**. Working in pairs, learners copy the place value chart onto paper or their whiteboard and solve the six calculations. Ask volunteers to explain how they worked out the answers writing them on the place value chart on **Slide 2**. Remind them that, like whole numbers, all the digits in a decimal number move one/two/three places to the left when multiplied by 10/100/1000. Ensure that learners understand that it is the digits and not the decimal point that move. Also, correct any misconceptions about 'adding zeros' in order to multiply by a power of 10. Confirm the answers as a class. (**a** 7, **b** 450, **c** 180, **d** 90·3, **e** 3425, **f** 716 230)
- **[T&T] [TWM.01/02]** Say: **I multiplied 5·78 and got the product 5780. What was the multiplier in the calculation?** Take responses and confirm the answer (1000). Ask: **How did you get the answer?** (each digit in 5·78 has moved three places to the left)
- 🗣 Ask: **What is 451·07 × 1000?** (451 070)
- Discuss the Let's learn section in the Student's Book.
- **[TWM.01/04]** Introduce the paired activity in the Student's Book, asking learners to work in pairs to write multiplication calculations that will give the specified answers. Choose learners to share their calculations and confirm they are correct.
- Discuss the Guided practice example in the Student's Book.

Practise 🔲 [TWM.01]

- Workbook

Title: Multiplying whole numbers and decimals by 10, 100 and 1000

Pages: 78–79

- Refer to Activity 1 from the Additional practice activities.

Apply 👥 💻

- Display **Slide 3**. Read the text and confirm the learners know what problem they need to solve. Ask: **How would you use your knowledge of multiplying by 1000 to find the combined mass of the rabbit and cat?** Remind learners of the relationship between grams and kilograms.

Review

- Write 782 × 100 = on the board. Ask learners to work out the answer, asking volunteers to explaining their method. Repeat for 89·34 × 1000 =.
- Say: **A number multiplied by 100 is 1234. What is the number?** (12·34) Discuss the methods used.

Assessment for learning

- Here is the number 485. Show me how to multiply it by 100.
- **[TWM.04]** Convince me that 2·18 multiplied by 1000 is 2180.

Same day intervention
Enrichment

- Extend multiplying two-place decimals to multiplying by 10 000, for example: 2·17 × 10 000. (21 700)

Number – Place value, ordering and rounding

Lesson 2: **Dividing whole numbers and decimals by 10, 100 and 1000**

Learning objective

Code	Learning objective
6Np.02	Use knowledge of place value to [multiply and] divide whole numbers and decimals by 10, 100 and 1000.

Resources

paper or mini whiteboard and pen (per pair)

Revise

Use the activity *10 or 100 times smaller* from Unit 10: *Place value, ordering and rounding decimals* in the Revise activities.

Teach 🔲 💻 📊 [TWM.01/03/04]

- Display **Slide 1**. Ask learners, working in pairs, to make a copy of the place value chart onto paper or their whiteboard and use it to solve the division calculations. Choose pairs to explain how they worked out the answers. Remind them that when a number is divided by 10/100, the digits move one/two places to the right. Confirm the answers as a class, writing them on the place value chart on **Slide 1**. (a 0·7, b 2·32, c 45·63, d 845·9)
- Display the **Function machine tool**, set up with 4820 and '÷ 10'. Ask: **What will the output be?** (482) Repeat for '÷ 100'. (48·2)
- [T&T][TWM.01] On the board, write: 53·8 ÷ 100 =, 131·9 ÷ 100 =. Ask: **How many decimal places do you expect the quotients of these division calculations to have? How do you know?** Take responses. Ask learners to solve the calculations and then confirm that each quotient has three decimal places. (0·538, 1·319)
- Write: 8045 ÷ 1000 =, 214 000 ÷ 1000 = and 45 789 ÷ 1000 =. Ask: **What happens to the digits in a number when you divide by 1000?** (confirm they move three places to the right) **Which of these division calculations is the odd one out? Why?** Take responses. Ask learners to calculate using a place value chart. Confirm the answers (8·045, 214, 45·789) and that 214 000 ÷ 1000 is the only calculation with a whole-number quotient.
- [TWM.03] Say: **Give me a division calculation where the divisor is 1000 and the quotient is a two-place decimal.** Take responses and confirm correct suggestions, for example, 4630 ÷ 1000. Ask: **What other questions of the form: 'Give me a division calculation where the divisor is [power of 10] and the quotient is a [x]-place decimal' can you think of?** Allow learners to pose their questions to the class.
- 🖱 Ask: **What is 10 304 divided by 1000?** (10·304)

- Discuss the Let's learn section in the Student's Book.
- [TWM.01/04] Introduce the paired activity in the Student's Book, asking learners to work in pairs to write division calculations that will give the specified answers. Choose learners to share their calculations and confirm they are correct.
- Discuss the Guided practice example in the Student's Book.

Practise 📘 [TWM.01]

- Workbook

Title: Dividing whole numbers and decimals by 10, 100 and 1000

Pages: 80–81

- Refer to Activity 1 (Variation) from the Additional practice activities.

Apply 👥 💻

- Display **Slide 2**. Read the text and confirm the learners know what problem they need to solve. Ask: **How would you use your knowledge of dividing by 1000 to find the combined volume of water poured from the two jugs?** Remind learners of the relationship between millilitres and litres.

Review

- Write 231·6 ÷ 100 = on the board. Ask learners to work out the answer, asking volunteers to explaining their method. Repeat for 4530 ÷ 1000 =.
- Say: **A number divided by 100 is 45·67. What is the number?** (4567) Discuss the methods used.

Assessment for learning

- Here is the number 8091. Show me how to divide it by 100.
- [TWM.04] Convince me that 4832 divided by 1000 is 4·832.

Same day intervention
Enrichment

- Extend dividing whole numbers to dividing by 10 000, for example: 45 300 ÷ 10 000. (4·53)

Lesson 3: **Rounding decimals to the nearest tenth**

Number – Place value, ordering and rounding

Learning objective

Code	Learning objective
6Np.04	Round numbers with 2 decimal places to the nearest tenth [or whole number].

Revise

Use the activity *Rounding targets* from Unit 10: *Place value, ordering and rounding decimals* in the Revise activities.

Teach [SB] [TWM.06]

- Review rounding numbers with one decimal place to the nearest whole number. Display the **Number line tool** set from 3 to 4, with increments of 0·1. On the board, write: 3·3, 3·5, 3·7. Set the tool pointer to 3·3. **[T&T]** Ask: **Which whole number is 3·3 nearest to?** (3) Say: **3·3 is rounded to 3 as this is its closest whole number.** Repeat for 3·5 (4), 3·7 (4).

- Highlight the tenths digits of the numbers on the board. Remind learners that when rounding a decimal number to the nearest whole number, the rounding digit is in the ones place. You study the digit that is one place to the right of the rounding digit, the tenths, and follow rules for rounding.

- On the board, write: 8·78. Say: **We are now going to round numbers to the nearest tenth. What do you think is the rounding digit?** Take responses. Establish that it is the digit in the tenths position. (tenths) Establish that when rounding to the nearest tenth, the rounding digit is the first decimal place. You study the digit that is one place to the right of the rounding digit, the hundredths, and follow rules for rounding. Choose a learner to circle the rounding digit (7). Ask: **What is the hundredths digit?** (8) **Should you round up or down?** (up) **How does the rounding digit change?** (from 7 to 8) **What happens to the hundredths digit?** (it becomes zero and the rounded number is 8·8)

- Repeat for 13·93. Ask: **What is the rounding digit?** (9) **Do you round up or down?** (down) **Why?** (the digit in the hundredths position is less than 5)

- Write: 5·46, 23·75, 19·02. Ask learners, working in pairs, to round each number to the nearest tenth. Choose pairs to explain their rounding strategy and confirm the rounded numbers (5·5, 23·8, 19).

- Write: 44·99. Say: **I think this number rounds to 44·0. Am I right?** Take responses. Praise any learner who explains that as the hundredths digit is 9, the digit in the tenths position needs to be rounded up. This means 9 resets to zero and the ones digit becomes 5, giving the rounded number 45. Draw a number line on the board marked out in tenths from 44·9 to 45 and show that the nearest multiple to 44·99 is 45.

- **[TWM.06]** Write: 6·35, 6·18, 6·32, 6·41, 6·23, 6·27. Ask: **How would you sort these numbers into**

three groups? Explain your groupings. Establish groupings based on rounding to the same number (6·23, 6·18 (6·2); 6·27, 6·32 (6·3); 6·35, 6·41 (6·4)).

- Ask: **What is 46·38 rounded to the nearest tenth?** (46·4)

- Discuss the Let's learn section in the Student's Book.

- Introduce the paired activity in the Student's Book. Ask: **What strategy did you use to find all the numbers? Is there a better method?** Take responses. Ask: **How many numbers are there?**

- Discuss the Guided practice example in the Student's Book.

Practise [WB]

- Workbook

Title: Rounding decimals to the nearest tenth

Pages: 82–83

- Refer to Activity 2 from the Additional practice activities.

Apply 👥 🖥

- Display **Slide 1**. Learners are asked to find the height of each zoo animal to the nearest tenth of a metre. Choose learners to explain how they rounded each number.

Review

- On the board, write: 14·28, 20·03, 45·66, 89·99. Ask learners to round each number to the nearest tenth. (14·3, 20, 45·7, 90)

Assessment for learning

- Explain to a partner how you would round a number with two decimal places to the nearest tenth.

- How do you round to the nearest tenth when the rounding digit is 5?

Same day intervention
Support

- Difficulty in rounding decimals is likely due to a lack of understanding of place values. To develop rounding skills, learners should be encouraged to put a number to be rounded in a place value chart. The place value headings will be a constant reminder of the value of the digits they are working with.

Lesson 4: **Rounding decimals to the nearest whole number**

Learning objective

Code	Learning objective
6Np.04	Round numbers with 2 decimal places to the nearest [tenth or] whole number.

Resources

paper or mini whiteboard and pen (per pair)

Revise

Use the activity *Rounding decimals* from Unit 10: *Place value, ordering and rounding decimals* in the Revise activities.

Teach SB [TWM.01/02/03]

- On the board, write: 14·36. Ask: **How do you round 14·36 to the nearest tenth?** Choose a learner to explain their strategy to the class. Expect: Round up to 14·4.

- **[TWM.03]** Say: **Let's round the same number to the nearest whole number. I expect that it will round in the same direction: up. Am I right?** Take responses. Expect learners to point out that the rounding digit is now 4, the digit in the ones position, and the digit to its right is less than 5. Therefore, you must round the number down to 14.

- On the board, write: 27·83, 3·56, 66·01, 104·29. Ask learners, working in pairs, to round each number to the nearest tenth, and to the nearest whole number, writing answers on paper or their whiteboard.

- Ask: **Which decimal rounds to the same number for both 'nearest tenth' and 'nearest whole number'?** Choose pairs to say their rounded numbers and confirm with the class (27·8, 28; 3·6, 4; 66, 66 (rounds the same); 104·3, 104).

- **[T&T] [TWM.02]** Say: **Find five other decimals that round to exactly the same number when rounded to the nearest tenth and nearest whole number. What do these numbers have in common?** Choose pairs to share their numbers. Confirm that these numbers have decimal digits between 0·1 and 0·4 or 0·95 and 0·99.

- ♫ Ask: **What is 45·35 rounded to the nearest whole number?** (45)

- Discuss the Let's learn section in the Student's Book.

- **[TWM.01]** Introduce the paired activity in the Student's Book. Choose learners to explain the strategy they used to choose each number.

- Discuss the Guided practice example in the Student's Book.

Practise WB

- Workbook

Title: Rounding decimals to the nearest whole number

Pages: 84–85

- Refer to Activity 2 (Variation) from the Additional practice activities.

Apply 👥 🖥

- Display **Slide 1**. Learners are asked to find the length of each vehicle to the nearest metre. Choose learners to explain how they rounded each number.

Review

- On the board, write: 6·37, 15·84, 38·25, 76·93. Ask learners to round each number to the nearest whole number. (6, 16, 38, 77)

Assessment for learning

- What is 59·51 rounded to the nearest whole number? (60)

- I rounded a two-place decimal to the nearest whole number and got 36. The number before rounding has 36 ones and 9 hundredths. What could the number be? (36·09, 36·19, 36·29, 36·39, 36·49)

Same day intervention

Support

- Difficulty in rounding decimals is likely due to a lack of understanding of place values. To develop rounding skills, learners should be encouraged to put a number to be rounded in a place value chart. The place value headings will be a constant reminder of the value of the digits they are working with.

Number – Place value, ordering and rounding

Additional practice activities

Activity 1

Learning objective
- Use knowledge of place value to multiply and divide whole numbers and decimals by 10, 100 and 1000.

What to do
- On the board, write: '39 to 390 000', '517 to 5 170 000', '7·2 to 72 000', '69·41 to 6 941 000'.
- Ask: **Using only combinations of '× 10', '× 100', '× 1000', what steps does it take to convert one number to the other?**
- Learners find solutions and discuss them with the class.
- They construct and write four new problems, swapping with another pair to solve.

Variation
On the board, write: '790 000 to 79', '801 000 to 8·01', '3 450 000 to 0·345'.

Ask: **Using only combinations of '÷ 10', '÷ 100', '÷ 1000', what steps does it take to convert one number to the other?**

Learners find solutions and discuss them with the class.

They construct and write four new problems, swapping with another pair to solve.

Activity 2

Learning objective
- Round numbers with 2 decimal places to the nearest tenth or whole number.

Resources
paper or mini whiteboard and pen (per learner); 1–6 dice or 1–6 spinner or 1–9 spinner from Resource sheet 5: 1–6 and 1–9 spinners (per pair); pencil and paper clip, for the spinner (per pair)

What to do
- Each learner has a whiteboard and pairs of learners have a 1–6 dice a 1–6 spinner or a 1–9 spinner.
- Pairs of learners work together to make a shopping list of five items to buy that each costs less than $10. They each list the items on paper or their whiteboard.
- They roll the dice or spin the spinner to give the number of dollars; they use the same digit for all prices.
- Then they roll the dice or spin the spinner twice for each item, to generate the number of cents

(tenths and hundredths digits), for example: $8.63, $8.24, $8.78, $8.45, $8.91. One learner in each pair records these prices alongside their items on their whiteboard.

- They order the set of prices from least to greatest and then round each number to the nearest tenth of a dollar. The other learner in the pair writes the rounded numbers on the other whiteboard. Remind learners that they should write the answers with a trailing zero in the hundredths column, for example, $8.40 not $8.4.
- The class pool together the whiteboards in two sets, showing rounded and unrounded figures, and work out which set of rounded numbers belong to which set of unrounded numbers.

Variation
Learners repeat the activity, rounding their five items to the nearest dollar.

Unit 11: Fractions (A)

Collins International Primary Maths Recommended Teaching and Learning Sequence Term 2, Week 4

Learning objectives

Code	Learning objectives
6Nf.01	Understand that a fraction can be represented as a division of the numerator by the denominator (proper and improper fractions).
6Nf.03	Use knowledge of equivalence to write fractions in their simplest form.
6Nf.04	Recognise that fractions, decimals (one or two decimal places) [and percentages] can have equivalent values.
6Nf.08	Understand the relative size of quantities to compare and order [numbers with one or two decimal places,] proper fractions with different denominators [and percentages,] using the symbols =, > and <.

Unit overview

In this unit, learners extend their knowledge of fractions. They understand that a fraction can be interpreted as a division of the numerator by the denominator and use this notion to convert a vulgar fraction (proper and improper fractions) to a decimal fraction. They understand that the quotient in a division calculation can always be expressed as a fraction. They learn that division notation and fraction notation mean the same thing and are interchangeable.

Learners use their knowledge of equivalence to write fractions in their simplest form by dividing the numerator and denominator by the highest common factor.

Learners also explore the equivalence between fractions and decimals up to two decimal places and compare and order proper fractions with different denominators.

Prerequisites for learning

Learners need to:
• understand unit and non-unit fractions
• understand the equivalence between decimals and fractions in halves, tenths and hundredths
• be able to recognise place value for decimal tenths and hundredths
• understand factors and multiples.

Vocabulary

numerator, denominator, decimal, highest common factor, simplify, equivalent fraction

Common difficulties and remediation

Misconceptions can arise when learners try to apply their understanding of whole numbers to comparing fractions. For example, they may say: '$\frac{5}{9}$ is greater than $\frac{5}{8}$ because 9 is greater than 8.' It is important to intervene where misunderstandings occur. Practical exercises, working with visual models, will help to develop their thinking in this area.

Learners should have a solid understanding of what fractions represent before moving on to comparing and ordering them. They need to understand that the denominator is the number of equal parts into which a whole is divided and the numerator is the number of parts being considered. Look for learners who assume the numerator cannot be greater than the denominator or that the larger the denominator, the larger the piece. They will require more work with concrete and then pictorial representations.

Supporting language awareness

Note terms that have more than one definition, particularly with meanings outside of mathematics. The term 'vulgar' implies that there is something incorrect about this type of fraction. Explain that although it is slightly misleading, 'vulgar' is the correct term for common fractions.

Promoting Thinking and Working Mathematically

In Lesson 2, learners convince (TWM.04) when they reason that only the highest common factor of two denominators will reduce a fraction to its lowest terms. Other common factors will simplify the fraction but the fraction can be reduced further.

In Lesson 3, learners improve (TWM.08) when they evaluate the models they have drawn to compare the relative sizes of two fractions.

Success criteria

Learners can:
• convert a vulgar fraction into a decimal fraction by dividing the numerator by the denominator
• simplify a fraction to its simplest form by identifying common factors in the numerator and denominator and dividing by the highest common factor
• recognise equivalence between fractions
• compare fractions with different denominators
• order a set of proper fractions with different denominators.

Number – Fractions, decimals, percentages, ratio and proportion

Lesson 1: **Fractions as division**

Learning objective

Code	Learning objective
6Nf.01	Understand that a fraction can be represented as a division of the numerator by the denominator (proper and improper fractions).

Resources

calculator (per learner); paper or mini whiteboard and pen (per learner)

Revise

Use the activity *Decimal fractions* from Unit 11: *Fractions (A)* in the Revise activities.

Teach 🆂🅱 💻 [TWM.05]

- On the board, write: $\frac{1}{4}$ and 0·25. **[T&T] [TWM.05]** Ask: **What is the connection between these two numbers?** Take responses. Confirm that the numbers have equivalent values, one being the fraction or decimal equivalent of the other. Ask: **How do you convert from a fraction to a decimal?** Remind learners that you divide the numerator by the denominator. Ask: **What is the connection between fractions and division?** Take responses. Establish that a fraction is basically a division problem. Explain that the bar of a fraction (vinculum) can be thought of as a division sign. To convert a fraction to a decimal, you divide the numerator by the denominator.

- On the board, write: $\frac{3}{4} = 3 \div 4 = 0\cdot75$. Establish that the calculation 3 divided by 4 gives the quotient 0·75; 0·75 is the decimal equivalent of the fraction $\frac{3}{4}$.

- Display and discuss **Slide 1**. On the board, write: $3 \div 5$. Ask: **Which fraction does this division represent?** ($\frac{3}{5}$). Provide learners with calculators. On the board, write: $\frac{4}{5}$. Ask: **What division gives us the decimal equivalent of this fraction?** ($4 \div 5$) Have learners use their calculators to convert the fraction to a decimal. (0·8)

- Write: $1\frac{1}{4}$. Ask: **What type of number is this?** (mixed number) **How do you represent this number as a division?** Elicit that you need to convert the mixed number to an improper fraction. Ask: **What is $1\frac{1}{4}$ as an improper fraction?** ($\frac{5}{4}$) Establish that the fraction can be represented as $5 \div 4$.

- Ask: **What fraction is represented by the division 50 ÷ 40?** ($\frac{50}{40}$) **Is this the only way to write this fraction?** Some learners may be able to simplify this answer to $\frac{5}{4}$ or $1\frac{1}{4}$.

- 🗣 Ask learners to explain to a partner what the fraction $\frac{3}{4}$ and the division $3 \div 4$ have in common.

- Discuss the Let's learn section in the Student's Book.

- Introduce the paired activity in the Student's Book. Learners write the amount of cake consumed at

the party as a fraction and express this fraction as a division.

- Discuss the Guided practice example in the Student's Book.

Practice 🆆🅱

- Workbook

Title: Fractions as division

Pages: 86–87

- Refer to Activity 1 from the Additional practice activities.

Apply 👥 💻

- Display **Slide 2**. Read the problem to learners. Ask them to state the fraction of juice drunk at the party and how this number can be expressed as a division.

Review

- On the board, write the following fractions in one row: $\frac{3}{5}, \frac{8}{7}, \frac{5}{3}, \frac{7}{8}$ and the following division problems in a second row: $5 \div 3, 8 \div 7, 7 \div 8, 3 \div 5$.

- Ask learners to record the numbers and match each fraction to the correct division problem.

Assessment for learning

- Give me a division that represents the fraction $\frac{7}{20}$. ($7 \div 20$)

- How would you represent the number $2\frac{3}{8}$ as a division? ($19 \div 8$)

Same day intervention
Support

- Look out for learners who see the denominator and the numerator as two separate whole numbers, recording a count of two different things. To remedy this 'double-count' view of fractions, support learners by providing practical exercises with part–whole models and developing a learner's relative size of fractions through visualising tasks.

- Provide sharing division exercises to explore the division meaning for fractions.

Lesson 2: Simplifying fractions

Learning objective

Code	Learning objective
6Nf.03	Use knowledge of equivalence to write fractions in their simplest form.

Resources

paper or mini whiteboard and pen (per learner)

Revise

Use the activity *Highest common factors* from Unit 11: *Fractions (A)* in the Revise activities.

Teach 🔲 💻 [TWM.04/07]

- Explain to learners that in this lesson, they will be simplifying fractions. On the board, write: $\frac{8}{20}$. **[T&T]** Say: **This fraction can be simplified. What do you think the phrase 'simplifying a fraction' means?** Take responses. Establish that the phrase means finding an equivalent fraction with the smallest denominator. You do this by finding the largest number that can divide both the numerator and denominator. Ask: **For the fraction $\frac{8}{20}$, what is the greatest number that can divide both the numerator and the denominator?** Agree: 4. Ask: **What do you call a number that can be divided into two different numbers, without leaving a remainder?** Agree that this is a common factor. Ask: **How do you work out the common factors of 8 and 20?** Remind learners that one method is to write out the factors of each number in separate rows and circle the factors common to both rows. **[TWM.04]** Ask: **Would you simplify a fraction to its lowest terms if you chose the first number common to both rows? If not, why?** Elicit that only the highest common factor simplifies the fraction to its lowest terms. Demonstrate that division by 2 reduces the fraction to $\frac{4}{10}$ but the fraction can be simplified further by dividing again by 2. A more efficient method is to divide by the highest common factor, 4. Ask: **What does dividing by 2 reduce the fraction to?** ($\frac{2}{5}$)
- Display and discuss **Slide 1**.
- On the board, write: $\frac{9}{12}$, $3\frac{12}{30}$. Ask learners to simplify each fraction to its lowest terms. For each, choose a learner to explain how they found the highest common factor for the numerator and denominator, and how they divided to reduce the fraction. ($\frac{3}{4}$, $3\frac{2}{5}$)
- On the board, write: $\frac{1}{2}$, $\frac{3}{4}$, $\frac{4}{5}$, $\frac{3}{10}$. Say: **These are examples of simplified fractions.** Ask: **What do you notice about the numerators and denominators?** Establish that the numerator and denominator of the reduced fraction have no common factors other than 1. Ask: **Which of the fractions on the board is a simplified form of $\frac{12}{15}$?** ($\frac{4}{5}$)

- 🗣 Say: **Explain why $\frac{4}{5}$ is equal in value to $\frac{8}{10}$.**
- Discuss the Let's learn section in the Student's Book.
- **[TWM.07]** Introduce the paired activity in the Student's Book, asking learners to determine if Zane has made a mistake when simplifying a fraction. Choose a learner to explain the error and how they would change Zane's thinking.
- Discuss the Guided practice example in the Student's Book.

Practice 📙

- Workbook

Title: Simplifying fractions

Pages: 88–89

- Refer to Activity 1 (Variation) from the Additional practice activities.

Apply 👥 💻

- Display **Slide 2**, which asks learners to solve a number problem involving simplifying a fraction. Invite them to explain how they solved the problem. Establish that the fraction $\frac{24}{56}$ can be reduced to $\frac{3}{7}$.

Review

- On the board, write: $\frac{21}{28}$, $\frac{22}{44}$, $\frac{56}{80}$. Ask learners to simplify then order the fractions, from smallest to greatest. ($\frac{1}{2}$, $\frac{7}{10}$, $\frac{3}{4}$)

Assessment for learning

- $\frac{54}{60}$ in its simplest form is $\frac{7}{10}$. Is this correct? How do you know?
- Give me three fractions that could be simplified to $\frac{7}{10}$.

Same day intervention
Support

- Learners need a secure understanding of factors in order to simplify fractions. Remind them of the meaning of a factor and model this concept by listing the factors of given numbers using the 'rainbow' method used in Stage 5.

Number – Fractions, decimals, percentages, ratio and proportion

Unit 11 Fractions (A)

Number – Fractions, decimals, percentages, ratio and proportion

Lesson 3: Comparing fractions with different denominators

Learning objectives

Code	Learning objective
6Nf.04	Recognise that fractions, decimals (one or two decimal places) [and percentages] can have equivalent values.
6Nf.08	Understand the relative size of quantities to compare [and order numbers with one or two decimal places,] proper fractions with different denominators [and percentages], using the symbols =, > and <.

Resources

mini whiteboard and pen (per learner)

Revise

Use the activity *Spin and match* (1) from Unit 11: *Fractions (A)* in the Revise activities.

Teach 💷 🖥 [TWM.08]

- On the board, write $\frac{5}{8}$ and $\frac{1}{4}$. Ask learners to work with a partner to draw a diagram to compare the two fractions. If they need help, prompt them to draw fraction circles or number lines. Choose pairs to share their drawings and explanations of how their diagrams show which is the greater fraction. **[TWM.08]** Invite learners to comment on the strengths and weaknesses of the different diagrams. Evaluate which models show the comparison the clearest and which ones need improvement. Confirm: $\frac{5}{8} > \frac{1}{4}$.

- Ask: **Is there a way you could compare the fractions without having to draw diagrams?** Take responses. Establish that the fractions $\frac{5}{8}$ and $\frac{1}{4}$ have different denominators and cannot be compared as easily as fractions with the same denominator. To compare fractions with unlike denominators, you convert them to equivalent fractions with the same denominator.

- Display **Slide 1**. Explain that in order to compare $\frac{1}{2}$ and $\frac{3}{8}$, both fractions need to have the same denominator. **[T&T]** Ask: **Which fraction should you convert and why?** Elicit that since you cannot convert $\frac{3}{8}$ to a fraction with a denominator of 2, you can only convert a half to a fraction with a denominator of 8. Discuss the slide and establish that you multiply both the numerator and denominator of $\frac{1}{2}$ to an equivalent fraction with a denominator of 8. Say and record: **Since $\frac{1}{2} = \frac{4}{8}$ and $\frac{4}{8}$ is greater than $\frac{3}{8}$, we know $\frac{1}{2} > \frac{3}{8}$.**

- On the board, write: $\frac{4}{5}$ and $\frac{9}{10}$. Ask: **Which is the greater fraction?** Ask learners to record the comparison using the symbol > or <. Have learners raise their whiteboards and confirm the answer ($\frac{9}{10}$). Choose a learner to explain how they compared the fractions. Confirm that you multiply $\frac{4}{5}$ by 2 to convert to $\frac{8}{10}$ and then compare the fractions. Since $\frac{8}{10}$ is less than $\frac{9}{10}$, you know that $\frac{4}{5}$ is less than $\frac{9}{10}$. Record: $\frac{4}{5} < \frac{9}{10}$.

- On the board, write: $\frac{1}{4}$ and $\frac{1}{5}$. Ask: **How could you use decimals to compare the fractions?** Take responses. Elicit that you could convert the numbers to decimals and compare their values. Ask: **What**

is $\frac{1}{4}$ as a decimal? (0·25) **What is $\frac{1}{5}$ as a decimal?** (0·2) **Which is greater?** (0·25) Record: $\frac{1}{4} > \frac{1}{5}$.

- On the board write $\frac{7}{10}$ and 0·6. Ask: **How would you use percentage to compare these numbers?** Take responses. Elicit that you could convert the numbers to percentages and compare their values. Ask: **What is $\frac{7}{10}$ as a percentage? What is 0·6 as a percentage? Which is greater?** Record: $\frac{7}{10} > 0·6$.

- On the board, write: $\frac{3}{4}$ and $\frac{7}{10}$. Ask learners to convert the fractions to decimals and write a statement that compares them. Choose a learner to state the answer and explain how they compared the numbers. Record: $\frac{3}{4}$ (0·75) $> \frac{7}{10}$ (0·7).

- Discuss the Let's learn section in the Student's Book.

- Introduce the paired activity in the Student's Book.

- Discuss the Guided practice example in the Student's Book.

Practice 📖

- Workbook

Title: Comparing fractions with different denominators

Pages: 90–91

- Refer to Activity 2 from the Additional practice activities.

Apply 👥 🖥

- Display **Slide 2**, which asks learners to solve a number problem involving comparison of two fractions. Invite learners to discuss the answer and solution.

Review

- On the board, write: $\frac{2}{5}$ and $\frac{3}{10}$, $\frac{5}{9}$ and $\frac{2}{3}$. Ask learners to identify the smaller fraction in each pair. ($\frac{3}{10}$, $\frac{5}{9}$).

Assessment for learning

- Use equivalent fractions to compare $\frac{5}{6}$ and $\frac{2}{3}$.
- Which is greater, $\frac{4}{5}$ or $\frac{9}{10}$? ($\frac{9}{10}$)

Same day intervention
Enrichment

- Learners compare fractions where one denominator is greater than 10, for example: **Which is greater, $\frac{4}{5}$ or $\frac{31}{40}$?** ($\frac{4}{5}$)

Lesson 4: **Ordering fractions with different denominators**

Learning objectives

Code	Learning objective
6Nf.04	Recognise that fractions, decimals (one or two decimal places) [and percentages] can have equivalent values.
6Nf.08	Understand the relative size of quantities to [compare and] order [numbers with one or two decimal places,] proper fractions with different denominators [and percentages], using the symbols =, > and <.

Resources

paper or mini whiteboard and pen (per learner)

Revise

Use the activity *Equivalent fractions* from Unit 11: *Fractions (A)* in the Revise activities.

Teach [SB]

- On the board, write: $\frac{5}{8}, \frac{1}{2}, \frac{7}{8}$ and $\frac{3}{4}$. **[T&T]** Ask: **How would you order these fractions, from least to greatest?** Take responses. Establish that the same methods for comparing fractions can be used to order them.
- Discuss converting to the same denominator. Ask: **Which denominator should you convert to and why?** Agree that you should convert to eighths. The greatest denominator is 8 and two of the fractions have denominators of 2 and 4, which are factors of 8. Have learners convert the four fractions to eighths and order them from least to greatest. Choose a learner to explain the conversions and state the order. Record: $\frac{5}{8}, \frac{1}{2} = \frac{4}{8}, \frac{7}{8}$ and $\frac{3}{4} = \frac{6}{8}$. Order: $\frac{1}{2} \left(\frac{4}{8}\right), \frac{5}{8}, \frac{3}{4} \left(\frac{6}{8}\right), \frac{7}{8}$.
- On the board, write: $\frac{7}{10}, \frac{3}{5}, \frac{3}{10}, \frac{1}{2}, \frac{2}{5}$. Ask learners to order the fractions, from least to greatest $\left(\frac{3}{5} = \frac{6}{10}, \frac{1}{2} = \frac{5}{10}, \frac{2}{5} = \frac{4}{10}\right)$. Order: $\frac{3}{10}, \frac{2}{5} \left(\frac{4}{10}\right), \frac{1}{2} \left(\frac{5}{10}\right), \frac{3}{5} \left(\frac{6}{10}\right), \frac{7}{10}$.
- On the board, write: $\frac{9}{10}$, 70% and 0·8. Ask: **How would you use percentages to order these numbers? What about decimals or fractions?** Take responses. Elicit that decimals, percentages and fractions are all interchangeable and can be converted into any of the three number forms. Confirm and record: $\frac{9}{10}$ > 0·8 > 70%.
- ⚐ Give learners the following problem: **Phoebe orders the fractions $\frac{2}{3}, \frac{7}{9}, \frac{1}{3}$. Which fraction is in the middle of the order?** $\left(\frac{2}{3}\right)$
- Discuss the Let's learn section in the Student's Book.

- Introduce the paired activity in the Student's Book, asking learners to work out how the children can be ordered by the fraction of points they score.
- Discuss the Guided practice example in the Student's Book.

Practice [WB]

- Workbook

Title: Ordering fractions with different denominators

Pages: 92–93

- Refer to Activity 2 (Variation) from the Additional practice activities.

Apply 👥 🖥

- Display **Slide 1**. Ask learners to order the bags of oranges by mass.

Review

- On the board, write: $\frac{1}{4}, \frac{1}{2}, \frac{3}{4}, \frac{5}{8}$. Ask learners to order the fractions, from greatest to least. $\left(\frac{3}{4}, \frac{5}{8}, \frac{1}{2}, \frac{1}{4}\right)$

Assessment for learning

- The fractions $\frac{5}{10}, \frac{3}{5}, \frac{9}{10}, \frac{4}{5}, \frac{7}{10}$ are placed in order. Which fraction is in the middle of the order? $\left(\frac{7}{10}\right)$
- Give me a fraction with a denominator of 4 that comes between $\frac{1}{2}$ and $\frac{7}{8}$. $\left(\frac{3}{4}\right)$

Same day intervention

Enrichment

- Provide tasks so that learners order fractions with related denominators where at least one denominator is greater than 10, for example: $\frac{3}{4}, \frac{11}{16}, \frac{5}{8}$.

Number – Fractions, decimals, percentages, ratio and proportion

Additional practice activities

Activity 1 👤 or 👥

Learning objectives

- Understand that a fraction can be represented as a division of the numerator by the denominator (proper and improper fractions).
- Use knowledge of equivalence to write fractions in their simplest form.

Resources

paper (per learner)

What to do

- On the board, write:

$$\frac{\Box}{5} = 1 \div \Box$$

$$\frac{3}{\Box} = \Box \div 8$$

$$\frac{\Box}{8} = 7 \div \Box$$

$$\frac{9}{\Box} = \Box \div 10$$

- Ask learners to fill in the missing numbers in the number statements.
- Learners then make their own missing number problems similar to the ones on the board. They swap with a partner to complete.

Variations

2 On the board, write a set of seven fractions: $\frac{1}{5}, \frac{1}{4}, \frac{2}{5}, \frac{1}{2}, \frac{7}{10}, \frac{3}{4}, \frac{4}{5}$.

Each learner multiplies the numerators and denominators of each fraction by a random number from 2 to 9 to produce a new list, for example, $\frac{3}{15}$ ($\frac{1}{5}$ by 3), $\frac{7}{28}$ ($\frac{1}{4}$ by 7), $\frac{10}{25}$ ($\frac{2}{5}$ by 5), $\frac{8}{16}$ ($\frac{1}{2}$ by 8), $\frac{42}{60}$ ($\frac{7}{10}$ by 6), $\frac{6}{8}$ ($\frac{3}{4}$ by 2), $\frac{32}{40}$ ($\frac{4}{5}$ by 8).

Learners then swap papers for their partner to reduce each fraction to its simplest form.

They return papers for marking.

3 Ask learners to write word problems for their partner to represent as a division and as a fraction, for example, 'Alisha divides a cake into five equal slices. Represent the size of each slice as a fraction of the whole and as a division'.

Activity 2 👤

Learning objective

- Understand the relative size of quantities to compare and order proper fractions with different denominators, using the symbols =, > and <.

Resources

paper or mini whiteboard and pen (per learner)

What to do [TWM.01/02]

- On the board write:
  'Halves greater than quarters'
  'Thirds greater than sixths'
  'Eighths greater than quarters'
  'Tenths greater than fifths'
  'Ninths greater than thirds'
- Learners work through each rule. They write a number statement that compares two fractions that follow the rule. For example, $\frac{2}{3} > \frac{1}{6}$ follows the rule: 'Thirds greater than sixths'.

- Name a rule and ask learners to share their statements. Confirm they are correct as a class.

Variation

2 On the board write:
'Halves, quarters, eighths'
'Thirds, ninths, thirds'
'Fifths, tenths, fifths'
'Halves, fifths, tenths'

For each set of denominators, learners write three fractions in order, from least to greatest. For example, for 'Fifths, tenths, fifths', learners might record the order: $\frac{2}{5} < \frac{7}{10} < \frac{4}{5}$.

Name a combination and ask learners to share their orders. Confirm they are correct as a class.

Unit 12: Fractions (B)

Collins International Primary Maths Recommended Teaching and Learning Sequence Term 2, Week 5

Learning objectives

Code	Learning objective
6Nf.02	Understand that proper and improper fractions can act as operators.
6Nf.05	Estimate, add and subtract fractions with different denominators.
6Nf.06	Estimate, multiply and divide proper fractions by whole numbers.

Unit overview

In this unit, learners extend their knowledge of fractions as operators to solving problems with improper fractions. Learners understand that the steps involved in finding an improper fraction of a number or quantity can involve different combinations of operations, for example, to find $\frac{3}{2}$ of 6, divide 6 by 2 then multiply by 3 or multiply by 3, then divide by 2.

In Lesson 2, learners investigate addition and subtraction of proper and improper fractions with different denominators. When adding fractions, learners express answers greater than 1 as an improper fraction and a mixed number, and reduce all fractions to their simplest form.

In Lessons 3 and 4 concrete, pictorial and abstract representations are used to support the multiplication and division of proper fractions by a whole number.

Prerequisites for learning

Learners need to:
• understand unit and non-unit fractions
• know how to calculate a unit fraction of an amount by dividing by the denominator of a fraction
• be able to multiply and divide unit fractions by a whole number.

Vocabulary

proper fraction, improper fraction, mixed number, numerator, denominator, lowest common multiple (LCM), common denominator, product, quotient

Common difficulties and remediation

To be able to add fractions with different denominators, learners need to have a strong understanding of equivalent fractions. Provide activities in which learners use a fraction wall, circular fraction tiles or other modelling methods to find equivalent fractions.

Look out for learners who do not convert fractions to an equivalent denominator before they add or subtract. If they do not understand that different denominators refer to different-sized fractions they may make errors, for example adding or subtracting numerators but retaining the larger denominator in the answer (e.g. $\frac{1}{4} + \frac{3}{8} = \frac{4}{8}$). Work with fraction walls and tiles will help learners understand that adding and subtracting fractions requires a common denominator.

This misconception can lead learners to make other errors, such as scaling the denominator without adjusting the numerator; for example, converting the problem $\frac{1}{5} + \frac{2}{10}$ to $\frac{1}{10} + \frac{2}{10}$. To remedy this, provide practical tasks where learners build equivalent fractions with tiles and fraction walls.

When multiplying fractions by a whole number, look out for learners who multiply the numerator and denominator. This misconception arises because learners do not know why we only multiply the numerator by the whole number. To correct this error, learners should change the whole number into a fraction and be encouraged to check their answer by writing the problem out as an addition.

Supporting language awareness

Encourage learners to write their own definition for new key words, for example: 'An improper fraction is a top-heavy fraction where the top number is bigger than the bottom number.'

Promoting Thinking and Working Mathematically

In Lesson 1, learners critique (TWM.07) when they suggest alternative strategies for finding a non-unit fraction of a quantity. They discuss how the steps involving multiplication and division can be switched.

In Lesson 2, learners conjecture (TWM.03) when they reason that any common multiple of a pair of fractions can be used as a common denominator. They understand, however, that the lowest common multiple is the most effective common denominator.

Success criteria

Learners can:
• calculate a non-unit fraction of an amount by dividing by the denominator and multiplying by the numerator of a fraction
• convert unlike fractions to fractions with the same denominators and add or subtract them
• multiply proper fractions by whole numbers by using fraction models and repeated addition
• divide proper fractions by whole numbers by using diagrams such as area models.

Number – Fractions, decimals, percentages, ratio and proportion

Unit **12** Fractions (B)

Number – Fractions, decimals, percentages, ratio and proportion

Lesson 1: **Fractions as operators**

Learning objective

Code	Learning objective
6Nf.02	Understand that proper and improper fractions can act as operators.

Resources

paper or mini whiteboard and pen (per learner) paper, for the Workbook (per learner)

Revise

Use the activity *Fractions of length measurements* from Unit 12: *Fractions (B)* in the Revise activities.

Teach [SB] ⬜ [TWM.03/07]

- On the board, write: $\frac{1}{5}$ of $60 = and, below it, write $\frac{3}{5}$ of $60 =. Ask: **If you divide $60 by 5, what will you find out?** Elicit that this is one fifth of $60 ($12). Write: $\frac{1}{5}$ of $60 = $12. Ask: **How will you find $\frac{3}{5}$ of $60?** Elicit that one-fifth must be multiplied by 3 to give three-fifths. Ask: **What is the answer?** ($36)

- On the board, write: $\frac{7}{5}$ of $60 =. Ask: **What is different about this question?** Take responses. Confirm that we are finding an improper fraction of an amount. Ask: **What calculations would give you the answer?** Agree that similar to a proper fraction, we find one-fifth of the amount and then multiply by 7 to find seven-fifths of the amount. Ask: **What is the answer?** ($84)

- [T&T] [TWM.07] Ask: **Can you think of a different way to find $\frac{7}{5}$ of $60?** Take responses. Ask learners to explain their strategy and confirm that it gives the same answer ($84). Ask questions to develop estimation skills, for example: **Would you expect the answer to be more than $100? How do you know?** Invite learners to comment on the methods suggested, asking them to say how efficient or 'easy' they consider them to be.

- Establish two other strategies. The first is to switch the order of division and multiplication: multiply by the numerator first and then divide by the denominator second. Ask learners to confirm that this works by performing the two calculations in this order. [TWM.03] Ask: **Why do you think this method works?** Take responses. Elicit that the order of division and multiplication does not matter. Demonstrate with examples: 5 ÷ 2 × 3 = is the same as 5 × 3 ÷ 2 =.

- Display **Slide 1**. Discuss the two methods of finding the fraction of the amount of milk given. Ask: **Which method is easier? Why?** Take responses. Discuss how method 1 might be preferable to method 2

when working with larger numbers. Ask: **For 500 ml of milk, what is $\frac{9}{5}$ of the amount?** Choose learners to explain the method they used (900 ml).

- 🗣 Say: **Give me three different methods for finding $\frac{5}{3}$ of 15 kg.** (25 kg)

- Discuss the Let's learn section in the Student's Book.

- Introduce the paired activity in the Student's Book, asking learners to find the amount of carrots sold.

- Discus the Guided practice example with learners.

Practice [WB]

- Workbook

Title: Fractions as operators

Pages: 94–95

- Refer to Activity 1 from the Additional practice activities.

Apply 👥 ⬜

- Display **Slide 2**, asking learners to find the best of three offers of fractional discounts on prices. Discuss their solutions. (a, $582)

Review

- On the board, draw a table with columns headed $\times \frac{4}{5}$, $\times \frac{7}{5}$, $\times \frac{11}{8}$ and rows labelled 40, 120, 320. Ask learners to copy and complete the table.

Assessment for learning

- Tilly wants to find $\frac{6}{5}$ of 30 kg. She decides to find $\frac{1}{5}$ of the amount and then multiply this number by 6. Show Tilly another method of finding $\frac{6}{5}$ of 30 kg.

- Show me that $\frac{7}{4}$ of $44 is $77.

Same day intervention
Enrichment

- Provide calculations with fractions that have larger numerators and denominators, for example: find $\frac{16}{17}$ of $153 ($144).

Lesson 2: **Adding and subtracting fractions**

Learning objective

Code	Learning objective
6Nf.05	Estimate, add and subtract fractions with different denominators.

Resources

paper or mini whiteboard and pen (per learner)

Revise

Use the activity *The answer is…* from Unit 12: *Fractions (B)* in the Revise activities.

Teach SB 🖥 [TWM.03]

- On the board, write: $\frac{1}{4} + \frac{2}{3} =$. **[T&T] [TWM.03]** Say: **These fractions have different denominators. How can you add them?** Take responses. Establish that since only like fractions can be added or subtracted, you first have to convert unlike fractions to equivalent fractions with the same denominator.

- Display **Slides 1** and **2**. Guide learners through the steps involved in converting fractions to equivalent fractions with the same denominator. Ask: **How do you find the lowest common multiple of two numbers?** Take responses. Remind learners that a good method is to write down several multiples of both the numbers and then identify the smallest common multiple among them. Identify that 12 is the LCM for 3 and 4. Explain that 12 is a common denominator for the two fractions.

- When converting to equivalent fractions, ask questions to develop understanding: **Why do we choose 12 for the common denominator? Why do we multiply the numerator and denominator of $\frac{1}{4}$ by 3?** (4 × 3 = 12) **Why do we multiply the numerator and denominator of $\frac{2}{3}$ by 4?** (3 × 4 = 12)

- On the board, write: $\frac{3}{5} - \frac{2}{7} =$. Say: **The fractions are unlike.** Ask: **How do you subtract them?** (convert to equivalent fractions with the same denominator) **Which common denominator should you use?** Give learners time to work out the answer and then take responses. Confirm the denominator: 35. Ask: **Why did you choose 35?** (LCM of 5 and 7) Ask: **What do you do next?** (convert the fractions to equivalent fractions with a denominator of 35) Ask: **How do you convert the first fraction, $\frac{3}{5}$ to a fraction with a denominator of 35?** Take responses. Prompt learners by asking how many 5s make 35. (7) Establish that both numerator and denominator need to be multiplied by 7. Ask: **What is the equivalent fraction?** ($\frac{21}{35}$) Record: $\frac{21}{35}$ below $\frac{3}{5}$.

  Ask: **How do you convert the second fraction, $\frac{2}{7}$?** Prompt learners by asking how many 7s make 35. (5) Establish that both the numerator and denominator need to be multiplied by 5. Ask: **What is the equivalent fraction?** ($\frac{10}{35}$) Record: $\frac{10}{35}$ below $\frac{2}{7}$. Explain that you can now subtraction like fractions. Ask: **What is $\frac{21}{35}$ subtract $\frac{10}{35}$?** Record: $\frac{3}{5} - \frac{2}{7} = \frac{11}{35}$.

- **[T&T] [TWM.03]** Return to the previous problem. Ask: **If instead we had decided that 70, a common multiple of 5 and 7, was the common denominator we should use, would this have given an incorrect answer?** Take responses. Work through the problem with 70 as the common denominator. Record: $\frac{3}{5} - \frac{2}{7} = \frac{42}{70} - \frac{20}{70} = \frac{22}{70}$. Explain that if you simplify $\frac{22}{70}$, you get $\frac{11}{35}$. Establish that the calculation leads to the same answer but involves a lot more work, multiplying larger numbers and simplifying the fraction.

- Write: $\frac{2}{15} + \frac{2}{3} =$, $\frac{8}{9} - \frac{5}{6} =$. Ask learners to solve the calculations with a partner. For each calculation, choose a pair to explain the steps involved. Ensure learners can reduce fractions to their simplest form.

- On the board, write: $\frac{11}{6} + \frac{4}{5} =$. Ask: **What is different about this problem?** Take responses. Agree that it includes an improper fraction. Explain that you solve it in the same way as previous examples. Repeat for: $\frac{9}{7} + \frac{5}{4}$ ($2\frac{15}{28}$) and $\frac{19}{10} - \frac{7}{4}$ ($\frac{3}{20}$).

- Discuss the Let's learn section in the Student's Book.
- Introduce the paired activity in the Student's Book.
- Discuss the Guided practice example.

Practice WB

- Workbook

Title: Adding and subtracting fractions

Pages: 96–97

- Refer to Activity 1 (Variation) from the Additional practice activities.

Apply 👥 🖥

- Display **Slide 3**. Discuss the problem. Choose learners to explain how they calculated the total.

Review

- Ask: **What is $\frac{7}{8} - \frac{4}{7}$?** ($\frac{7}{8} - \frac{4}{7} = \frac{49}{56} - \frac{32}{56} = \frac{17}{56}$)

Assessment for learning

- How many fifteenths is the difference between $\frac{16}{5}$ and $\frac{2}{3}$? (38)

Same day intervention

Enrichment

- Provide problems that have higher lowest common multiples, for example: $\frac{9}{7} + \frac{14}{8}$ or $\frac{22}{8} - \frac{4}{9}$ ($\frac{170}{56} = \frac{85}{28} = 3\frac{1}{28}$; $\frac{166}{72} = \frac{83}{36} = 2\frac{11}{36}$).

Number – Fractions, decimals, percentages, ratio and proportion

Lesson 3: **Multiplying fractions by whole numbers**

Learning objective

Code	Learning objective
6Nf.06	Estimate, multiply [and divide] proper fractions by whole numbers.

Resources

mini whiteboard and pen (per learner)

Revise

Use the activity *Multiplying fractions* from Unit 12: *Fractions (B)* in the Revise activities.

Teach 📘 💻 [TWM.03/07]

- On the board, write: $\frac{3}{4} \times 3 =$. **[TWM.07] [T&T]** Ask: **How would you model this multiplication?** Learners work in pairs to draw a diagram that models the calculation. Choose pairs to share their models. Engage the class in discussing the advantages and disadvantages of the models shared and how they could be improved.
- Display **Slide 1**. Referring to the model and related calculation, establish that, since multiplication is repeated addition, $\frac{3}{4} \times 3$ is equivalent to the addition $\frac{3}{4} + \frac{3}{4} + \frac{3}{4}$ ($\frac{9}{4}$ or $2\frac{1}{4}$).
- Ask: **What is $\frac{3}{8} \times 5$?** Ask questions to develop estimation skills, for example: **Would you expect the answer to be less or more than $2\frac{1}{2}$? How do you know?** Give pairs time to devise models for the calculation and share them with the class. Confirm the answer. ($\frac{15}{8}$ or $\frac{17}{8}$).
- Ask learners to answer the three multiplication calculations on **Slide 1**. Give pairs time to devise models for the calculations and share them with the class. Confirm the answers. (**a** $\frac{10}{5}$ or 2 **b** $\frac{18}{10}$ or $1\frac{4}{5}$ **c** $\frac{20}{6}$ or $3\frac{1}{3}$)
- On the board, write: $\frac{3}{3} \times 15 =$. Say and record on the board: **Shane solves this problem by drawing 15 models of three-thirds. He adds each fraction together to find the sum, $\frac{45}{3}$. He then divides 45 by 3 and finds that the answer is 15. [TWM.03]** Ask: **How could Shane have changed the question to make it much easier?** Take responses. Explain that three-thirds are equivalent to 1. 15 is effectively multiplied by 1 and so the product is 15. Establish that if the multiplier in a calculation is a fraction with a numerator equal to its denominator, then the multiplier is effectively 1 and the multiplicand remains unchanged.
- 🔁 Ask: **What is the product of $\frac{5}{8} \times 4$ as a mixed number reduced to its simplest form?** ($2\frac{1}{2}$)
- Discuss the Let's learn section in the Student's Book.
- **[TWM.04]** Introduce the paired activity in the Student's Book, asking learners to prove that four pieces that are each three-eighths of a pizza are equivalent to $1\frac{1}{2}$ whole pizzas.

- Discuss the Guided practice example in the Student's Book.

Practice 📒 [TWM.04]

- Workbook

Title: Multiplying fractions by whole numbers

Pages: 98–99

- Refer to Activity 2 from the Additional practice activities.

Apply 👥 💻

- Display **Slide 2**. Talk through the problem with learners. Learners calculate the combined fraction of the four bars eaten by the children. Choose a pair of learners to explain how they solved the problem.

Review

- On the board, write: $\frac{1}{4}, \frac{2}{5}, \frac{3}{8}$. Learners work in pairs to draw two connected function machines labelled: × 2 and × 3. They estimate the outputs when the numbers on the board are passed through both machines, first the × 2 function and then the × 3 function. Have learners write each output as both an improper fraction and a mixed number, reducing the fraction to its simplest form. How close were their estimates to the outputs? ($\frac{6}{4}$ ($1\frac{1}{2}$), $\frac{12}{5}$ ($2\frac{2}{5}$), $\frac{18}{8}$ ($2\frac{1}{4}$))
- Repeat for different multipliers.

Assessment for learning

- Toby says, 'When a proper fraction is multiplied by a whole number, the product is always greater than 1.' Is Toby correct? How do you know? (not always, for example, $\frac{3}{8} \times 2 = \frac{6}{8}$)
- How many sevenths are there in the product of $\frac{3}{7} \times 4$? (12) How would you write the product as a mixed number? ($1\frac{5}{7}$)

Same day intervention
Enrichment

- Provide problems with fractions that have larger numerators and denominators and larger whole numbers, for example: $\frac{27}{4} \times 13$ ($\frac{351}{4}$ or $87\frac{3}{4}$).

Number – Fractions, decimals, percentages, ratio and proportion

Lesson 4: **Dividing fractions by whole numbers**

Learning objective

Code	Learning objective
6Nf.06	Estimate, [multiply] and divide proper fractions by whole numbers.

Resources

ruler (per pair); paper or mini whiteboard and pen (per pair)

Revise

Use the activity *Fraction stories* from Unit 12: *Fractions (B)* in the Revise activities.

Teach 📖 🖥 [TWM.01/02/04/07/08]

- On the board, write: $\frac{2}{3} \div 4 =$. Ask learners how they might model the division of a unit fraction by a whole number. Take responses.
- Display **Slide 1**. Carefully explain how an area model can be used to represent a division problem.
- Write: $\frac{3}{4} \div 5 =$. Ask questions to develop estimation skills, for example: **Would you expect the answer to be less or more than $\frac{3}{4}$? How do you know?** Provide rulers. Ask learners to work with a partner to draw an area model for the fraction $\frac{3}{4}$. Ask them to hold up their models and confirm they are an effective representation. Expect them to have divided a rectangle into four equal rows or columns and shaded three of these areas. Ask: **How would you use your model to show $\frac{3}{4}$ divided by 5?** Give pairs time to adjust their diagrams. Expect them to have drawn straight parallel lines to divide their models into 20 equal areas. Draw the area model on the board. Ask: **Which part of the diagram shows us the answer, the quotient of three-quarters divided by 5?** Invite a learner to the board to identify one-fifth of $\frac{3}{4}$. Remind learners that dividing by 5 is the same as multiplying by $\frac{1}{5}$. Ask: **What is the answer?** $\left(\frac{3}{20}\right)$
- Repeat the previous activity for the question: $\frac{2}{5} \div 6 =$. $\left(\frac{2}{30} \text{ or } \frac{1}{15}\right)$
- [TWM.07/08] Write: $\frac{4}{5} \div 5 =$. On the board, draw a rectangle divided into five rows and shade four of the rows to model $\frac{4}{5}$. Draw four vertical lines to model division by 5. This creates a 5 by 5 array of 25 equal parts with 20 parts shaded. [T&T] [TWM.01/02] Ask: **I have shaded 20 equal parts of the area diagram. This must mean that four-fifths divided by 5 is $\frac{20}{25}$. Is that correct?** Take responses and establish the error: 20 equal parts of the diagram represents the fraction $\frac{20}{25}$ which is equivalent to $\frac{4}{5}$. No division has actually taken place. Establish that you have to find the shaded parts in one column to find $\frac{20}{25}$ divided by 5. $\left(\frac{4}{25}\right)$
- ♫ Ask learners to draw a diagram to model the division $\frac{5}{8} \div 3$. Ask: **What is the answer?** $\left(\frac{5}{24}\right)$
- Discuss the Let's learn section in the Student's Book.

- [TWM.04] Introduce the paired activity in the Student's Book, asking learners to prove that if $\frac{5}{6}$ of a pizza is divided into four equal slices, then each slice represents $\frac{5}{24}$ of the whole pizza.
- Discuss the Guided practice example in the Student's Book.

Practice 📗 [TWM.04]

- Workbook

Title: Dividing fractions by whole numbers

Pages: 100–101

- Refer to Activity 2 (Variation) from the Additional practice activities.

Apply 👥 🖥

- Display **Slide 2**. Ask learners to work out how to share the lemonade between the five children. Choose a pair of learners to explain how they worked out the answer.

Review

- On the board, write: $\frac{3}{4}, \frac{2}{5}, \frac{4}{9}$. Learners work in pairs to draw two connected function machines labelled: ÷ 2 and ÷ 3. They estimate the outputs when the numbers on the board are passed through both machines, first the ÷ 2 function and then the ÷ 3 function. Have learners write each output, reducing the fraction to its simplest form. How close were their estimates to the outputs? $\left(\frac{3}{24} \left(\frac{1}{8}\right), \frac{2}{30} \left(\frac{1}{15}\right), \frac{4}{54} \left(\frac{2}{27}\right)\right)$
- Repeat for different divisors.

Assessment for learning

- Lara says, 'When a proper fraction is divided by a whole number greater than 1, the quotient is always smaller than the dividend.' Is Lara correct? (yes) How do you know?
- What is the quotient of $\frac{3}{7} \div 6$ reduced to its simplest form? $\left(\frac{1}{14}\right)$

Same day intervention

Enrichment

- Provide problems with fractions that have larger numerators and denominators, for example: $\frac{7}{18} \div 3$. $\left(\frac{7}{54}\right)$

Number – Fractions, decimals, percentages, ratio and proportion

Additional practice activities

Activity 1

Learning objectives
- Understand that proper and improper fractions can act as operators.
- Estimate, add and subtract fractions with different denominators.

Resources
2–9 spinner made from Resource sheet 3: Spinner (1) (per pair); pencil and paper clip for the spinner (per pair); paper or mini whiteboard and pen (per learner)

What to do
- Learners take turns to spin the spinner and multiply the number spun by 100 to create a three-digit multiple of 100. This number is the 'quantity'.
- One player plays as 'quarters'; the other as 'fifths'.
- The 'quarters' player spins again and the number spun forms the numerator for that number of quarters.
- The 'fifths' player spins again and the number spun forms the numerator for that number of fifths.
- Players then work out their fraction of the quantity. The player with the higher fraction of the quantity wins the equivalent in points.
- The game continues for five rounds. The winner is the player with the greater score.

Variation
Learners take turns to spin the spinner four times. The four digits are used to write a fraction calculation. For example, if they spin 5, 3, 6, 7 they write $\frac{5}{3} + \frac{6}{7} =$.

They calculate the sum and the learner with the greater total scores a point.

They repeat the four spins, but this time they write a subtraction calculation with a positive difference. For example, if they spin 3, 6, 9, 4 they could write $\frac{9}{4} - \frac{2}{3} =$. Allow learners to swap the numbers if they get them the wrong way around.

They calculate the difference and the learner with the smaller difference scores a point.

The winner is the player with the higher score after three rounds. Each round is one addition and one subtraction calculation.

Activity 2 ▪ or ▪▪

Learning objective
- Estimate, multiply and divide proper fractions by whole numbers.

Resources
paper (per learner)

What to do
- Learners write a fraction word problem about a pizza divided into a fraction.
- They describe a scenario based on a group of children eating pizza. Each child is given a pizza of which they consume the same non-unit fraction, for example: 'There are three children. Each child is given a pizza of which they eat two-fifths. What fraction of the three pizzas have the children eaten altogether?' ($\frac{2}{5} \times 3 = \frac{6}{5} = 1\frac{1}{5}$)

- Encourage them to be imaginative.
- They swap papers with a partner to solve. They return the papers for marking.

Variation
Change the description of the scenario to a division problem, for example: 'Three children share $\frac{2}{3}$ of a pizza. What fraction of the whole pizza will each get?' ($\frac{2}{3} \div 3 = \frac{2}{9}$)

Unit 13: Percentages

Collins International Primary Maths Recommended Teaching and Learning Sequence Term 2, Week 6

Learning objectives

Code	Learning objective
6Nf.04	Recognise that fractions, decimals (one or two decimal places) and percentages can have equivalent values.
6Nf.07	Recognise percentages (1%, and multiples of 5% up to 100%) of shapes and whole numbers.
6Nf.08	Understand the relative size of quantities to compare and order numbers with one or two decimal places, proper fractions with different denominators and percentages, using the symbols =, > and <.

Unit overview

In this unit, learners recognise percentages of shapes focusing on 1% and multiples of 5%. They make the connection between fractions, decimals and percentages, for example $30\% = 0.3 = \frac{3}{10}$ (learners colour three columns rather than counting individual squares), $75\% = 0.75 = \frac{3}{4}$ (learners colour a half plus a quarter of the grid).

Learners complete exercises where they use the equivalence between fractions, decimals and percentages to calculate percentages of whole numbers.

Contextual problems are introduced in Lesson 3, which require learners to work out percentage increases and decreases in prices. Other examples explore the relationship between percentages, fractions and angles of sectors in a pie chart.

Learners extend their knowledge of comparing and ordering fractions, decimals and percentages.

Note

Lesson 3 introduces the concept of pie charts. If learners have not yet covered this in Unit 25, ensure that time is spent exploring what they are and how they work.

Prerequisites for learning

Learners need to:
- be able to explain percentage as 'the number of parts in every hundred'
- know how to use a 100 grid or counters to model a percentage
- recognise that 50% is a half, 25% is a quarter, 10% is a tenth and 1% is a hundredth
- be able to find simple percentages of shapes.

Vocabulary

per cent (%), percentage, fraction, decimal, denominator, discount, pie chart, angle, quantity

Common difficulties and remediation

Look out for learners who view a percentage as a number rather than part of an amount, or who have yet to make a sound link with fractions. Also, look out for learners who have made a link with fractions but use the value of the percentage as the denominator, for example, thinking 15% is equivalent to $\frac{1}{15}$ and subsequently finding 15% of 200 by attempting to divide 200 by 15. To remediate, provide models of percentages as fractions, saying 'out of a hundred' while drawing the fraction line (vinculum). Similarly, provide models and examples of percentages as fractions. This helps learners to make links with problem solving and finding a percentage of any amount, not just an amount out of a hundred.

Supporting language awareness

Whenever discussing percentage, it is important to say what the percentage is relevant to, stating the whole amount – what 100% is. For example, '30% of learners travel to school by car' does not mean the same as the context-specific: 'In a school of 900 learners, 30% travel to school by car.'

Provide learners with frequent opportunities to experience percentages in real-life contexts to give more meaning to their calculations and promote understanding.

Promoting Thinking and Working Mathematically

In Lesson 1, learners convince (TWM.04) when they demonstrate how to quickly show 55% of squares on a 100 grid by shading half then adding 5, rather than counting individual squares.

In Lesson 2, learners improve (TWM.08) when they discuss the efficiency of different strategies for finding a multiple of 5% of an amount.

Learners characterise (TWM.05) in Lesson 3 when they use the percentage results of a survey to calculate the angles of sectors on a pie chart.

Success criteria

Learners can:
- use a 100 grid or counters to model 1% and multiples of 5%
- use the equivalence between fractions, decimals and percentages to calculate percentages of whole numbers and quantities
- compare and order percentages of whole numbers and quantities.

Number – Fractions, decimals, percentages, ratio and proportion

Unit **13** Percentages

Number – Fractions, decimals, percentages, ratio and proportion

Lesson 1: **Percentages of shapes**

Learning objectives

Code	Learning objective
6Nf.04	Recognise that fractions, decimals (one or two decimal places) and percentages can have equivalent values
6Nf.07	Recognise percentages (1%, and multiples of 5% up to 100%) of shapes and whole numbers.

Resources

100 grid from Resource sheet 14: Waffle diagrams (per pair)

Revise

Use the activity *Out of $100* from Unit 13: *Percentages* in the Revise activities.

Teach 🔲 📊 [TWM.04]

- Display the **Number square tool**. Explain to learners that in this lesson, they will revise percentages. Ask: **What is a percentage?** Take responses. Remind learners that when you consider parts out of 100, you can use a special type of fraction called a percentage. A percentage is a fraction with the denominator 100. Instead of writing the denominator, you use the percentage sign.
- Shade 1 square. On the board, write: per cent, %. Ask: **What percentage of squares is shaded?** Confirm and write: 1%. Shade seven squares. Ask: **What percentage of squares is shaded?** Confirm and write: 7%.
- Shade the squares in three columns on the grid. Ask: **What fraction and percentage is this?** Agree $\frac{30}{100}$, 30% and record the fraction on the board. Ask: **What is 30% as a decimal?** (0·3)
- Choose a learner to colour 70% of the squares on the grid. Expect them to shade seven columns rather than counting individual squares. Ask: **What is this percentage as a decimal?** (0·7) **As a fraction?** ($\frac{70}{100}$, $\frac{7}{10}$)
- Choose a learner to colour 25% of the squares on the grid. Expect them to shade two and a half columns rather than counting individual squares. Ask: **What is this percentage as a decimal?** (0·25) **As a fraction?** ($\frac{1}{4}$) [T&T] [TWM.04] Ask: **Why is the fraction 1 over 4?** Take responses. Remind learners that $\frac{25}{100}$ is $\frac{1}{4}$ when reduced to its simplest form (by dividing both numerator and denominator by 25).
- Shade 75 squares on the grid. Ask: **What fraction and percentage is this?** Agree $\frac{75}{100}$, 75%. Ask: **What is this percentage as a decimal?** (0·75) **As a fraction?** ($\frac{3}{4}$) [T&T] Ask: **Why is the fraction 3 over 4?** Take responses. Remind learners that $\frac{75}{100}$ is $\frac{3}{4}$ when reduced to its simplest form (by dividing both numerator and denominator by 25).
- Shade five squares on the grid. Ask: **What fraction and percentage is this?** Agree 5%. Ask: **How do you write this as a fraction?** ($\frac{5}{100}$) **Can it be simplified?** (yes, $\frac{1}{20}$) Repeat for multiples of 5%, for example: 15%, 35%, 45%.
- [TWM.03] Ask: **If you were given a 100 grid and asked to represent the fraction 55% by shading squares as quickly as possible, how**

would you do it? Take responses. Establish that the quickest way would be to shade half (50%) of the squares and then five more.
- Ask a learner to shade 65%. Expect them to shade six and a half columns rather than count individual squares. Ask: **What is this percentage as a fraction?** ($\frac{65}{100}$) **Can it be simplified?** (yes, $\frac{13}{20}$)
- 📖 Ask: **How many columns of the grid would you shade to represent 95%?** (9 and a half)
- Discuss the Let's learn section in the Student's Book and address any misunderstandings.
- Provide each pair with a 100 grid from Resource sheet 14: Waffle diagrams and introduce the paired activity. Choose a learner to explain how they shaded the squares in a 100 grid to represent 85%.
- Discuss the Guided practice example with learners.

Practice 📓

- Workbook

Title: Percentages of shapes

Pages: 102–103

- Refer to Activity 1 from the Additional practice activities.

Apply 👥 🖥️

- Display **Slide 1**, asking learners to find the percentage expressed by the number of red pencils in a tray (36%).

Review 📊

- Display the **Number square tool** set as a blank 100 grid. Shade in different multiples of five squares on the grid. Ask learners to find both the total fraction and the percentage shaded. Repeat for different amounts of squares.

Assessment for learning

- 55 out of 100 learners prefer the colour red. How would you show this as a percentage?
- 7 out of 20 fruits in a box are apples. What percentage of apples is this? (35%)

Same day intervention

Support

- Give learners plenty of opportunities to model percentages with 100 grids and manipulatives, such as cubes.

Lesson 2: **Percentages of whole numbers (1)**

Learning objectives

Code	Learning objective
6Nf.04	Recognise that fractions, decimals (one or two decimal places) and percentages can have equivalent values
6Nf.07	Recognise percentages (1%, and multiples of 5% up to 100%) of shapes and whole numbers.

Resources

mini whiteboard and pen (per learner)

Revise

Use the activity *Spin and match (2)* from Unit 13: *Percentages* in the Revise activities.

Teach 📘 📊 🖥️

- Display the **Number square tool** set up as blank six columns by ten rows. Present a context in which the grid represents 60 seeds planted by a gardener in a garden plot. Ask: **How would you find 25% of 60 seeds?** Take responses. If learners find the question difficult then provide a hint: convert the percentage to a fraction. Ask: **What is the fraction?** ($\frac{1}{4}$) Ask: **What is a quarter of 60?** Choose a learner to come to the board to shade a quarter of the squares. Ask: **How many is this?** (15) **How would you write this as a calculation?** Have learners write the calculation on their whiteboards. Ask them to hold up their whiteboards to confirm they are correct. Expect: $\frac{1}{4} \times 60$ or $60 \div 4$.
- Display **Slide 1**. Discuss how to find 20% of $40 both pictorially and as a calculation. Ensure that learners know that 20% is equivalent to $\frac{1}{5}$.
- On the board, write a table with columns headed 25%, 40%, 50%, 75%, 80% and rows headed $10, $20, $60 and $80. Ask learners to work with a partner and copy and complete the table. They find the percentage of each amount of money. Pick a cell in the table and choose a pair of learners to explain how they found the percentage. Repeat for other cells.
- On the board, write: What is 15% of 4 kg? **[T&T]** Ask: **How would you answer this question?** Take responses. Choose learners to explain the strategy they would use. Invite responses from the class, asking learners to comment on how efficient they think the strategy is and how it could be improved. Ask: **What is 15% as a fraction?** ($\frac{15}{100}$) **Can it be reduced?** ($\frac{3}{20}$) **The calculation is now $\frac{3}{20}$ of 4. How do you solve this?** Expect: $4 \div 20 \times 3$ or $3 \times 4 \div 20 = 0\cdot6$ kg. Alternatively, discuss conversion of 4 kg to 4000 g and calculating $\frac{3}{20}$ ($3 \times 4000 \div 20 = 600$ g). Some learners may have chosen to find the sum or difference of partial percentages, for example: 10% + 5% or 20% – 5%.
- Repeat for other percentage problems: 35% of 8 litres (2·8ℓ), 85% of 6 km (5·1 km).
- Ask: **If I have a block of wood that is 20 centimetres long and I cut a piece from it that is 5 centimetres long, what percentage of the block did I cut?** Expect learners to use their knowledge

of equivalence to find that 5 cm is $\frac{1}{4}$ of 20 and therefore, 25%. Repeat for other questions, for example: **If I have 500 ml of juice and I drink 200 ml, what percentage did I drink?** ($\frac{200}{500} = \frac{2}{5} = 40\%$)

- 🖐️ Ask: **How would you find 35% of 40 m?** (14 m)
- Discuss the Let's learn section in the Student's Book and address any misunderstandings.
- Introduce the paired activity, asking learners to explain how they calculated the amount Daisy has left to spend ($3.90). If learners struggle to understand the problem ask: **What percentage of the amount does Daisy NOT spend? How can you use this to calculate the amount left?**
- Discuss the Guided practice example with learners.

Practice 📒 🖥️

- Workbook

Title: Percentages of whole numbers (1)

Pages: 104–105

- Refer to Activity 1 (Variation) from the Additional practice activities. Use **Slide 2** APA1: Variation

Apply 👥

- On the board, write the statement: 'All percentages can be found as the sum of a set of smaller percentages.' Learners investigate, providing evidence to support their answers.

Review

- Draw a table with headings: Amount, 5%, 35%, 85%. In different rows under 'Amount' write: $1400, $2800 and $4200. Ask learners to complete the table by finding the percentages of each amount.

Assessment for learning

- What is 75% of 140 g? (105 g)
- 1% of $1900 is $19 and 10% is $190. How would you use this to find 31%?

Same day intervention

Enrichment

- Some learners may be ready to investigate other percentage-fraction equivalences, for example: $12\cdot5\% = \frac{1}{8}$, $37\cdot5\% = \frac{3}{8}$. Provide questions where learners calculate with these percentages, for example: **What is 12·5% of $40?** ($5)

Number – Fractions, decimals, percentages, ratio and proportion

185

Unit **13** Percentages

Number – Fractions, decimals, percentages, ratio and proportion

Lesson 3: **Percentages of whole numbers (2)**

Learning objectives

Code	Learning objective
6Nf.04	Recognise that fractions, decimals (one or two decimal places) and percentages can have equivalent values
6Nf.07	Recognise percentages (1%, and multiples of 5% up to 100%) of shapes and whole numbers.

Resources

paper or mini whiteboard and pen (per learner)

Revise

Use the activity *Spin and match (3)* from Unit 13: *Percentages* in the Revise activities.

Teach 📖 🖥 [TWM.05]

- Display **Slide 1**. Discuss the meaning of the word 'discount' as an amount taken off the usual price of something. Ask learners to say where they have seen the word used. Establish that discounts are often given as percentages. Equally percentages may be used to indicate a rise in price. Point to the surfboard and ask learners what the new price will be. Explain that if an item is initially priced at $100 and the price increases by 10%, then the increase is $10 and the new price will be $110.

- Ask learners to work in pairs to work out the new price of each item after the percentage rise or discount given. Choose each item in turn and ask a pair of learners to explain how they calculated the new price (beach shorts: $40, bag: $45, sandals: $4.50, sunglasses: $140, radio: $130).

- Display **Slide 2** and introduce pie charts as a chart that uses pie slices to show relative sizes of data and that the chart is divided into sectors, where each sector shows the relative size of each value. Also explain how the entire pie represents 100 percent of a whole, while the pie slices represent portions of the whole. Explain that the pie chart was drawn by Lucy using the results of a class survey that asked the question: 'What is your favourite flavour of fruit juice?' **[T&T] [TWM.05]** Ask: **How did Lucy decide on the angle of each slice (sector) of the pie chart?** Take responses. Establish that Lucy knows that there are 360 degrees in a circle. She uses the percentages to calculate the angle made by each sector. Ask: **What angle does the 'orange juice' sector make at the centre of the circle?** Give learners time to work out the answer and then choose a pair to explain the method they used. Ask: **How did you find 40% of 360°?** Some learners may have calculated $\frac{40}{100} \times 360°$ or $\frac{4}{10} \times 360°$. Ask: **How could you have simplified the calculation?** Establish that 40% is equivalent to $\frac{2}{5}$. Record: $\frac{2}{5} \times 360 = 2 \times 360 \div 5 = 720 \div 5 = 144°$.

- Choose each flavour in turn and ask a pair of learners to explain how they calculated the angle of the sector (pineapple: 36°, grapefruit: 90°, apple: 72°, grape: 18°) Confirm the answers by checking that the sum of the angles is 360° (144 + 36 + 90 + 72 + 18 = 360).

- 💬 Ask: **What would you rather have, 30% of $90 or 70% of $40? Why?** (30% of $90)

- Discuss the Let's learn section in the Student's Book and address any misunderstandings.

- Introduce the paired activity in the Student's Book, asking learners to work out the new price of the car after a 15% increase ($5577.50).

- Discuss the Guided practice example with learners.

Practice 📒

- Workbook

Title: Percentages of whole numbers (2)

Pages: 106–107

- Refer to Activity 2 from the Additional practice activities.

Apply 👥 🖥

- Display **Slide 3**. Explain the problem. Choose learners to explain how they found the price of the watch after two discounts and a price rise.

Review

- Ask learners to work in pairs. They each draw a layout similar to **Slide 3**, showing an item with its original price and then three changes in price. These could be two discounts and a rise, or two rises and a discount. Learners swap papers to calculate the new price. They return their paper for marking.

Assessment for learning

- A sofa is on sale for $3400. The shop owner places a 30% discount sticker on the sofa. How much does it cost now? ($2380)

- 35 out of 100 people prefer to play cricket. If you were to draw a pie chart for this data, what angle would you give for the 'cricket' sector? (126°)

Same day intervention
Support

- When calculating discounted prices, some learners may calculate the value of the discount but fail to use it to find the new price. Set up a classroom shop and involve learners in real-life transactions giving them opportunities to calculate discounts, for example '20% off'. This will help them to understand that the value of the discount is usually much smaller than the price.

Lesson 4: **Comparing percentages**

Learning objective

Code	Learning objective
6Nf.08	Understand the relative size of quantities to compare and order numbers with one or two decimal places, proper fractions with different denominators and percentages, using the symbols =, > and <.

Resources

paper or mini whiteboard and pen (per learner)

Revise

Use the activity *Higher or lower* from Unit 13: *Percentages* in the Revise activities.

Teach [TWM.01/02]

- Remind learners that percentages can be compared like any number as long as they refer to the same amount.

- Ask: **Which is the larger amount, 30% of 1 litre or 25% of 1 litre?** (30%) **[T&T]** Ask: **How did you decide the larger amount?** (30% > 25%) Establish that it is very easy to compare percentages when they refer to the same quantity.

- Ask similar questions that compare: percentages and fractions of quantities, for example: **Which is the larger amount, 30% of 1kg or $\frac{1}{4}$ of 1kg? (30%) How do you know? Take responses. Establish** that we can compare numbers in different forms, for example percentages and fractions, by converting them to the same form. Ask: **What should we convert to, a percentage or a fraction?** Elicit that as percentages are very easy to compare it would be an efficient strategy to convert to this form. Ask: **What is a $\frac{1}{4}$ as a percentage?** (25%) On the board, write 30% [ ] 25%. Invite a student to the board to write the symbol < or > in the box. Agree: 30% > 25% and therefore 30% of 1 kg > $\frac{1}{4}$ of 1 kg.

- Ask: **Which is the smaller amount, 0·5 of 1 metre or 40% of 1 metre?** (40% of 100 cm). **How do you know?** Invite a student to the board to explain how they decide which number is smaller. Expect them to convert the numbers to the same form, for example percentages, recording: 50% (0·5) of 1 metre > 40%. Confirm 40% is the smaller number.

- On the board, write: $\frac{3}{4}$ of 1 minute, 80% of 1 minute or 0·7 of 1 minute. Ask: **How should we order these numbers, from smallest to largest?** Take responses. Establish that similar to how we compared two numbers in different forms, we order three numbers in different forms by converting to the same form. Give learners time to work out the answer and then choose a learner to explain how they decided the order. Explain that we convert to the same form, percentage for example. Record: 75% ($\frac{3}{4}$) of 1 minute, 80% of 1 minute, 70% (0·7) of 1 minute. Therefore the order is 0·7, ¾, 80%.

- Ask learners questions where they have to order fractions, decimals and percentages, for example: $\frac{2}{5}$ of 1 km, 0·3 of 1 km, 25% of 1 km (25% of 1 km < 0·3 (30%) of 1 km < $\frac{2}{5}$ (40%) of 1 km)

- [TWM.01/02] Say: **Give me a percentage that would come in between 0·5 and $\frac{7}{10}$ (60%) What would you rather have $\frac{4}{5}$, 70% or 0·9 of a cake? (0·9)**

- Introduce the paired activity in the Student's Book, asking learners to work out which number is greater: 75%, 0·8 or $\frac{7}{10}$? (0·8)

- Discuss the Guided practice example with learners.

Practice [TWM.01]

- Workbook

  Title: Comparing percentages

  Pages: 108–109

- Refer to Activity 2 (Variation) from the Additional practice activities.

Apply

- Display **Slide 2**. Discuss the problem. Choose learners to explain how they ordered the children by the distance they jumped.

Review

- Ask learners to work in pairs. They each write three 'Would you rather...' questions, for example: 'Would you rather have 25% of a sticker collection, $\frac{1}{5}$ of the collection or 0·1 of the collection?' Learners swap papers to answer the problems and then return them for marking.

Assessment for learning

- Four children recorded their scores in a video game. Kara scored 75% of the total points, Archie scored 0·8 of the total points, Lucy scored $\frac{3}{5}$ and Toby scored $\frac{7}{10}$. Order the scores, from highest to lowest (Archie, Kara, Toby, Lucy).

Same day intervention

Support

- Learners need to understand that 50% is equivalent to one half or 0·5, 25% is equivalent to one quarter or 0·25, 10% is equivalent to one tenth or 0·1 and 1% is equivalent to one hundredth or 0·01. Learners who experience difficulty understanding percentages and their fraction and decimal equivalents in the early stages will need extended practical experience in working with models or number lines, exploring the relationship between equivalent forms that express part of a whole.

Number – Fractions, decimals, percentages, ratio and proportion

Additional practice activities

Activity 1

Learning objectives
- Recognise that fractions, decimals and percentages can have equivalent values.
- Recognise percentages of shapes and whole numbers.

Resources
number cards: multiples of 5 from 5 to 100 from Resource sheet 2: 0–100 number cards (per pair); Resource sheet 14: Waffle diagrams (per learner)

What to do
- Learners work in pairs.
- Each learner takes a card from the top of the number card pile without their partner seeing. They shade in that number of squares on one of their 100 grids from Resource sheet 14: Waffle diagrams.
- Learners then swap grids and describes the percentage represented by the shaded squares. They also write the percentage as a fraction, reduced to its simplest form.

- Learners then return grids and confirm whether or not their partner has correctly identified the percentage and corresponding fraction.
- The activity continues for three further rounds with learners using a different 100 grid from Resource sheet 14: Waffle diagrams for each round.

Variations 🖥
1 Learners work with multiples of 10% only.

2 Display **Slide 2** (APA 1: Variation). Ask learners to calculate the percentages and decide on the better option.

Answers: A; B; C; A.

Activity 2

Learning objectives
- Recognise that fractions, decimals (one or two decimal places) and percentages can have equivalent values.
- Recognise percentages (1%, and multiples of 5% up to 100%) of shapes and whole numbers.
- Understand the relative size of quantities to compare and order numbers with one or two decimal places, proper fractions with different denominators and percentages, using the symbols =, > and <.

Resources
Resource sheet 15: Multiples of 5% cards (per pair); Resource sheet 16: Multiples of 20 cards (per pair)

What to do
- Shuffle each set of cards and place them face down in two seperate piles. The first pile is the percentage 'discount' cards and the second pile is prices in dollars.
- Each learner takes a card from each pile. They calculate the percentage discount of the price. They then subtract the discount from the price to give a 'discounted' price.

- Learners compare discounted prices. The learner with the lowest price scores a point.
- The winner is the learner with the higher score after five rounds.

Variation
2 Learners repeat the activity and then each learner orders their discounted prices, from lowest to highest. They score points according to this order: 1 point for the lowest score, 2 points for the next highest score… up to the maximum number of points for the highest score.

Unit 14: Addition and subtraction of decimals

Collins International Primary Maths Recommended Teaching and Learning Sequence Term 3, Week 3

Learning objective

Code	Learning objective
6Nf.09	Estimate, add and subtract numbers with the same or different number of decimal places.

Unit overview

Learners revise mental methods for addition and subtraction. They use place value counters to add and subtract whole numbers and decimals by partitioning, and a number line for counting on and back. Learners estimate, add and subtract numbers with the same or different number of decimal places.

Mental strategies are developed for addition and subtraction of tenths with two digits, or hundredths less than 1. Learners practise using the compensation strategy for addition when one of the decimals has eight or nine tenths or hundredths, and for subtraction when the minuend or subtrahend has eight or nine tenths or hundredths for subtraction, for example: $3.45 - 0.9 = 3.45 - 1 + 0.1 = 2.45 + 0.1 = 2.55$ or $6.9 - 0.43 = 7 - 0.43 - 0.1 = 6.57 - 0.1 = 6.47$.

Learners apply the formal method of column addition and subtraction to working with numbers with up to five digits and with up to three decimal places.

Guided examples are provided, together with practical activities that require learners to use appropriate strategies to solve problems. In so doing, they learn that some strategies are more effective and efficient than others, depending on the calculation.

Prerequisites for learning

Learners need to:
- be able to use effective, efficient and appropriate strategies when adding and subtracting decimals
- be able to use known facts and knowledge of adding two-digit numbers to add pairs of tenths and hundredths
- know how to use column addition and subtraction to add or subtract decimals with one or two places, aligning digits and decimal points correctly.

Vocabulary

place value, regroup, addend, subtrahend, trailing zero

Common difficulties and remediation

As calculations become more complex, some learners may struggle to use known number facts and place value to add or subtract mentally. Provide these learners with guided examples, modelled with Base 10 equipment or other manipulatives, such as place value counters.

Link each practical step to a recorded step. Encourage learners to estimate to help find their answers, while acknowledging that estimating is a difficult skill to master – praise learners who are willing to try it.

For work with decimals, provide fraction tiles to allow learners to regroup tenths into wholes. Learners who lack a sound knowledge of place value will continue to make mistakes with column addition or subtraction. Teach them to estimate so that they can recognise when they have made an error. When teaching the algorithm, it is useful to reinforce the principles of place value used in the operation.

Learners are more likely to connect the word 'regroup' to what they can see is actually happening rather than learn it as a rule.

Supporting language awareness

Adapt language when it becomes a barrier to learning; for example, use simpler terms to describe the strategy of decomposition: 'exchanging a one for 10 tenths'.

Promoting Thinking and Working Mathematically

In Lesson 1, learners convince (TWM.04) when they suggest an alternative strategy for adding two numbers mentally. In Lesson 2, learners conjecture (TWM.03) when they suggest the reason why a zero with no value is inserted in front and at the end of a number in a column addition.

Success criteria

Learners can:
- use effective, efficient and appropriate strategies when adding and subtracting decimals with the same or different numbers of decimal places
- use column addition and subtraction to add or subtract decimals with up to three decimal places, aligning digits and decimal points correctly.

Number – Fractions, decimals, percentages, ratio and proportion

Unit 14 Addition and subtraction of decimals

Lesson 1: **Adding decimals (mental strategies)**

Learning objective

Code	Learning objective
6Nf.09	Estimate, add [and subtract] numbers with the same or different number of decimal places.

Resources

calculator (per learner)

Revise

Use the activity *Quick sums* from Unit 14: *Addition and subtraction of decimals* in the Revise activities.

Teach [SB] 🖥 📊 [TWM.04/08]

- On the board, write 4·7 + 0·64 = . Ask: **How would you work this out?** Take responses.
- Display **Slide 1** and the **Place value counters tool**. Use the tool to model the calculation. Ensure that learners understand that addition begins with the smallest place value (hundredths) but since there are no hundredths in the augend the addition moves to the tenths. Explain that the addition of the tenths requires regrouping: the exchange of 13 tenths for 1 one and 3 tenths. The sum is 5·34.
- Refer to the first number line on **Slide 1** demonstrating the 'counting on' strategy: adding 0·64 in two steps (+ 0·6, + 0·04). **[TWM.04] [T&T]** Ask: **Are there any other methods to add the two numbers mentally?** Take responses. Together, identify another regrouping strategy: 4·7 + 0·64 = 4·7 + 0·3 + 0·34 = 5 + 0·34 = 5·34. Confirm this using a pictorial approach.
- Repeat the above strategies for 7·6 + 0·87 = 8·47. Choose learners to model the calculation on the **Place value counters tool**. Ensure they understand that 6 tenths and 8 tenths (0·06 + 0·08) are exchanged for 1 one and 4 tenths, and the one is carried to the ones column.
- On the board, write: 5·7 + 0·58 =. Ask learners to solve the calculation using a mental strategy. Choose a learner to explain the strategy they used and confirm the answer (6·28). Ask learners to use a calculator to check the answer to their calculation.
- Write: 86 + 79 =. Ask: **How would you solve this calculation mentally?** Take responses. Establish that as 79 is a near multiple of 10, you can use a compensation strategy.
- **[TWM.08]** Ask: **What is a compensation strategy?** Choose a learner to explain the method and then ask: **Is this a good explanation?** Take responses from the class. Refine the definition by demonstrating the strategy: 86 + 79 = 86 + 80 – 1 = 165. Write: 0·86 + 0·79 =. Ask: **How can you apply this strategy to decimal addition?** Refer to the second number line on **Slide 1** to demonstrate and establish that this is similar to the whole-number calculation, but the numbers are just ten times smaller.

- Write: 0·46 + 0·88 =. Ask: **How would you use a compensation strategy to add these numbers?** Take responses. Establish that 0·88 can be rounded to 0·9 and the answer reduced by 0·02. Record: 0·46 + 0·9 – 0·02 = 1·36 – 0·02 = 1·34, comparing it to the whole-number problem: 46 + 88 = 134 where the numbers are 100 times bigger.
- 🖐 Ask: **How would you solve 0·77 + 0·38?** Describe the strategy you would use. (1·15)
- Discuss the Let's learn section in the Student's Book.
- Introduce the paired activity in the Student's Book. Discuss the different strategies. Which method do learners consider the most efficient and efficient?
- Discuss the Guided practice example in the Student's Book.

Practise [WB] [TWM.07]

- Workbook

Title: Adding decimals (mental strategies)

Pages: 110–111

- Refer to Activity 1 from the Additional practice activities.

Apply 👥 🖥

- Display **Slide 2**. Once complete, ask pairs of learners to explain the strategies they used to add the numbers.

Review

- On the board, write: 2·7. Ask learners to call out a two-place decimal less than 1, which they add, using a mental strategy. They then round the number to the nearest tenth. They continue adding and rounding until the total reaches 10.

Assessment for learning

- What is the sum of 7·6 and 0·83? (8·43)
- Convince me that 0·89 doubled is 1·78.

Same day intervention
Enrichment

- Introduce additions where both numbers are three- or four-digit decimals with two decimal places and where the solution can be found using a compensation strategy, for example: 7·6 + 5·99 or 26·3 + 18·98.

Lesson 2: **Adding decimals (written methods)**

Learning objective

Code	Learning objective
6Nf.09	Estimate, add [and subtract] numbers with the same or different number of decimal places.

Resources

paper or mini whiteboard and pen (per learner); calculator (per learner)

Revise

Use the activity *Find the sum* from Unit 14: *Addition and subtraction of decimals* in the Revise activities.

Teach 🆂🅱 🖵 [TWM.03/06]

- Display **Slide 1**. Remind learners that when additions get more complicated, involve larger numbers, or numbers with lots of digits, that a formal written method may be more effective and efficient than a mental strategy. Take learners through the expanded written method. Explain that methods are identical to those used for whole numbers, except for the decimal point. **[T&T] [TWM.03]** Ask: **Why have zeros with no value been placed in front and at the end of some of the numbers in the addition?** Take responses. Explain that to avoid mistakes when adding numbers with different numbers of decimal places, it helps to make sure the columns are neat. You fill out the columns on the right with trailing zeros (zeros inserted after the final decimal digit that do not change the value of a number) and on the left with leading zeros (zeros inserted before the first digit of a number that do not change the value of a number).
- Take learners through the formal written method on the slide. Ask questions to develop understanding, for example: **[TWM.06] In which columns has regrouping taken place and why? Why have small '1' digits been written below the answer line in the ones and tenths columns?**
- Write: 5·776 + 68·89 =. Ask: **What is a good estimate for the calculation?** Remind learners that rounding is a good strategy for estimation. Take responses. Record: 6 + 70 = 76 (or 5·8 + 69 = 74·8). Ask learners to solve the problem, in pairs, with one learner using the expanded written method and the other using the formal written method. They check they have the same answer and compare it with the estimate (74·666). Ask learners to use a calculator to check the answer to their calculation.
- 🔁 Say: **35·28 litres of water is poured into a barrel containing 8·963 litres. How much water is now in the barrel?** (44·243 ℓ)
- Discuss the Let's learn section in the Student's Book.

- Introduce the paired activity in the Student's Book, asking learners to find the total amount of sand in each container and order the containers by the mass of sand they contain.
- Discuss the Guided practice example in the Student's Book.

Practise 🆆🅱

- Workbook

Title: Adding decimals (written methods)

Pages: 112–113

- Refer to Activity 1 (Variation) from the Additional practice activities.

Apply 👥 🖵

- Display **Slide 2** and read the text to the class. Give learners time to complete the task and then ask: **How did you work out each combination of pieces?**

Review

- On the board, write: 18·76. Ask learners to call out numbers less than 10 with three decimal places, which they add to a running total. Give them one minute to find the answer, using a written strategy. Addition continues until the sum reaches 50.

Assessment for learning

- Show me how to add 58·65 and 7·761 using the expanded written method.
- What is 44·88 kg add 4·888 kg? (49·768 kg)

Same day intervention
Support

- By Stage 6, most learners should be encouraged to use the formal written method for calculations they cannot do using a mental strategy. Pictorial representations and the expanded written method should only be taught or consolidated in order to ensure that learners fully understand the formal written procedure, and the effectiveness and efficiency of the method.

Number – Fractions, decimals, percentages, ratio and proportion

Unit 14 Addition and subtraction of decimals

Number – Fractions, decimals, percentages, ratio and proportion

Lesson 3: **Subtracting decimals (mental strategies)**

Learning objective

Code	Learning objective
6Nf.09	Estimate[, add] and subtract numbers with the same or different number of decimal places.

Resources

calculator (per learner)

Revise

Use the activity *Quick subtractions* from Unit 14: *Addition and subtraction of decimals* in the Revise activities.

Teach [SB] [💻] [📊] [TWM.03]

- On the board, write: 5·7 – 3·64 = . **[T&T] Ask: How would you work this out?** Take responses.
- Display **Slide 1** and the **Place value counters tool**. Use the tool to model the calculation shown. Ensure that learners understand that subtraction begins with the smallest place value (hundredths). **[TWM.03] Ask: How do you take away 4 hundredths when there are no hundredths in 5·7?** Establish that for subtraction to proceed, 1 tenth needs to be exchanged for 10 hundredths. Make the exchange and then ask: **What is the subtraction in the hundredths column now?** (10 hundredths – 4 hundredths = 6 hundredths) **What is the subtraction in the tenths column?** (6 tenths – 6 tenths = 0 tenths) **What is the subtraction in the ones column?** (5 – 3 = 2) Confirm the answer: 2·06.
- Refer to the first number line on **Slide 1** demonstrating the: 'counting back' strategy: taking away 3·64 in three steps (– 3, – 0·6, – 0·04).
- Repeat the above strategies for 8·2 – 5·43 = 2·77. Choose learners to model the calculation on the **Place value counters tool**. Ensure they understand that 1 tenth must be exchanged for 10 hundredths to allow subtraction to proceed in the hundredths column (10 hundredths – 3 hundredths = 7 hundredths); and 1 one must be exchanged for 10 tenths to allows subtraction to proceed in the tenths column (10 + 1 tenths – 4 tenths = 7 tenths). Complete the subtraction in the ones column (7 – 5 = 2) and confirm the answer: 2·77. Draw an empty number line on the board and demonstrate the counting back strategy: 8·2 – 5 – 0·4 – 0·03 = 2·77.
- On the board, write: 9·5 – 4·85 =. Choose a learner to explain the strategy they would use and confirm the answer (4·65). Ask learners to use a calculator to check the answer to their calculation.
- Refer to the second number line on **Slide 1** and discuss how finding the difference by counting on is a useful strategy when the two numbers are close together. Write: 8·664 – 8·62 = and ask learners to solve the calculation. (0·044)
- Refer to the third number line on **Slide 1** and discuss how a compensation strategy can be used to solve 7·28 – 0·9 =. Ask: **Why do you need to add 0·01 in the second step of the calculation?** (to compensate for subtracting 1 from 7·28 which is 0·01 too many)
- Write: 6·7 – 2·39 =. Ask: **How would you use a compensation strategy to subtract this number?** Take responses. Establish that 2·39 can be rounded to 2·4, subtract 6·7 – 2·4 = 4·3 and then add 0·01 to compensate. Record: 6·7– 2·4 + 0·01 = 4·3 + 0·01 = 4·31.
- Discuss the Let's learn section in the Student's Book.
- Introduce the paired activity in the Student's Book. Discuss the different strategies for solving the subtraction problem. Which is the most efficient?
- Discuss the Guided practice example.

Practise [WB] [TWM.04]

- Workbook

Title: Subtracting decimals (mental strategies)

Pages: 114–115

- Refer to Activity 2 from the Additional practice activities.

Apply [👥] [💻]

- Display **Slide 2**. Once complete, ask: **How did you work out the height of each balloon above the ground?**

Review

- On the board, write: 9·6 and explain that this is the total. Call out a number less than 10 with two decimal places, which learners subtract from the total, using a mental strategy. They then round the number to the nearest tenth. They continue subtracting and rounding until the total reaches 0.

Assessment for learning

- Convince me that 9·5 subtract 5·28 is 4·22.
- Which of the following calculations could be best solved using a compensation strategy: 8·6 – 4·26 or 6·8 – 2·49? Why?

Same day intervention
Enrichment

- Introduce subtractions where both numbers are less than 100 with one or two decimal places and where the solution can be found using a compensation strategy.

Lesson 4: **Subtracting decimals (written methods)**

Learning objective

Code	Learning objective
6Nf.09	Estimate[, add] and subtract numbers with the same or different number of decimal places.

Resources

paper or mini whiteboard and pen (per learner); calculator (per learner)

Revise

Use the activity *Find the difference* from Unit 14: *Addition and subtraction of decimals* in the Revise activities.

Teach 〔SB〕 ⬜ [TWM.04]

- Display **Slide 1**. Remind learners that when subtractions get more complicated, involve larger numbers, or numbers with lots of digits, that a formal written method may be more effective and efficient than a mental strategy. Also remind learners that methods are identical to those used for whole numbers, except for the decimal point. Take learners through the first formal written method on the slide. Emphasise the need to estimate by rounding first. Also, point out where a trailing zero is inserted to neaten up a column. **[T&T]** Ask questions to develop understanding, for example: **[TWM.04] In which column has regrouping taken place and why?**

- Repeat for the second subtraction on the slide: 62·354 – 28·54. Again, discuss the estimation strategy and the insertion of a trailing zero. Ask: **In which columns has regrouping taken place?**

- Write on the board: 5·8 – 2·57 =. Have learners work in pairs to estimate and solve the calculation, with one learner calculating using the formal written method and the other using a mental method. They check that they have the same answer and compare it with the estimate. Ask learners to use a calculator to check the answer to their calculation (3·23).

- Repeat for 75·687 – 37·86 =. Ask: **How would you calculate a good estimate for the difference?** Expect and record: 76 – 38 = 38. Have learners calculate the answer using the formal written method. Choose a pair to come to the board and demonstrate the solution. (37·827)

- ▱ Say: **A car uses 16·64 litres of fuel on a journey. If the car's tank had 54·387 litres to begin with, how much fuel is in the tank now?** (37·747 ℓ)

- Discuss the Let's learn section in the Student's Book.

- Introduce the paired activity in the Student's Book, asking learners to copy the table and fill in the missing numbers. Ask: **What calculations did you need to complete this? Were they all subtractions?** Confirm the answers as a class

- Discuss the Guided practice example in the Student's Book.

Practise 〔WB〕

- Workbook

Title: Subtracting decimals (written methods)

Pages: 116–117

- Refer to Activity 2 (Variation) from the Additional practice activities.

Apply 👥 ⬜

- Display **Slide 2** and read the text to the class. Give learners time to complete the task and then ask: **How did you work out each combination of ribbon and scissor cuts?**

Review

- On the board, write: 98·765. Ask learners to call out numbers less than 100 with two decimal places, which they subtract from a running total. Give them one minute to find the answer, using a written strategy. Subtraction continues until the answer reaches 0.

Assessment for learning

- Show me how to subtract 18·72 from 43·556 using the formal written method.

- What is 9·8 km subtract 5·37 km? (4·43 km)

Same day intervention

Support

- By Stage 6, most learners should be encouraged to use the formal written method for calculations they cannot do using a mental strategy. Pictorial representations and the expanded written method should only be taught or consolidated in order to ensure that learners fully understand the formal written procedure, and the effectiveness and efficiency of the method.

Number – Fractions, decimals, percentages, ratio and proportion

Number – Fractions, decimals, percentages, ratio and proportion

Additional practice activities

Activity 1

Resources
paper (per learner)

What to do
• On the board, write: 'Logan: $8.60 and $0.83', 'Maisie: $0.76 and $0.89', 'Jamie: $8.40 and $0.76' and 'Rashida: $0.86 and $0.78'.
• Say: **Four children went to the shop where they were allowed to buy two items each.**
• Ask pairs of learners to calculate the amount each child spent in total and order the amounts, from greatest to least.
• They discuss their solutions to confirm they are correct. (Logan: $9.43; Jamie: $9.16; Maisie: $1.65; Rashida: $1.64)

Variation
On the board, write: 'Zoe: 4·257 m and 57·68 m', 'Baisha: 3·687 m and 56·77 m', 'Toby: 5·863 m and 55·96 m' and 'Florence: 6·765 m and 54·66 m'.

Say: **Four children took a walk in two stages, one long walk and one short walk.**

Ask learners to calculate the combined distance of each child's walk and then put the children in order of the length of their walk, from shortest to longest.

They discuss their solutions to confirm they are correct. (Baisha: 60·457 m, Florence: 61·425 m, Toby: 61·823 m, Zoe: 61·937 m)

Activity 2

Resources
two sheets of paper (per pair)

What to do
• Ask pairs of learners to think of a word that is related to mathematics that contains no repeat letters, for example: 'minus', 'equal' or 'decimal'.
• They write each letter of the word down the left side of a piece of paper (Paper A). They copy the word on to a second piece of paper (Paper B) but with the letters in a random order.
• Learners write a subtraction calculation alongside each letter in the word on Paper A. The calculations should be those solved by one or more of the mental subtraction strategies they have studied in this unit.

Learners must make sure that all the answers to the calculations are different.
• They solve each calculation and write the answer alongside the matching letter on Paper B. Once the answers are written, the letters on Paper A are erased.
• Pairs then swap Papers A and B with another pair. The pairs solve each subtraction on Paper A in turn and find the letter that matches the answer on Paper B. They use the answers to build and decode the secret word.

Variation
Learners write calculations that are solved by one or more of the written subtraction strategies they have studied.

Unit 15: Multiplication of decimals

Collins International Primary Maths
Recommended Teaching and
Learning Sequence Term 3, Week 5

Learning objective

Code	Learning objective
6Nf.10	Estimate and multiply numbers with one or two decimal places by 1-digit and 2-digit whole numbers.

Unit overview

In this unit, learners revise multiplication of numbers with one decimal place by one-digit whole numbers. They are then introduced to multiplication of numbers with two decimal places. In Lessons 3 and 4, learners extend their skills to multiplying by two-digit whole numbers. Decimals include no more than four digits and answers have no more than two decimal places.

Pictorial approaches to whole-number multiplication are extended to decimals, with learners using number facts, partitioning and the grid method to help them calculate. Once secure in these processes, learners move on to multiplying by using written strategies, specifically partitioning and the extended written method.

Emphasise the importance of estimation throughout the unit. The placing of the decimal point can be checked by estimating the size of the answer expected.

Prerequisites for learning

Learners need to:
- be able to mentally multiply tenths less than one by a one-digit number
- know how to use the grid method to multiply one- or two-digit numbers with one decimal place by a one-digit whole number
- know how to use partitioning or the expanded written method to multiply one- or two-digit numbers with one decimal place by a one-digit whole number.

Vocabulary

place value, grid method, product, partial product, partition, expanded written method, short multiplication, long multiplication

Common difficulties and remediation

Without a secure understanding of decimals, learners may operate on both sides of the decimal point independently; for example, mistakenly answering $8 \times 34 \cdot 7 = 272 \cdot 56$. They need to understand that decimal places are connected. Develop this understanding by providing concrete approaches; for example, using a place value chart and counters to model multiplication and regrouping.

If learners have difficulty placing the decimal point in the answer, provide plenty of opportunities to practise regrouping. For TO.th × O calculations, regrouping will take place across the tenths, ones and tens columns.

Estimating the product will help verify that the placement of the decimal point is correct, and that the answer is reasonable.

Supporting language awareness

Throughout this unit, learners will benefit from working with visual models while multiplying. Consistently use and be precise with mathematical language. Use the language '74 hundredths' (instead of point seven-four) when referring to multiplication and decimals. This level of precision in vocabulary will provide learners with the opportunity to hear the connections between decimals and fractions.

Promoting Thinking and Working Mathematically

Opportunities to develop all four pairs of TWM characteristics are provided throughout the unit.

In Lesson 1, learners generalise (TWM.02) when they identify the distributive property of multiplication as the rule of number that allows the partitioning of a multiplicand and the multiplication of the parts by the multiplier.

In Lesson 2, learners convince (TWM.04) when they prove that $0 \cdot 8 \times 70 = 8 \times 7$.

In Lesson 3, learners improve (TWM.08) when they suggest alternative strategies for a HTO.t × TO calculation.

Success criteria

Learners can:
- mentally multiply tenths less than one by a one-digit number or a multiple of 10
- use place value to link multiplying by a multiple of 10 with multiplying by a ones digit that is 10 times less, e.g. $72 \cdot 34 \times 50$ with $72 \cdot 34 \times 5$, i.e. as 50 is 10 times larger than 5, so the answer will be 10 times larger
- use the grid method to multiply one- or two-place decimals by one- or two-digit whole numbers
- use partitioning or the expanded written method to multiply one- or two-place decimals by one- or two-digit whole numbers
- use formal methods of short and long multiplication to multiply numbers with one or two decimal places by one- or two-digit whole numbers.

Number – Fractions, decimals, percentages, ratio and proportion

Unit 15 Multiplication of decimals

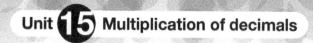

Number – Fractions, decimals, percentages, ratio and proportion

Lesson 1: Multiplying decimals by 1-digit whole numbers (1)

Learning objective

Code	Learning objective
6Nf.10	Estimate and multiply numbers with one or two decimal places by 1-digit [and 2-digit] whole numbers.

Resources

paper or mini whiteboard and pen (per learner); paper, for the Workbook (per learner)

Revise

Use the activity *Multiplication function machines* from Unit 15: *Multiplication of decimals* in the Revise activities.

Teach [SB] 🖵 📊 [TWM.02]

- Write on the board: $0.7 \times 4 =$. **[T&T]** Ask: **How would you solve this calculation? Is there a related fact you can use?** Elicit that you could use a whole number fact and adjust by place value. Say: **Since you know that $7 \times 4 = 28$, you also know that $0.7 \times 4 = 2.8$.** (10 times smaller) Confirm the product by using a pictorial approach. Display **Slide 1** and talk through the first example. Display the **Place value counters tool**. Choose a learner to use the tool to model the multiplication: $0.7 \times 4 = 4$ rows of 7 tenths (2.8).
- Display **Slide 2**. Ask: **What is a good estimate for this calculation?** (28; 3.5×8) Ask: **How could you use place value counters to model this calculation?** Choose a learner to demonstrate on the **Place value counters tool**. Expect them to partition the multiplicand $(3 + 0.4)$ and multiply both parts by 8. **[TWM.02]** Ask: **Which property of multiplication is being applied here?** (distributive) Write the new calculation on the board: $3.4 \times 8 = 3 \times 8 + 0.4 \times 8$ and ask a learner to use the tool to complete each multiplication: $3 \times 8 = 8$ rows of 3 ones (24); $0.4 \times 8 = 8$ rows of 4 tenths (3.2). Add the partial products to give the answer $(24 + 3.2 = 27.2)$. Compare with the estimate and confirm.
- Point to the grid method solution for $3.4 \times 8 =$ on **Slide 2** and remind learners of the steps involved.
- Write $6.6 \times 8 =$ on the board. With learners working in pairs, they find an estimate for the calculation. Then learners calculate the product, one by drawing a place value counter model and the other using the grid method. They confirm both solutions give the same answer and compare with the estimate (52.8).
- Display **Slide 3**. Ask: **What is a good estimate for this calculation?** (240; 40×6) Choose a learner to use the **Place value counters tool** to model the multiplication. Compare with the estimate and confirm (253.8).
- Point to the grid method solution for 42.3×6 and guide learners through the steps involved.
- Say: **We are now going to use place value counters and the grid method to multiply a greater range of decimals by one-digit numbers.** Display **Slide 4, 5, 6 or 7.** Take learners through the two methods (place value counter model, grid method). Then refer to the 'Your turn' calculation, and as before, learners work in pairs to estimate then calculate the answer with one learner using a place value counter model, and the other using the grid method. They then compare answers and their estimate.
- Repeat above choosing other types of calculations from **Slides 4, 5, 6 or 7.**
- 🔄 Say: **Maisie draws a line 47.8 mm in length. Luke draws a line four times longer. How long is Luke's line?** (191.2 mm)
- Discuss the Let's learn section in the Student's Book.
- Introduce the paired activity in the Student's Book.
- Discuss the Guided practice example.

Practise [WB]

- Workbook

Title: Multiplying decimals by 1-digit whole numbers (1)

Pages: 118–119

- Refer to Activity 1 from the Additional practice activities.

Apply 👥 🖵

- Display **Slide 8**. Learners are given the mass of an orange and asked to work out the combined mass of eight identical oranges. Choose a pair of learners to explain how they calculated the combined mass.

Review

- Write on the board different HTO·t × O and TO·th × O calculations to be solved using the grid method. In each example, make one mistake. Ask learners to identify and correct each mistake.

Assessment for learning

- Show me two different methods for calculating 534.8×7. (3743.6)
- A chair has a mass of 78.37 kg. What is the combined mass of 6 chairs? (470.22 kg)

Same day intervention
Support

- The strategies introduced for multiplication of decimal numbers carry their own difficulties. A good way to combat these is to show learners examples where mistakes have been made and ask them to spot the errors.

Lesson 2: **Multiplying decimals by 1-digit whole numbers (2)**

Learning objective

Code	Learning objective
6Nf.10	Estimate and multiply numbers with one or two decimal places by 1-digit [and 2-digit] whole numbers.

Resources

paper or mini whiteboard and pen (per learner)

Revise

Use the activity *Grid method (1)* from Unit 15: *Multiplication of decimals* in the Revise activities.

Teach 🆂🖵 [TWM.04]

- Write on the board: 0·8 × 70 =. **[T&T]** Ask: **How would you solve this calculation? Is there a related fact you could use?** Elicit that you could use a whole number fact and adjust by place value. Say: **Since you know that 8 × 7 = 56, you also know that 0·8 × 70 = 56. [TWM.04]** Ask: **How would you prove that 0·8 × 70 = 8 × 7?** Take responses. On the board, show the decomposition and reordering of terms: 0·8 × 70 = 0·8 × 7 × 10 = 0·8 × 10 × 7 = 8 × 7.
- Write: 0·9 × 60 =, 0·6 × 80 =. Ask learners to solve the calculations using related facts (0·9 × 60 = 9 × 6 = 54; 0·6 × 80 = 6 × 8 = 48).
- Say: **We are now going to use the grid method to multiply a greater range of decimals by multiples of 10.** Display **Slide 1, 2, 3, 4 or 5**. Take learners through the grid method and then the multiplication of the answer by 10 to arrive at the answer to the calculation.
- Then refer to the 'Your turn' calculation, and ask learners to work in pairs to calculate the answer.
- Repeat above choosing other types of calculations from **Slides 1, 2, 3, 4 or 5**.
- Say: **We are now going to use written methods to solve examples of the different multiplication forms we have been working with.** Display **Slide 6, 7, 8, 9 or 10**. Take learners through the partitioning method, expanded and formal written method. Emphasise the place value link between whole number multiplications and decimal multiplication and the adjustment factors of ÷ 10 or ÷ 100. Also, point out the use of brackets for detailing the calculation of partial products.
- Then, as before, refer to the 'Your turn' calculation, and ask learners to work in pairs to calculate the answer.
- Repeat above choosing other types of calculations from **Slides 6, 7, 8, 9 or 10**.
- Say: **We are now going to use written methods to solve examples of the different multiplication forms we have worked with where the multiplier is a multiple of 10.** Display **Slide 11, 12, 13, 14 or 15**. Take learners through the partitioning and expanded written method. Emphasise the place

value link between whole number multiplications and decimal multiplication and the adjustment factors of ÷ 10 or ÷ 100.
- Then, as before, refer to the 'Your turn' calculation, and ask learners to work in pairs to calculate the answer.
- Repeat above choosing other types of calculations from **Slides 11, 12, 13, 14 or 15**.
- 🗎 Say: **A swimming pool has a capacity of 246·7 litres. What would be the capacity of a pool seven times the size? Calculate using a written method.** (1726·9 litres)
- Discuss the Let's learn section in the Student's Book.
- Introduce the paired activity in the Student's Book.
- Discuss the Guided practice example.

Practise 🆆🅱

- Workbook

Title: Multiplying decimals by 1-digit whole numbers (2)

Pages: 120–121

- Refer to Activity 1 (Variation) from the Additional practice activities.

Apply 👥🖵

- Display **Slide 16**. Choose a pair of learners to explain how they calculated the height of the actual rocket.

Review

- Write on the board 487·6 × 4 = and 82·36 × 30 =. Ask learners to work out the answer to each calculation. Then ask individual learners to explain their method. Discuss with the class how the first calculation can be solved using short multiplication and the second using the expanded written method.

Assessment for learning

- Show me two different methods for calculating 63·27 × 6. (379·62)
- A meteor travels 563·2 km every minute. How far does it travel in 8 minutes? (4505·6 km)

Same day intervention

Enrichment

- In pairs, learners devise four word problems similar to Apply. They swap papers with other pairs to solve, then return them for checking.

Number – Fractions, decimals, percentages, ratio and proportion

197

Lesson 3: **Multiplying decimals by 2-digit whole numbers (1)**

Number – Fractions, decimals, percentages, ratio and proportion

Learning objective

Code	Learning objective
6Nf.10	Estimate and multiply numbers with one or two decimal places by [1-digit and] 2-digit whole numbers.

Resources

paper or mini whiteboard and pen (per learner); paper, for the Workbook (per learner)

Revise

Use the activity *Related facts* from Unit 15: *Multiplication of decimals* in the Revise activities.

Teach 📖 🖥 [TWM.08]

- Say: **We are now going to multiply decimals by two-digit whole numbers.** Display **Slide 1**. Ask: **How would you use a rounding strategy to estimate the answer?** Take responses. Establish a good strategy is to round the whole number multiplier: $0.6 \times 30 = 18$.

- Take learners through the grid method and point out the use of a trailing zero to neaten up the columns when adding the partial products.

- Then refer to the 'Your turn' calculation, and ask learners to work in pairs to calculate the answer. Remind learners to first estimate the answer. Once learners have worked out the answer, choose a pair of learners to come to the board to demonstrate their solution. Confirm the answers as a class.

- Repeat above choosing other types of calculations from **Slides 2, 3, 4, 5 or 6**.

- **[T&T] [TWM.08]** On the board, write: 453.8×37. Say: **I can solve this calculation by multiplying 4538×370 and then dividing the answer by 10. Is this a more efficient strategy? If not, why?** Take responses and invite learners to suggest other possible strategies. Evaluate them as a class and decide if they provide a quicker method of calculation.

- 🖉 Say: **A submarine descends 3·78 metres every minute. How far will it have descended in 33 minutes?** (124·74 m)

- Discuss the Let's learn section in the Student's Book.

- Introduce the paired activity in the Student's Book. Ask a pair of learners to demonstrate how they used the grid method to find how far the mountaineer will have climbed in the time given.

- Discuss the Guided practice example in the Student's Book.

Practise 📝

- Workbook

Title: Multiplying decimals by 2-digit whole numbers (1)

Pages: 122–123

- Refer to Activity 2 from the Additional practice activities.

Apply 👥 🖥

- Display **Slide 7**. Learners are given the length of a model ship and asked to work out the length of an actual ship, which is 63 times longer. Choose a pair of learners to explain how they calculated the length of the actual ship.

Review

- Write on the board $56.27 \times 17 =$ and $345.3 \times 44 =$, and ask learners to solve both calculations using the grid method. (956·59, 15 193·2)

Assessment for learning

- Which has the greater product, 47.8×72 or 51.4×67? (51.4×67)

- Toby drives 168·7 km. What would be the distance of a journey 41 times further? (6916·7 km)

Same day intervention
Enrichment

- Ask learners to solve word problems that involve multiplying decimals by two-digit whole numbers, for example: **The mass of a small carton of juice is 63·37 g. What is the mass of 34 cartons?** (2154·58 g)

Lesson 4: **Multiplying decimals by 2-digit whole numbers (2)**

Learning objective

Code	Learning objective
6Nf.10	Estimate and multiply numbers with one or two decimal places by [1-digit and] 2-digit whole numbers.

Resources

paper or mini whiteboard and pen (per learner)

Revise

Use the activity *Grid method (2)* from Unit 15: *Multiplication of decimals* in the Revise activities.

Teach ▣ 💻

- Display **Slide 1**. Take learners through the partitioning method and point out the use of brackets for detailing the calculation of partial products.
- Then refer to the 'Your turn' calculation, and ask learners to work in pairs to calculate the answer. Remind learners to first estimate the answer.
- Repeat above choosing other types of calculations from **Slides 2, 3, 4, 5 or 6**. Take learners through the partitioning method, expanded written method and long multiplication method solutions, pointing out the use of brackets for detailing the calculation of partial products. Emphasise the place value link between whole number multiplications and decimal multiplication and the adjustment factors of ÷ 10 or ÷ 100.
- Then, as before, refer to the 'Your turn' calculation, and ask learners to work in pairs to estimate and calculate the answer. Once learners have worked out the answer, choose a pair of learners to come to the board to demonstrate their solution. Confirm the answers as a class.
- 🗣 Say: **A pipe leaks 237·8 litres of water every minute. How much water will the pipe have leaked after 53 minutes?** (12 603·4 litres)
- Discuss the Let's learn section in the Student's Book.
- Introduce the paired activity in the Student's Book. Ask a pair of learners to demonstrate how they use the formal written method to find the bird's adjusted altitude.
- Discuss the Guided practice example in the Student's Book.

Practise � [TWM.07]

- Workbook

Title: Multiplying decimals by 2-digit whole numbers (2)

Pages: 124–125

- Refer to Activity 2 (Variation) from the Additional practice activities.

Apply 👥 💻

- Display **Slide 7**. Learners are given details of how much faster a jet travels than a car. Choose a pair of learners to explain how they calculated the speed of the jet.

Review

- Write on the board 287·6 × 14 = and 64·38 × 46 =. Ask learners to work out the answer to each calculation. Then ask individual learners to explain their method. Discuss with the class how the first calculation can be solved using the expanded written method and the second to be solved using long multiplication. (4026·4, 2961·48)

Assessment for learning

- Show me two different methods for calculating 85·33 × 63. (5375·79)
- A child is born with a mass of 3·78 kg. In 12 years time, her mass is 13 times heavier. What is her mass now? (49·14 kg)

Same day intervention
Enrichment

- Ask learners to work in pairs to devise four word problems similar to the one described in the Apply section. They swap papers with other groups and return them for checking.

Number – Fractions, decimals, percentages, ratio and proportion

Additional practice activities

Activity 1

Learning objective
• Estimate and multiply numbers with one or two decimal places by 1-digit whole numbers.

Resources
two sets of 1–9 digit cards from Resource sheet 13: 0–9 digit cards (per pair); paper (per learner)

What to do
• Pairs of learners shuffle the digit cards and place them face down in a pile.
• Each learner takes five cards and arrange these into a TO.th × O multiplication.
• They multiply the two numbers together using any preferred method and add the product to a running total.

• After each round, the digit cards are returned to the pile and the cards are reshuffled. Learners continue drawing new cards each round.
• The winner is the player with the higher score after six rounds.

Variations
1 Learners work with digits 1–3 only for the TO.th multiplicand and 2–3 only for the multiplier, removing the requirement to regroup ('carry').

2 Each player begins with 1000 points and subtracts the product of the multiplication from this number. The winner is the player with the lower score after six rounds.

Activity 2

Learning objective
• Estimate and multiply numbers with one or two decimal places by 2-digit whole numbers.

Resources
two sets of 1–9 digit cards from Resource sheet 13: 0–9 digit cards (per pair); paper (per learner)

What to do
• Pairs of learners shuffle the digit cards and place them face down in a pile.
• Each learner takes six cards and arranges these into a HTO.t × TO multiplication.
• They multiply the two numbers together using any preferred method and add the product to a running total.

• After each round, the digit cards are returned to the pile and the cards are reshuffled. Learners continue drawing new cards each round.
• The winner is the player with the higher score after six rounds.

Variation
2 Each player begins with 1000 points and subtracts the product of the multiplication from this number. The winner is the player with the lower score after six rounds.

Unit 16: Division of decimals

Collins International Primary Maths
Recommended Teaching and
Learning Sequence Term 3, Week 6

Learning objective

Code	Learning objective
6Nf.11	Estimate and divide numbers with one or two decimal places by whole numbers.

Unit overview

In this unit, learners use their knowledge of division facts to perform mental calculations for related facts involving decimals. They find estimates and use the expanded written method and formal written method of short division to divide decimals with one or two decimal places by a one- or two-digit number. For divisions of the form TO.th ÷ O and TO.th ÷ TO, examples are used that are easy for learners to recognise and calculate, i.e. the dividend can easily be recognised as a multiple of the divisor. For example: 20·05 ÷ 5; 23·46 ÷ 23 (instead of 23·56 ÷ 23 as 56 is not multiple of 23).

Learners are encouraged to look carefully at the numbers in the calculation and decide on the most efficient, effective and appropriate strategy, including using partitioning to calculate mentally.

Emphasise the importance of estimation throughout the unit. The placing of the decimal point can be checked by estimating the size of the expected answer.

Prerequisites for learning

Learners need to:
- be able to partition decimal numbers with up to two decimal places
- be able to recall multiplication and division facts for multiplication tables up to 10 × 10
- understand the effect of dividing a number by 10 and 100
- know how to use the expanded written method of division.

Vocabulary

partition, expanded written method, short division

Common difficulties and remediation

Learners now have a set of strategies for the division of a whole number by a whole number, including examples when there is a remainder. The aim of this unit is to introduce strategies for dividing a decimal by a whole number. Ensure learners notice that the steps in the process are similar to whole number division when using partial quotients in the expanded written

method or the division algorithm for short division. For decimal division, learners need to be secure in their understanding of place value and where the decimal point goes in the quotient. They should be made aware that dividing a decimal by a whole number gives a quotient smaller than either number.

Supporting language awareness

Throughout this unit, learners will benefit from working with visual models while dividing. Consistently use and be precise with mathematical language. Use the language '74 hundredths' (instead of point seven-four) when referring to division and decimals. This level of precision in vocabulary will provide learners with the opportunity to hear the connections between decimals and fractions.

Promoting Thinking and Working Mathematically

Opportunities to develop all four pairs of TWM characteristics are provided throughout the unit.

In Lesson 1, learners conjecture (TWM.03) when they suggest alternative strategies to calculate 66 ÷ 3 = 22 using their knowledge of place value.

In Lesson 2, learners convince (TWM.04) when they identify that when dividing the '9' digit in 98 by 4 they are actually working with a dividend of 9 tens not 9 ones and must write the quotient in the tens column of the answer line, not the ones column.

In Lesson 3 (Additional practice activity), learners classify (TWM.06) when they identify which numbers in a set can be divided by 3 to give quotients with exactly two decimal places, and which numbers can be divided by 4.

Success criteria

Learners can:
- make a reasonable estimate for the answer to a calculation
- use partitioning or the expanded written method to divide decimal numbers by one- or two-digit whole numbers
- use short division to divide decimal numbers by one- or two-digit whole numbers.

Number – Fractions, decimals, percentages, ratio and proportion

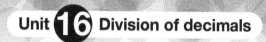
Lesson 1: **Dividing one-place decimals by whole numbers (1)**

Number – Fractions, decimals, percentages, ratio and proportion

Learning objective

Code	Learning objective
6Nf.11	Estimate and divide numbers with one [or two] decimal place[s] by whole numbers.

Resources

paper or mini whiteboard and pen (per learner); paper, for the Workbook (per learner)

Revise

Use the activity *Estimate and divide* from Unit 16: *Division of decimals* in the Revise activities.

Teach 🆂🅱 💻 [TWM.03]

- Write on the board: $6{\cdot}6 \div 3 =$. **[T&T]** Ask: **How would you answer this calculation? Is there a related fact you can use?** Elicit that you could use a whole number fact and adjust by place value. Say: **Since you know that $66 \div 3 = 22$, you also know that $6{\cdot}6 \div 3 = 2{\cdot}2$** (10 times smaller). **[TWM.03]** Ask: **Is there another mental method you could use to answer this calculation?** Take responses. Establish that a partitioning strategy could be used. Record: $6{\cdot}6 \div 3 = (6 \div 3) + (0{\cdot}6 \div 3) = 2 + 0{\cdot}2 = 2{\cdot}2$. Ask: **Which property of number allows us to split the problem in this way?** (distributive rule)
- On the board, write: $4{\cdot}6 \div 2 =$, $9{\cdot}9 \div 3 =$, $4{\cdot}8 \div 4 =$. Learners work in pairs: one learner solves each calculation using a 'related fact' strategy and the other uses a partitioning strategy. The pair confirm they have the same answer each time. ($2{\cdot}3$, $3{\cdot}3$, $1{\cdot}2$)
- Display **Slide 1**. Ask: **What is a good estimate for the quotient?** Take responses. Establish that since the dividend is just over double the divisor, they should expect the answer to be close to 2.
- Explain that you are now going to use the expanded written method for decimal divisions. Ask: **Why do you solve $86 \div 4$ not $8{\cdot}6 \div 4$?** Explain that it is easier to work with a whole number dividend than a decimal one. Say: **For this division, multiply $8{\cdot}6$ by 10 to make the dividend 86 and then adjust the quotient by dividing by 10.**
- Guide learners through the steps involved in using the expanded written method to solve $86 \div 4$. **[T&T]** **[TWM.03]** Ask questions to develop understanding: **Why do you record '$- 80$' below '86'? What do you do with the remainder '2'?** Explain that you express the remainder as a fraction ($\frac{2}{4}$), simplify and convert it to a decimal ($0{\cdot}5$) to give the quotient $21{\cdot}5$. Ask: **What is the final step?** Agree that you need to divide $21{\cdot}5$ by 10 to give the answer $2{\cdot}15$.
- Ask learners to work in pairs to solve the 'Your turn' calculation ($8{\cdot}4 \div 8 = 1{\cdot}05$).

- Say: **We are now going to use the expanded written method to divide other decimals by a one-digit whole number.**
- Display and discuss **Slide 2**, then **Slide 3**. Emphasise how each dividend is converted to a whole number by multiplying by 10 and the quotient adjusted by dividing by 10. Ask learners to work in pairs to solve each 'Your turn' calculation. ($47{\cdot}4 \div 3 = 15{\cdot}8$, $375{\cdot}6 \div 6 = 62{\cdot}6$).
- Discuss the Let's learn section in the Student's Book.
- Introduce the paired activity in the Student's Book. Ask a pair of learners to explain how they worked out the mass of each chair.
- Discuss the Guided practice example.

Practise 🆆🅱

- Workbook

Title: Dividing one-place decimals by whole numbers (1)

Pages: 126–127

- Refer to Activity 1 from the Additional practice activities.

Apply 👥 💻

- Display **Slide 4**. Choose a pair of learners to explain how they calculated the answer.

Review

- Write on the board $224{\cdot}4 \div 3 =$ and $848{\cdot}4 \div 4 =$. Ask learners to decide which one of the calculations could be solved using a mental method and which one is best solved using a written method. Learners solve the problems using the appropriate method.

Assessment for learning

- Show me two methods for calculating $606{\cdot}9 \div 3$.
- Four tomatoes have a mass of $191{\cdot}2$ g. If all tomatoes have the same mass, what is the mass of one tomato? ($47{\cdot}8$ g)

Same day intervention
Support

- Learners need to have instant recall of division facts. Allow learners who find memorising difficult time to study representations of division (arrays, groups of objects) and to spot patterns, using 100 squares and the multiples themselves.

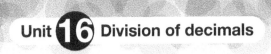

Lesson 2: Dividing one-place decimals by whole numbers (2)

Learning objective

Code	Learning objective
6Nf.11	Estimate and divide numbers with one [or two] decimal place[s] by whole numbers.

Resources

paper or mini whiteboard and pen (per learner)

Revise

Use the activity *Quick division* from Unit 16: *Division of decimals* in the Revise activities.

Teach [SB] 🖥 [TWM.04]

• Display **Slide 1**. Explain that you are now going to use the short division method to solve decimal division calculations. Ask: **What is a good estimate for the quotient?** Take responses. **[TWM.04]** Ask: **Why do you solve 98 ÷ 4, not 9·8 ÷ 4?** Remind learners that it is easier to work with a whole number dividend than a decimal one. Say: **For this division, multiply 9·8 by 10 to make the dividend 98 and then adjust the quotient by dividing by 10.**

• Work through the calculation, reminding learners of the steps in short division, comparing and contrasting it to the expanded method. Follow the completed example on **Slide 1** but write out the entire calculation alongside as you move through the steps. Explain that, to set out the problem correctly, you place the divisor outside the long division bar. You place the dividend inside the long division bar. The quotient is recorded above the division bar.

• Begin the division. Ask: **How many 4s go into 9?** (2) Write '2' on the answer line in the tens column. **[T&T] [TWM.04]** Ask: **Why do we record the '2' in the tens column when we ask the question: 'how many 4s go into 9?'** Take responses. Explain that since the 9 digit in 98 has a value of 90, we are really asking how many 4s go into 90 (20). Say: **This leaves a remainder of 10.** Write a small '1' to the left of the next digit in the dividend (8) to make 18. Ask: **How many 4s go into 18?** (4) Write '4' on the answer line in the ones column. Say: **This leaves a remainder of 2. We write this on the answer line to give the quotient 28 r $\frac{2}{4}$ with the remainder expressed as a fraction of the divisor.** (4) Ask learners how to simplify $\frac{2}{4}$ and convert the answer to a decimal ($\frac{1}{2}$, 24·5). Ask: **Is this the answer to the calculation 9·8 ÷ 4?** (no) Establish that you still need to adjust for multiplying the dividend by 10 by dividing the quotient by 10. Say: **So, 9·8 ÷ 4 = 2·45.**

• Ask learners to work in pairs to solve the 'Your turn' calculation (8·4 ÷ 3 = 2·8).

• Say: **We are now going to use the short division to divide other decimals by a one-digit whole number.** Display and discuss **Slide 2**, then **Slide 3**. Emphasise how each dividend is converted to a whole number by multiplying by 10 and the answer adjusted by dividing by 10. Ask learners to work in pairs to solve each 'Your turn' calculation.

• 🔁 Say: **Lucy divides 254·2 ml of blackcurrant juice equally between four glasses. How much juice is in each glass?** (63·55 ml)

• Discuss the Let's learn section in the Student's Book.

• Introduce the paired activity in the Student's Book. Ask a pair of learners to explain how they worked out the length of each of hop.

• Discuss the Guided practice example.

Practise [WB]

• Workbook

Title: Dividing one-place decimals by whole numbers (2)

Pages: 128–129

• Refer to Activity 1 (Variation) from the Additional practice activities

Apply 🖥 👥

• Display **Slide 4**. Choose a pair of learners to explain how they calculated the answer.

Review

• Write on the board 369·6 ÷ 3 = and 456·2 ÷ 4 =. Ask learners to decide which one of the calculations could be solved using a mental method and which one is best solved using a written method. Learners solve the problems using the appropriate method.

Assessment for learning

• Show me two methods for calculating 474·6 ÷ 3. (158·2)

• A skier travels at a constant speed and covers 525·6 m in 6 minutes. What is her speed in metres per minute? (87·6 m/min)

Same day intervention
Enrichment

• Ask learners to create eight numbers that are all less than 1000 with one decimal place. The first number should be divisible by 2, the second by 3, the third by 4, and so on, up to the last number divisible by 9. In each case, the quotient must have only one decimal place.

Number – Fractions, decimals, percentages, ratio and proportion

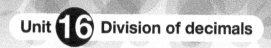

Unit 16 Division of decimals

Lesson 3: **Dividing two-place decimals by whole numbers (1)**

Learning objective

Code	Learning objective
6Nf.11	Estimate and divide numbers with [one or] two decimal places by whole numbers.

Resources

paper or mini whiteboard and pen (per learner); sheet of paper, for the Workbook (per learner)

Revise

Use the activity *Division stories* from Unit 16: *Division of decimals* in the Revise activities.

Teach [SB] 🖥 [TWM.03]

- Write on the board: 4·88 ÷ 4 =. **[T&T]** Ask: **How would you solve this calculation? Is there a related fact you can use?** Elicit that you could use a whole number fact and adjust by place value. Say: **Since you know that 488 ÷ 4 = 122, you also know that 4·88 ÷ 4 = 1·22 (100 times smaller). [TWM.03]** Ask: **Is there another mental method you could use to solve the calculation?** Take responses. Establish that a partitioning strategy could be used. Record: 4·88 ÷ 4 = (4 ÷ 4) + (0·88 ÷ 4) = 1 + 0·22 = 1·22.
- On the board, write: 6·46 ÷ 2 =, 9·69 ÷ 3 =, 8·04 ÷ 4 =. Learners work in pairs: one learner solves each calculation using a 'related fact' strategy and the other uses a partitioning strategy. The pair confirm they have the same answer each time (3·23, 3·23, 2·01).
- Display **Slide 1.** Say: **We are now going to divide numbers with two decimal places by whole numbers.** Ask: **What is a good estimate for the quotient?** Take responses. Establish that since the dividend is just over three times the divisor, you should expect the answer to be close to 3.
- Ask: **Why do you solve 942 ÷ 3 not 9·42 ÷ 3?** Remind learners that it is easier to work with a whole number dividend than a decimal one. Say: **For this division, multiply 9·42 by 100 to make the dividend 942 and then adjust the quotient by dividing by 100.**
- Guide learners through the steps involved in using the expanded written method to solve 942 ÷ 3 =. **[T&T] [TWM.03]** Ask questions to develop understanding: **Why do you record '– 900' below '942'? Why do you record '42' below '– 900'?** Ask: **You have the quotient 314. Is this the answer to 9·42 ÷ 3?** Agree that you need to adjust the quotient 314 by dividing by 100 to give the answer 3·14.
- Ask learners to work in pairs to solve the 'Your turn' calculation (8·28 ÷ 4 = 2·07).
- Say: **We are now going to use the expanded written method to divide other decimals by one- and two-digit whole numbers.**

- Display and discuss **Slide 2**, then **Slide 3**. Emphasise how each dividend is converted to a whole number by multiplying by 100 and the answer adjusted by dividing by 100. Ask learners to work in pairs to solve each 'Your turn' calculation. (27·18 ÷ 3 = 9·06, 48·32 ÷ 16 = 3·02).
- 💧 Say: **A tap leaks at a constant rate. If the tap leaks 48·54 litres every 6 minutes, how much water is lost every minute?** (8·09 *l*)
- Discuss the Let's learn section in the Student's Book.
- Introduce the paired activity in the Student's Book. Ask a pair of learners to explain how they worked out how much each person receives.
- Discuss the Guided practice example in the Student's Book.

Practise [WB]

- Workbook

Title: Dividing two-place decimals by whole numbers (1)

Pages: 130–131

- Refer to Activity 2 from the Additional practice activities.

Apply 👥 🖥

- Display **Slide 4.** Choose a pair of learners to explain how they calculated the equal share of the putty.

Review

- On the board write: $4.24, $8.44, $36.12, $48.28. Ask learners to find a quarter of each amount using the expanded written method.

Assessment for learning

- Show me two methods for calculating 64·48 ÷ 8.
- Find a third of 27·18 litres. (9·06 ℓ)

Same day intervention

Enrichment

- Ask learners to determine different ways in which 9·72 litres of juice could be divided equally between two or more glasses (up to nine glasses). In each case, the amount in the glass must be a number with no more than two decimal places.

Lesson 4: **Dividing two-place decimals by whole numbers (2)**

Learning objective

Code	Learning objective
6Nf.11	Estimate and divide numbers with [one or] two decimal places by whole numbers.

Resources

paper or mini whiteboard and pen (per learner)

Revise

Use the activity *Compare the methods* from Unit 16: *Division of decimals* in the Revise activities.

Teach ▣ ▯ [TWM.04]

- Display **Slide 1**. Explain that you are now going to use the short division method to solve division calculations where the dividend has two decimal places. Ask: **What is a good estimate for the quotient?** Take responses. **[TWM.04]** Ask: **Why do you solve 624 ÷ 3, not 6·24 ÷ 3?** Remind learners that it is easier to work with a whole number dividend than a decimal one. Say: **For this division, multiply 6·24 by 100 to make the dividend 624 and then adjust the quotient by dividing by 100.**

- Work through the calculation, reminding learners of the steps in the short division method, comparing and contrasting it to the expanded written method. Follow the completed example on **Slide 1** but write out the entire calculation alongside as you move through the steps.

- Begin the division. Ask: **How many 3s go into 6?** (2) Write '2' on the answer line in the hundreds column. **[T&T] [TWM.04]** Ask: **Why do we record the '2' in the hundreds column when we ask the question: 'how many 3s go into 6?'** Take responses. Explain that since the digit 6 in 624 has a value of 600, you are really asking how many 3s go into 600 (200). Ask: **How many 3s go into 2?** (0) **This leaves a remainder of 2.** Write a small '2' to the left of the next digit in the dividend (4) to make 24. Ask: **How many 3s go into 24?** (8) Write '8' in the answer line in the ones column. Say: **The quotient is 208. Is this the answer to the calculation 6·24 ÷ 3?** (no) Establish that you still need to adjust for multiplying the dividend by 100 by dividing the quotient by 100 (208 ÷ 100 = 2·08). Say: **Therefore, 6·24 ÷ 3 = 2·08.**

- Ask learners to work in pairs to solve the 'Your turn' calculation (8·36 ÷ 4 = 2·09).

- Say: **We are now going to divide other decimals by a one- and two-digit whole number.** Display and discuss **Slide 2**, then **Slide 3**. Take learners through the solutions for each division form. Emphasise how each dividend is converted to a

whole number by multiplying by 100 and the answer adjusted by dividing by 100. Ask learners to work in pairs to solve each 'Your turn' calculation. (81·45 ÷ 9 = 9·05, 65·39 ÷ 13 = 5·03).

- 🗩 Say: **Prices in a shop are reduced by one twelfth. How much is taken off a suitcase priced at $84.36?** ($7.03)

- Discuss the Let's learn section in the Student's Book.

- Introduce the paired activity in the Student's Book. Ask a pair of learners to explain how they worked out the length of each of piece of ribbon.

- Discuss the Guided practice example in the Student's Book.

Practise ▣

- Workbook

Title: Dividing two-place decimals by whole numbers (2)

Pages: 132–133

- Refer to Activity 2 (Variation) from the Additional practice activities.

Apply ▣▣ ▯

- Display **Slide 4**. Discuss the problem. Choose a pair to explain how they decided which number of equal sections of material is possible.

Review

- On the board write: $9.12, $12.36, $24.33, $39.18. Ask learners to use short division to find a third of each amount. ($3.04, $4.12, $8.11, $13.06)

Assessment for learning

- Show me two different methods for calculating 35·77 ÷ 7. (5·11)
- Find an eighth of 48·72 kg. (6·09 kg)

Same day intervention

Enrichment

- Some learners may wish to calculate HTO.th ÷ O calculations, for example, 369·27 ÷ 3.

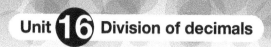

Additional practice activities

Activity 1

Learning objective
- Estimate and divide numbers with one decimal place by whole numbers.

Resources

paper (per learner)

What to do [TWM.04]

- Draw two connected function machines on the board, the first labelled '÷ 3' and the second '÷ 4'.
- Write the inputs on the left side of the first machine: 259·2, 278·4, 316·8.

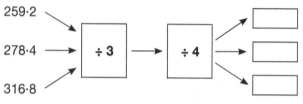

- Learners pass the inputs through both machines using the expanded written method to calculate the quotient at each stage.

- Confirm the answers: 21·6, 23·2, 26·.4.
- Say: **You divided each input number by 3, and then by 4. Is there another way of solving the problem?** Take responses. Praise learners who suggest dividing by 12 in one stage rather than by 3 and then by 4 in two stages.

Variations

2 Use a different set of inputs: 331·2, 403·2, 573·6.

Learners pass the inputs through both machines using the short division method to calculate the quotient at each stage. (27·6, 33·6, 47·8)

3 Challenge learners to divide by 3, then by 4, then by 3 again. Ask: **Is there another way of solving the problem?** (divide by 36)

Activity 2

Learning objective
- Estimate and divide numbers with two decimal places by whole numbers.

Resources

paper (per learner)

What to do [TWM.06]

- On the board write: 32·32, 9·33, 15·27, 24·12, 21·24, 8·36.
- Ask learners to work in pairs to sort the numbers into dividends that give a quotient with two decimal places when divided by 3, and those that give a quotient with two decimal places when divided by 4. Say: **One of the numbers will go in both lists. Which number is it?**
- Learners sort the numbers and then complete the divisions (divided by 3 or 4) using the expanded written method. 24·12 is the number that goes in both lists.

Variation [TWM.06]

2 On the board, write: 49·21, 42·18, 28·42, 36·66, 30·.24, 63·77.

Ask learners to work in pairs to sort the numbers into dividends that give a quotient with two decimal places when divided by 6, and those that give a quotient with two decimal places when divided by 7.

Learners sort the numbers and then complete the divisions (divided by 6 or 7) using short division.

Unit 17: Proportion and ratio

Collins International Primary Maths
Recommended Teaching and
Learning Sequence Term 3, Week 7

Learning objectives

Code	Learning objective
6Nf.12	Understand the relationship between two quantities when they are in direct proportion.
6Nf.13	Use knowledge of equivalence to understand and use equivalent ratios.

Unit overview

In this unit, learners extend their knowledge of ratio and proportion. They revise their understanding of the term 'proportion' as a comparison between part and whole and use simple fractions to describe proportional relationships. The concepts are applied to a variety of real-world situations, including scaling of recipe ingredients.

Learners understand what is meant by 'in proportion' and the concept of 'direct proportion' is developed through examples such as the scaling of models. They understand that when one quantity increases (or decreases) the other quantities increase (or decrease) in the same ratio.

Learners review the term 'ratio' and explore further examples of comparison between two quantities. Contextual examples and problems are provided where learners are asked to work out new and unknown quantities based on an existing (known) ratio.

Prerequisites for learning

Learners need to:
- know how to use fractions to describe and estimate a simple proportion
- understand ratio as a means of comparing part to part
- be able to use simple ratios in context.

Vocabulary

proportion, in every direct proportion, scale (scaling), scale factor, ratio, for every, to every, equivalent ratio

Common difficulties and remediation

Ratio and proportion can be difficult concepts for learners to grasp. The subjects can be confusing because the terms are frequently used interchangeably in many aspects of daily life. The words 'ratio' and 'proportion' actually represent two quite different ways of looking at the same thing. It is important to make frequent use of familiar phrases that help support development of the concepts involved. Use of the phrase 'for every' helps to introduce ratio. For example: 'For every red bead there are three yellow beads.' Explain that ratio compares one part to another.

After the introduction of ratio, and having completed lessons on proportion, it is advisable to use practical examples to clarify the difference between the two terms. An effective way to illustrate the difference is to use beads on a string. For example, show a string comprising a repeating pattern of one red bead for every three yellow beads. You could say: **One in every four beads is red, so the fraction of red beads is $\frac{1}{4}$.** Explain that you are comparing one part to the whole and that this is called proportion. Ask: **You can see that three in every four beads are yellow, so what proportion is yellow?** $(\frac{3}{4})$

Supporting language awareness

At every stage, learners require mathematical vocabulary to access questions and problem-solving exercises. If appropriate, when a new key word is introduced, ask learners to write a definition in their books, drawing a box around it for emphasis. Encourage learners to write the definition in their own words, for example: 'proportion compares a part with the whole' and 'ratio compares one part with another part'. Sketching a diagram alongside will help embed the definition.

Promoting Thinking and Working Mathematically

Opportunities to develop all four pairs of TWM characteristics are provided throughout the unit.

In Lesson 1, learners convince (TWM.04) when they prove that a scale factor can make a quantity smaller.

In Lesson 2, learners specialise (TWM.01) when they identify how to change the dimensions of a picture of a robot and still maintain proportion.

Learners classify (TWM.02) in Lesson 3 when they find other ratios are that are equivalent to the ratio 24:8.

Success criteria

Learners can:
- use direct proportion to increase or decrease quantities
- use direct proportion to find unknown values and solve problems
- use equivalent ratios to solve problems involving proportional reasoning.

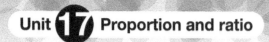

Unit 17 Proportion and ratio

Lesson 1: **Direct proportion (1)**

Learning objective

Code	Learning objective
6Nf.12	Understand the relationship between two quantities when they are in direct proportion.

Resources

interlocking cubes (per group)

Revise

Use the activity *Shape proportions* from Unit 17: *Proportion and ratio* in the Revise activities.

Teach SB [TWM.01/04]

- Revise the term 'proportion'. Construct a tower from 20 cubes, making a repeating pattern of three blue cubes and two red cubes. **[T&T]** Ask: **How would you describe the pattern?** Prompt learners to use the phrase 'in every'. Take responses. Establish the form of the pattern. Say: **In every five cubes, there are three blue cubes and two red cubes.** Remind learners that the phrase 'in every' means that you are talking about proportion – how the parts compare to the whole. Reduce the tower to five cubes, comprising three blue and two red. Ask: **Think of the whole as five cubes, so the cubes are parts of the whole. What fraction of the whole are the blue cubes?** ($\frac{3}{5}$) **What fraction are the red cubes?** ($\frac{2}{5}$)
- Remind learners that you can replace the word 'fraction' with 'proportion'. The proportion of blue cubes is $\frac{3}{5}$; the proportion of red cubes is $\frac{2}{5}$. Also, remind learners that the relationship between the blue and red cubes can be expressed as a ratio, 3:2.
- Organise learners into groups and provide them with trays of interlocking cubes. Ask them to construct the five-cube blue and red tower previously described. **[TWM.01]** Ask: **How would you double the number of cubes but maintain the same ratio, three blue cubes to two red cubes?** Have learners construct towers of ten cubes. Choose a pair to demonstrate their tower and explain how they maintain the same ratio. Ask: **How many blue cubes are there in total?** (6) **How many red cubes?** (4) On the board, write: $\frac{6}{10}$ and $\frac{4}{10}$. **[TWM.04]** Say: **These are now the proportions of blue and red cubes but they look different from the fractions we began with, $\frac{3}{5}$ and $\frac{2}{5}$. Are they wrong? Have the proportions changed?** Take responses. Praise learners who explain that the proportion has not changed. $\frac{6}{10}$ and $\frac{3}{5}$, and $\frac{4}{10}$ and $\frac{2}{5}$, are equivalent fractions: $\frac{6}{10}$ can be simplified to $\frac{3}{5}$, and $\frac{4}{10}$ can be simplified to $\frac{2}{5}$. Ask: **Start with your five-cube tower and triple the number of cubes. How many blue cubes are there?** (9) **What proportion is this?** ($\frac{9}{15}$) **Has the proportion changed?** (no) **How do you know?** ($\frac{9}{15} = \frac{3}{5}$) Repeat for the red cubes ($\frac{6}{15} = \frac{2}{5}$) Repeat for a tower of 20 cubes.

- Line up the different blue-red towers of five cubes, ten cubes, 15 cubes and 20 cubes. Explain that the numbers of red and blue cubes in all four towers are in direct proportion. Say: **If you double one quantity, you double the other quantity; if you multiply one quantity by four, you increase the other quantity by four times.**
- **[TWM.04]** Ask: **Can multiplying a quantity by a number make the quantity smaller? How would you prove this?** Take responses. Establish that if you multiply by a fraction, this results in a smaller quantity.
- Discuss the Let's learn section in the Student's Book.
- Introduce the paired activity in the Student's Book.
- Discuss the Guided practice example.

Practise WB

- Workbook

Title: Direct proportion (1)

Pages: 134–135

- Refer to Activity 1 from the Additional practice activities.

Apply 👥 🖥

- Display **Slide 1**. Discuss the problem. Choose a pair of learners to explain how they worked out the diameter of the wheel on the actual car.

Review

- Choose several learners to form a queue at the front. Make proportion statements about the queue's composition. For example: **Two in every five people in the queue are girls. Three-fifths are boys.** Include some statements that are incorrect and ask the class to show whether they think each statement is true or false.

Assessment for learning

- If you get six questions in a spelling test right and there are ten questions altogether, what proportion did you answer correctly? ($\frac{3}{5}$)
- A model train and a real train are in proportion. The train is 50 times longer than the model. If the real train has a carriage of length 18·5 m, what is the length of the model carriage?

Same day intervention

Support

- Stress that proportion compares one part to the whole and is often expressed as a fraction, or by using the phrase 'in every'.

Lesson 2: **Direct proportion (2)**

Learning objective

Code	Learning objective
6Nf.12	Understand the relationship between two quantities when they are in direct proportion.

Revise

Use the activity *Scaling triangles* from Unit 17: *Proportion and ratio* in the Revise activities.

Teach [SB] 🖥 [TWM.01/04]

• Display **Slide 1**. Read and discuss the text. **[T&T] [TWM.01/04]** Ask: **What dimensions, width and height, could Carrie change the robot to and still maintain proportion?** Take responses. Write the pairs of dimensions on the board, for example, 24 cm and 32 cm, 36 cm and 48 cm. Ask: **What do all of these pairs of numbers have in common?** Help learners to answer the question by writing the numbers as ratios including the original dimensions, 12:16, 24:32, 36:48. Take responses. Establish that all the ratios are in proportion and can be reduced to 12:16. Ask: **Can the ratio be reduced further? How?** Confirm that the ratio can be simplified to 3:4.

• Return to **Slide 1**. Ask: **What could this simplified ratio help you to do?** If learners require help, ask: **How could you use this ratio to reduce the size of the robot on the screen?** Establish that the robot could be reduced to dimensions of 3 cm by 4 cm and still maintain proportion. Ask: **Could you make the robot larger but still smaller than the original size. How?** Take responses. Establish that you could double or triple the ratio to get the dimensions 6 cm by 8 cm, or 9 cm by 12 cm.

• Display **Slide 2**. **[T&T]** Ask: **How would you change the amount of each ingredient when baking for eight people?** Take responses. Confirm that to increase the amounts for four people to eight people requires a doubling of the amounts. Choose a learner to list the amount of each ingredient required for eight people (1200 g flour, 800 g sugar, 960g butter, 8 eggs).

• Ask: **How would you change the quantity of each ingredient for just two people? Would you divide or multiply? If so, by which number?** (divide by 2 or multiply by $\frac{1}{2}$) **What quantities are required?** (300 g flour, 200 g sugar, 240 g butter, 2 eggs) Ask: **How much of each ingredient is required for one person/12 people/20 people?**

• 🖐 Ask: **A photocopy is in proportion to the original. If the original is 20 cm by 30 cm, what sizes can the photocopies be?**

• Discuss the Let's learn section in the Student's Book.

• Introduce the paired activity in the Student's Book. Ask a pair of learners to explain how they calculated

the number of pieces of homework Fred completes in 8·5 months.

• Discuss the Guided practice example in the Student's Book.

Practise [WB]

• Workbook

Title: Direct proportion (2)

Pages: 136–137

• Refer to Activity 1 (Variation) from the Additional practice activities.

Apply 👥 🖥

• Display **Slide 3**. Discuss the problem. Ask learners to explain how they calculated the price of each tin of paint. Confirm the answers as a class.

Review

• Say: **To make five banana muffins I need 3 bananas, 4 teaspoons of baking powder and 2 cups of flour, plus some other ingredients. What amount of each ingredient is required to make 15 muffins?** (9 bananas, 12 teaspoons of baking powder, 6 cups flour)

Assessment for learning

• I buy 4 packets of stickers for $6. How many packets will I get for $24? (16)

• A photograph of a dog is made 8 times larger. If the enlarged photo has a width of 288 cm, what was the width of the original photo? (36 cm)

Same day intervention
Support

• Learners can find word problems challenging and need reminding of how to approach them for the best results. Read the question very carefully and check that they are clear what they need to find out. Look at the facts given in the question and decide how to use them to find the answer. Remember there may be more than one stage. Note down the results at each stage. When they have found an answer, make sure that it fits the original information and answers the question.

Number – Fractions, decimals, percentages, ratio and proportion

Unit 17 Proportion and ratio

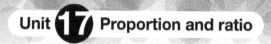

Number – Fractions, decimals, percentages, ratio and proportion

Lesson 3: **Equivalent ratios (1)**

Learning objective

Code	Learning objective
6Nf.13	Use knowledge of equivalence to understand and use equivalent ratios.

Resources

2D shapes (per pair); paper or mini whiteboard and pen (per learner)

Revise

Use the activity *Simplifying ratios* from Unit 17: *Proportion and ratio* in the Revise activities.

Teach 📖 📊 🖥 [TWM.02/08]

- Remind learners that ratio is a way of comparing numbers and quantities, part to part.
- Display the **Pattern tool** showing a pattern of six squares and two triangles. Ask: **How many squares are there compared with triangles?** Revise the expressions 'for every' and 'to' to compare the two parts. Say: **There are six squares for every two triangles. You can write this as 6:2.** Remind learners that the colon (:) is the ratio symbol and is read as 'to'. **[T&T]** Ask: **Is this the simplest way to write the ratio?** Take responses. Confirm that the ratio can be simplified to 3:1. On the board, write: 24:8. Ask: **Is this the same ratio? If so, how do you know?** Establish that 24:8 is an equivalent ratio to 3:1. Multiplying both sides of the ratio 3:1 by 8 gives 24:8. **[TWM.02]** Ask: **What other ratios are equivalent to 24:8 where the values in the ratio are greater than 24 and 8?** Expect and record: 48:16, 96:32, and so on. Ask: **What about if both values in the ratio are multiplied by a decimal, for example, 1·5? What ratio does this give?** Take responses. Confirm that 36:12 is also an equivalent ratio.
- Provide trays of 2D shapes. Ask learners to make shape sequences using two shapes and express these as a ratio. Ask questions such as: **If there were twice/3 times/5 times as many squares, how many circles would you need?** Ask learners to record these as equivalent ratios.
- Say: **A builder uses sand and cement. He uses three bags of sand for every two bags of cement. If there are ten bags of cement, how many bags are sand?** Ask learners to work out this amount in their pairs. Confirm the answer (15) and choose pairs to explain the strategy they used to work out the amount of sand. **[TWM.08]** Ask other learners to comment on the efficiency of each method and make suggestions as to how the method could be improved.
- Display **Slide 1**. Discuss the use of a table to work out equivalent ratios and finding unknown amounts.

Ask learners to copy and complete the second table. Choose pairs to explain how they found the unknown amounts.

- 🗣 Say: **Orange paint is made in a ratio of seven cups of red paint and four cups of yellow paint. If 21 cups of red paint are used to make a tin of orange paint, how many cups of yellow paint are needed?** (12)
- Discuss the Let's learn section in the Student's Book.
- Introduce the paired activity in the Student's Book. Ask a pair of learners to explain how they worked out the number of children that are wearing shoes.
- Discuss the Guided practice example in the Student's Book.

Practise 📒

- Workbook

Title: Equivalent ratios (1)

Pages: 138–139

- Refer to Activity 2 from the Additional practice activities.

Apply 👥 🖥

- Display **Slide 2**. Discuss the problem. Ask learners to explain how they worked out the number of sheep for each set of cows.

Review [TWM.06]

- Write the following ratios on the board and ask learners to identify the odd one out: 27:15, 90:50, 54:30, 9:5, 34:20, 63:35. (34:20)

Assessment for learning

- Give me three equivalent ratios to 6:5.
- The ratio of squares to triangles in a pattern is 11:6. If there are 24 triangles, how many squares will there be? (44)

Same day intervention
Support

- Encourage learners to create tables or double number lines of equivalent ratios. Both are useful for visualising and comparing ratios.

Lesson 4: **Equivalent ratios (2)**

Learning objective

Code	Learning objective
6Nf.13	Use knowledge of equivalence to understand and use equivalent ratios.

Resources

paper or mini whiteboard and pen (per learner)

Revise

Use the activity *Equivalent fraction chains* from Unit 17: *Proportion and ratio* in the Revise activities.

Teach 🆂🅱 💻 [TWM.07]

- Say: **Lewis plants flowers in a garden plot. For every 15 daffodils, he plants 21 roses. If Lewis plants 63 roses, how many daffodils does he plant? [T&T]** Ask learners to work out this amount in their pairs. Confirm the answer (45) and choose pairs to explain the strategy they used to work out the number of daffodils planted. **[TWM.07]** Ask other learners to comment on the efficiency of each method and make suggestions as to how to improve it.
- Display **Slide 1**. Discuss the use of a table to work out equivalent ratios and finding unknown amounts. Ask learners to copy and complete the second table. Choose pairs to explain how they found the unknown amounts. The final column in the table is a challenging question. Praise learners who simplify the ratio to 5:7 before multiplying. (5:7 × 2 = 10:14)
- Display **Slide 2**. Read out the problem. Give learners time to work out the answer using the table format. Confirm the answer (30 cups of apple juice for 24 cups of orange juice) and choose a pair to explain the strategy they used to work out the amounts. Ask learners to copy and complete the table. Choose pairs to explain how they found the unknown amounts.
- 💬 Say: **Connor makes cheese and tomato sandwiches for a party. He uses two slices of bread for every three slices of cheese. How many slices of bread does he use for every 27 slices of cheese?** (18)
- Discuss the Let's learn section in the Student's Book.
- Introduce the paired activity in the Student's Book. Ask a pair of learners to explain how they worked out the number of chocolate muffins.
- Discuss the Guided practice example in the Student's Book.

Practise 🅦🅑

- Workbook

Title: Equivalent ratios (2)

Pages: 140–141

- Refer to Activity 2 (Variation) from the Additional practice activities.

Apply 👥 💻

- Display **Slide 3**. Discuss the problem. Ask learners to explain how they worked out the number of each size of T-shirt for sale.

Review

- Ask learners to make a repeating shape pattern with moons and stars in the ratio 6:5, for example. Working in pairs, learners ask questions about each other's patterns, for example: 'If you had three times the number of moons, how many stars would you need?' Choose pairs to share their questions and answers with the class.

Assessment for learning

- Give me three equivalent ratios to 7:6.
- The ratio of dogs to cats is 12:7. If there are 35 cats, how many dogs will there be? (60)

Same day intervention
Support

- Look out for learners who add rather than multiply when increasing in a ratio or proportion. For example, given a flapjack recipe that includes two tablespoons of syrup for three people, and asked to work out how many tablespoons they would need for 12 people, although learners realise there are nine extra people, they add either nine tablespoons or 18 (two tablespoons each). They do not see that the recipe is for four times as many people, or recognise the ratio of 4:1, giving eight tablespoons. To remediate this problem, allow learners to take part in practical cooking activities and give them opportunities to practise calculations involving ratio and proportion.

Number – Fractions, decimals, percentages, ratio and proportion

Additional practice activities

Activity 1

Learning objective
- Understand the relationship between two quantities when they are in direct proportion.

Resources
interlocking cubes in two colours (per pair); 1–10 number cards from Resource sheet 2: 0–100 number cards (per pair); paper (per pair)

What to do [TWM.04]
- Learners shuffle the number cards and place them face down in a pile.
- Each learner takes a card, then working together they use the cards to make a proper fraction that represents a proportion, for example, $\frac{5}{8}$ (5 in every 8).

- Learners choose the smallest number of cubes (≥ 20) that are in this proportion, i.e. 24 cubes: 15 blue and 9 red.
- Learners record the proportion and the number of cubes in each colour.
- They repeat until all 10 cards have been used.

Variation [TWM.04]
Learners take turns to create a shopping scenario, for example: 'I buy 8 chocolate bars for $12. How many bars can I buy for $24? For $48? For $3?' Learners respond and confirm the answers. (16, 32, 2)

Activity 2

Learning objective
- Use knowledge of equivalence to understand and use equivalent ratios.

Resources
1–12 number cards from Resource sheet 2: 0–100 number cards (per pair); paper (per pair)

What to do
- Learners shuffle the number cards and place them face down in a pile.
- Each learner takes two cards.
- The cards represent the ratio of cups of blue paint to cups of red paint to make purple paint, for example, 4: 7 (4 cups of blue paint to 7 cups of red paint).
- Each learner then picks a third card that they keep secret from their partner. They multiply one of the values in their partner's ratio and state the new amount, for example, for the ratio 4:7, and a multiplier of 5, they might say: 'How many cups of blue paint is needed for 35 cups of red paint?' The answer is 20 (4 × 5).

- Repeat for different cards and different ratios.
- If time allows, collect and reshuffle the cards and repeat the activity.

Variation
- The cards represent the ratio of sunflower seeds to tomato seeds in a garden plot, for example, 5:4 (sunflower : tomato).
- Learners then pick a third card that they keep secret from their partner. They multiply one of the values in their partner's ratio and state the new amount, for example, for the ratio 5: 4, and a multiplier of 3, they might say: 'How many sunflower seeds are needed for 24 tomato seeds?' The answer is 30 (5 × 6).
- Repeat for different cards and different ratios.
- If time allows, collect and reshuffle the cards and repeat the activity.

Unit 18: 2D shapes and symmetry

Collins International Primary Maths Recommended Teaching and Learning Sequence Term 1, Week 6

Learning objectives

Code	Learning objective
6Gg.01	Identify, describe, classify and sketch quadrilaterals, including reference to angles, symmetrical properties, parallel sides and diagonals.
6Gg.02	Know the parts of a circle: - centre - diameter - radius - circumference.
6Gg.08	Identify rotational symmetry in familiar shapes, patterns or images with maximum order 4. Describe rotational symmetry as 'order x'.
6Gg.11	Construct circles of a specified radius or diameter.

Unit overview

Learners are introduced to quadrilaterals, including parallelogram, rhombus, trapezium and kite. They learn to identify, describe, classify and sketch quadrilaterals, including reference to angles, symmetrical properties, parallel sides and diagonals.

Practical activities are used to introduce learners to parts of a circle: centre, diameter, radius and circumference. The relationship between diameter and radius is elicited and the rule d = 2 × r introduced. The use of compasses is demonstrated and learners practise drawing circles of a specified radius or diameter.

Learners investigate rotational symmetry in familiar shapes, patterns or images with maximum order 4. The order of rotational symmetry is introduced visually and learners explore this, using tracing paper. They learn to describe rotational symmetry as 'order x'.

In the Collins International Primary Maths teaching and learning sequence, it is recommended that this unit is taught prior to Unit 19: 3D shapes. However, some schools may prefer to change the order of these two units and teach Unit 19 before teaching this unit.

Prerequisites for learning

Learners need to:
- know the properties of a square and a rectangle and the terms 'parallel' and 'perpendicular'
- understand the meaning of the term 'symmetry' and be able to find symmetry in shapes and patterns.

Vocabulary

quadrilateral, congruent, rectangle, square, rhombus, parallelogram, kite, trapezium, diagonal, parallel, perpendicular, bisect, circle, centre, radius, radii, diameter, circumference, compass, separation distance, rotational symmetry, order

Common difficulties and remediation

When working with shape, learners should be given opportunities to explore physical models and not just pictorial representations to help determine a shape's properties.

Introduce learners to a range of different examples and models of each type of quadrilateral while reminding them that they all have four sides and two diagonals.

Look out for learners who identify a square turned 45° as a 'diamond'. Point out that there is no geometric shape named 'diamond'. Explain, with modelling, that the shape is either a square that has been rotated or a rhombus. For this reason, it is important that when learners are asked to compare and contrast quadrilaterals and their properties, squares are drawn tilted and in an upright position.

Supporting language awareness

Note that we use the words 'diameter' and 'radius' in two different ways, as an object and as a measurement. A line segment from the centre of a circle to any point on the circle is a radius of the circle and considered an object. There are an infinite number of line segments. In contrast, the length of such a segment is the radius of the circle and is considered a measurement. There is only one value for any particular circle.

Promoting Thinking and Working Mathematically

In Lesson 1, learners classify (TWM.06) when they identify the differences in properties between two shapes.

In Lesson 4, learners convince (TWM.04) when they consider the smallest possible order of symmetrical rotation.

Success criteria

Learners can:
- use properties to identify, describe, classify and sketch a parallelogram, rhombus, trapezium and kite
- use compasses to draw circles
- name the parts of a circle
- use the rule d = 2 × r to calculate the diameter or the radius of a circle
- identify rotational symmetry in shapes, patterns and images
- describe rotational symmetry as 'order x'.

Geometry and Measure – Geometrical reasoning, shapes and measurements

Lesson 1: **Quadrilaterals**

Learning objective

Code	Learning objective
6Gg.01	Identify, describe, classify and sketch quadrilaterals, including reference to angles, symmetrical properties, parallel sides and diagonals.

Resources

Geoboard and bands, or other shape-building equipment (per learner); trays of quadrilaterals (one tray of each type): rectangles (squares, rectangles), parallelograms, rhombuses, kites, trapeziums (per group); Resource sheet 17: Square dot paper (per pair); sorting hoops (per pair)

Revise

Use the activity *Quadrilaterals* from Unit 18: *2D shapes and symmetry* in the Revise activities.

Teach 📖 💻 [TWM.01/05/06]

- Display **Slide 1**. Explain that any polygon with four sides is called a quadrilateral. Say: **Quadrilaterals have very specific names and, in some cases, more than one name.**
- Arrange learners into five groups and give each group one of the five trays of quadrilaterals. Distribute Geoboards and bands. Each group makes four shapes that would be classified as their allocated quadrilateral. **[T&T]** Ask: **Does this shape have any pairs of parallel lines/right angles/congruent sides (same length)? Why does this shape belong in this category?** Explain that 'congruent' means two figures are exactly the same, even if they are in different positions. When two line segments have the same length, they are congruent. When two figures have the same shape and size, they are congruent.
- Bring the class together. **[TWM.06]** Ask: **Who has been working with a shape that has pairs of adjacent sides that are equal?** (kites) Invite learners to display their shapes.
- Invite the trapezium group forward. Ask: **What property distinguishes a trapezium from all other quadrilaterals?** (one pair of parallel sides) Challenge the rectangle group to transform one of their shapes into a trapezium.
- Invite the parallelogram group forward. Ask: **What property distinguishes a parallelogram from other quadrilaterals?** (opposite sides are parallel and equal in length; opposite angles are equal) Challenge them to transform their shape into a rhombus.
- Distribute square dot paper. Ask learners, working in pairs, to draw six shapes on the paper: square, rectangle, parallelogram, rhombus, kite and trapezium. Have them draw the diagonals of each shape. Name a shape and choose learners to describe the properties of the diagonals. Establish that the diagonals of a square are of equal length and perpendicular to each other. Introduce the word 'bisect' (to divide into two exactly equal parts) and explain that the diagonals of a square bisect each other. **[TWM.01]** Ask: **How do the diagonals of**

a rectangle differ from a square? How are they similar? Agree that diagonals are of equal length and bisect each other. They differ from a square in that they are not perpendicular. Display **Slide 2** to establish the properties of the diagonals and symmetry for the different types of quadrilaterals.

- 🖽 **Consider a square and a kite. Give me two similarities and two differences.**
- Discuss the Let's learn section in the Student's Book.
- **[TWM.05]** Introduce the paired activity in the Student's Book, then discuss Guided practice.

Practise 📒 [TWM.06]

- Workbook

Title: Quadrilaterals

Pages: 142–143

- Refer to Activity 1 from the Additional practice activities.

Apply 👥 💻 [TWM.04]

- Display **Slide 3**.

Review 📊

- Display the **Geoboard tool** showing a parallelogram and rhombus, side by side. Say: **A rhombus is a special kind of parallelogram, with four equal sides.** Ask learners what other shapes would be considered parallelograms. Agree that squares and rectangles are also parallelograms.
- Using the trays of rectangles (squares and rectangles) and rhombuses, and the sorting hoops, sort the shapes according to whether the diagonals are the same or different lengths. Ask: **What did you discover?** (diagonals of a rhombus are not congruent, unlike in squares and rectangles)

Assessment for learning

- I have a mix of quadrilaterals in my tray. How would I identify a rhombus? A kite?

Same day intervention
Support

- Ensure learners can identify parallel and perpendicular lines as they will need to know this when classifying polygons.

Lesson 2: **Parts of a circle**

Learning objective

Code	Learning objective
6Gg.02	Know the parts of a circle: – centre – diameter – radius – circumference.

Resources

large pair of compasses (per class); Resource sheet 18: Circle (per learner); ruler (per learner)

Revise

Use the activity *Drawing a circle* from Unit 18: *2D shapes and symmetry* in the Revise activities.

Teach [SB] 💻 [TWM.06]

- Use a large pair of compasses to draw a circle on the board. Do not discuss the mechanics of how to use a compass at this stage as this will be covered in the next lesson. If a large pair of compasses is not available, then draw with a chalk or pencil attached to a string anchored at a central point. **[T&T]** Ask: **What are the properties of a circle?** Choose learners to share these properties with the class.

- Display **Slide 1**. Discuss the different parts of a circle and then demonstrate how to mark these on the circle on the board. Mark random points on the circumference and label the centre. Establish that although only some points have been marked, it is possible to continue marking points all around the shape. Explain that a circle is a set of points that are the same distance from a given point, called the centre, and that there are an infinite number of these.

- Invite a volunteer to the board and give them a ruler. They measure and record the distance from the centre to a point on the circumference. Invite volunteers to the board to measure from the centre to other points on the circle. Agree that the distance is always the same.

- Using a ruler, draw a line from the centre of the circle to the circumference and label the line 'radius'. Explain that the distance from the centre point to any point on a circle is called the radius of the circle.

- Distribute copies of Resource sheet 18: Circle and rulers. Ask learners to draw a radius of the circle then to hold up their sheets and confirm that they have correctly positioned a radius.

- Place a ruler on the board so that the edge passes through the centre of the circle. Draw a diameter of the circle. Label the line 'diameter'. Explain that a diameter is any line that joins two points on the circle and passes through the centre of the circle. Reinforce the definition by drawing other diameters of the circle. Draw a line that does not pass through the centre. Ask: **Is this a diameter?** (no) **Why?**

- Ask learners to use a ruler to draw and label a diameter of the circle on the resource sheet.

- **[TWM.06]** Ask: **What is the relationship between the diameter and the radius?** Establish that since the radius is halfway across the circle just to the

centre, and the diameter is the entire way across, the diameter is twice as long as the radius. On the board, write: d = 2 × r. Explain that d and r are variables, where d represents diameter and r represents radius.

- Label the radius '12 cm'. Ask: **What is the diameter of the circle? How do you know?** Agree that as the diameter is double the radius, the diameter is 24 cm.

- Label the diameter '54 cm'. Ask: **What is the radius? How do you know?** Agree that since the radius is half the diameter, the radius of the circle is 27 cm.

- 🖉 Ask: **The diameter of a circle is 92 metres. What is the radius?** (46 m)

- Discuss the Let's learn section in the Student's Book.

- Introduce the paired activity in the Student's Book, then discuss the Guided practice example.

Practise [WB]

- Workbook

Title: Parts of a circle

Pages: 144–145

- Refer to Activity 1 (Variation) from the Additional practice activities.

Apply 👥 💻

- Display **Slide 2**. Ensure learners understand that the dimensions of both circles are needed to work out the depth at which the water wheel sits below the surface. (4 m)

Review

- Display **Slide 1** again and recap the parts of a circle, asking volunteer learners to offer a definition of each part, including the relationship between radius and diameter.

Assessment for learning

- Describe to your partner the relationship between the radius and the diameter of a circle.

Same day intervention
Support

- Some learners think that a circle has only one radius and diameter. Intervene to show that an infinite number of radii (teach learners the plural) and diameters can be drawn.

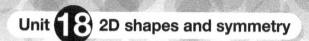

Geometry and Measure – Geometrical reasoning, shapes and measurements

Lesson 3: **Constructing circles**

Learning objective

Code	Learning objective
6Gg.11	Construct circles of a specified radius or diameter.

Resources

paper (per learner); compass (per learner); ruler (per learner); cardboard (optional) (per learner); tape (optional) (per learner)

Revise

Use the activity *Circles in a row* from Unit 18: *2D shapes and symmetry* in the Revise activities.

Teach ⬛ 🖥 [TWM.01/05]

- Display **Slide 1**. Discuss the steps involved in using a compass to draw a circle.
- Provide learners with paper, a compass and ruler, and if appropriate cardboard and tape (see below). Hold up a compass. Ask learners to follow the steps to load a pencil into the compass correctly: Insert the pencil into the cam lock and tighten the screw. Ensure that the tip of the pencil is at the same height as the sharp point of the compass when the compass is closed. Learners adjust the pencil height and tighten the cam screw once the pencil is at the correct height.
- Provide instructions for preparing the paper for drawing: Place a sheet of paper on top of cardboard to prevent damage to the table surface and to keep the needle point from slipping as the pencil is rotated. Demonstrate that you turn the knob on top of the compass to draw the circle.
- Ask learners to practise pressing the needle point firmly on to the paper and turning the compass using the handle. Learners may wish to tape the cardboard down to prevent it from spinning.
- Give learners time to draw circles of various sizes and become familiar with adjusting the compass to obtain a larger separation.
- Once learners have familiarised themselves with correct use of the compass, ask them to draw circles with a specific dimension, radius or diameter. For example, ask them to use a ruler to set the separation distance of the compass to 10 cm, and draw a circle. Ensure that learners place the needle point at 0 when measuring the separation distance.
- [TWM.01] [T&T] Ask: **I want to draw a circle with a diameter of 8 cm. What separation should the compass be set to?** Give learners time to consider the problem, then accept answers. Confirm that a separation of 4 cm is required as this is the radius that will give a diameter of 8 cm.
- [TWM.05] Ask learners to mark the centre of a circle they have drawn, the point where the needle of the compass was placed. Have them mark three points on the circle. Ask: **What do all three points have in common?** Take responses. Establish that the points are all the same distance from the centre of the circle.

Ensure learners understand that a circle is a set of points that are the same distance from the centre. To reinforce this concept, ask them to draw three points B, C and D, that are exactly 3 cm from a point A. Choose learners to explain their solution. Establish that drawing a circle with a centre A and radius of 3 cm gives a range of points that are equidistant (3 cm) from the centre.

- Repeat for lines of various lengths representing radii and diameters.
- 🗣 Say: **Give me the instructions for how to use a compass to draw a circle with a radius of 5 cm.** Ensure that answers including measuring the width of the compass to 5cm, and drawing with a steady hand.
- Discuss the Let's learn section in the Student's Book.
- Introduce the paired activity in the Student's Book, then the Guided practice example.

Practise 🔲

- Workbook

Title: Constructing circles

Pages: 146–147

- Refer to Activity 2 from the Additional practice activities.

Apply 👥 🖥

- Display **Slide 2**. Ensure learners understand the diagram they need to construct. Choose pairs of learners to demonstrate their drawings.

Review

- Ask learners to construct a circle with a radius of 6 cm. Ask: **What is the diameter of the circle?** (12 cm)
- Ask learners to construct a circle with a diameter of 18 cm. Ask: **What is the radius of the circle?** (9 cm)

Assessment for learning

- I have drawn a circle with a radius of 19 cm. What is the diameter of the circle? (38 cm)
- I use a compass to draw a circle with a diameter of 17 cm. What separation distance do I set the compass to? (8·5 cm)

Same day intervention
Support

- Left-handed learners may find it easier to reverse the direction of the compass rotation.

Lesson 4: **Rotational symmetry**

Learning objective

Code	Learning objective
6Gg.08	Identify rotational symmetry in familiar shapes, patterns or images with maximum order 4. Describe rotational symmetry as 'order x'.

Resources

tray of 2D shapes (per group); split pins (per pair); scissors (per pair); two pieces of A4 card (per pair); Resource sheet 36: Symmetrical letters (per pair)

Revise

Use the activity *Rotating objects* from Unit 18: *2D shapes and symmetry* in the Revise activities.

Teach 🔲 🖥 [TWM.04]

• Display **Slide 1**. Choose each shape in turn and describe a full rotation of the shape about its centre. **[T&T] Ask: How many times does this shape appear exactly like it did before the rotation?** Take responses. Explain that if at any point the object appears exactly like it did before the rotation, then the object has rotational symmetry.

• Distribute trays of 2D shapes, split pins, scissors and card. Working in pairs, learners select a shape and draw around it on both pieces of card. Next, they cut out the shape from one of the pieces of card and lay it over its outline on the A4 card. They fix the cut-out shape to the A4 card by pushing a split pin through the centre of both shapes.

• Learners turn the shape a full rotation, noting the number of times the shape fits into its original position. Ask learners to share the number of times this happens. Explain that this number is the 'order of symmetry' – or for short, 'order'. For example, a square has rotational symmetry of order 4. Ask learners to identify the order of rotational symmetry for a rectangle (2) and an equilateral triangle (3).

• **[TWM.04] Ask: What is the smallest order of rotational symmetry?** Take responses. Establish that this would be an order of 2. Ask: **Why isn't it possible to have a shape or object that has rotational symmetry of the order 1?** Explain that an order of 1 would mean a shape has to be fully rotated for it to appear as it did before the rotation. This would mean it had no rotational symmetry at all.

• 🖉 Say: **Draw two shapes that have rotational symmetry of order 4.**

• Discuss the Let's learn section in the Student's Book.

• Introduce the paired activity in the Student's Book, asking learners to work out the order of rotational symmetry for each shape.

• Discuss the Guided practice example in the Student's Book.

Practise 📕

• Workbook

Title: Rotational symmetry

Pages: 148–149

• Refer to Activity 2 (Variation) from the Additional practice activities.

Apply 👥 🖥

• Display **Slide 2**, give learners a copy of resource sheet 36 and read the text to the class. Give learners time to investigate the letters for rotational symmetry. Choose pairs to say the order for each letter. Confirm the answers as a class.

Review

• Have learners find four shapes/objects from around the class that have rotational symmetry. They arrange the shapes in order of rotational symmetry.

Assessment for learning

• What is the order of rotational symmetry for a rhombus? How do you know? (2)

• Name a shape that has rotational symmetry of order 4.

Same day intervention

Enrichment

• Provide grid paper and ask learners to rule lines to divide the grid into four quadrants. They shade cells in each quadrant to create a pattern that has rotational symmetry of order 4.

Geometry and Measure – Geometrical reasoning, shapes and measurements

Additional practice activities

Activity 1 👥 ▲2

Learning objectives
- Identify, describe, classify and sketch quadrilaterals, including reference to angles, symmetrical properties, parallel sides and diagonals.
- Know the parts of a circle: centre, radius, diameter, circumference.

Resources
Resource sheet 17: Square dot paper (per learner); ruler (per learner); tracing paper (per learner); Resource sheet 18: Circle (per learner) (for variation)

What to do
- Each learner draws the following shapes on Resource sheet 17: Square dot paper: square, rectangle, parallelogram, rhombus, kite, trapezium.
- They use a ruler to mark the diagonals of each shape.
- They place a piece of tracing paper over the dot paper and trace the diagonals of each shape.
- Learners swap tracing papers with their partner and complete the quadrilaterals from the diagonals. They name each shape and comment on the properties of each pair of diagonals.

Variations [TWM.05]
▲2 Provide each learner with a copy of Resource sheet 18: Circle. They use a ruler to mark three lines from one side of the circle to the other only one of which is the true diameter.

They label the lines A, B and C. They also mark three lines X, Y and Z, only one of which is the true radius.

They swap papers with their partner who has to say which of the lines represents the radius and diameter of the circle.

3️⃣ Learners draw only one diagonal of each shape on the dot paper. They write the name of the shape underneath and ask their partner to locate a possible position of the other diagonal and complete the shape.

Activity 2 👤 and 👥 ▲2

Learning objectives
- Construct circles of a specified radius or diameter.
- Identify rotational symmetry in familiar shapes, patterns or images with maximum order 4. Describe rotational symmetry as 'order x'.

Resources
compass (per learner); ruler (per learner); tray of regular and irregular 2D shapes (per pair) (for variation)

What to do
- On the board, write: Circle A: Diameter 8 cm; Circle B: Radius 5 cm; Circle C: Diameter: 6 cm; Circle D: Radius 6 cm.
- Ask learners to make a concentric circle pattern. They mark a dot in the centre of a piece of paper.

Next, they use a compass to draw circles A to D that all share the same centre.
- Ask learners to write the letter of each circle in ascending order of diameter. (C, A, B, D)

Variation [TWM.06]
▲2 Distribute trays of 2D shapes to pairs of learners. Ask them to sort the shapes according to the order of rotational symmetry: a group of order 2, a group of order 3, and a group of order 4.

Unit 19: 3D shapes

Learning objectives

Code	Learning objective
6Gg.04	Identify, describe and sketch compound 3D shapes.
6Gg.06	Identify and sketch different nets for cubes, cuboids, prisms and pyramids.

Unit overview

This unit introduces compound 3D shapes – structures built from two or more component shapes. For example, an octahedron formed from two triangular-pyramids (tetrahedrons) or an L-shape formed from two cuboids, one fixed at 90 degrees to the other.

Learners are given opportunities to construct their own compound shapes. They name the component shapes and describe the compound shape formed. They identify and count the faces, vertices and edges on the 'new' shape. Learners are asked to sketch (rather than accurately draw) the shapes on squared or triangular dot paper.

In Lessons 3 and 4, learners revise and extend knowledge of nets. They are reminded of which nets will/will not produce a cube, cuboids, prisms and pyramids. They are given opportunities to make their own nets and use them to describe the properties of 3D shapes. Learners sketch nets on squared or triangular dot paper.

In the Collins International Primary Maths teaching and learning sequence, it is recommended that this unit is taught after Unit 18: 2D shapes and symmetry. However, some schools may prefer to change the order of these two units and teach this unit prior to teaching Unit 18.

Prerequisites for learning

Learners need to:
- be able to identify and describe 3D shapes
- be able to visualise 3D shapes from 2D drawings
- be able to visualise, from the front and from the side, 2D representations of 3D shapes made with interlocking cubes
- use knowledge of the properties of prisms and pyramids to identify and draw different nets.

Vocabulary

compound 3D shape, component shape, net

Common difficulties and remediation

Look out for learners who find it difficult to identify nets that will form a cube. This is typical of learners who have had insufficient experience of visualising 3D shapes, particularly how they can be opened up and folded again. Consequently, they find it difficult to 'see' how a net can be folded to make a 3D shape.

Provide extended opportunities to draw and make nets of shapes, including those that do not create a complete 3D shape. Learners should experience unfolding a range of packaging, such as tubes and prism-shaped packages.

Supporting language awareness

Encourage frequent usage of mathematical language to help embed vocabulary. For example, model the use of the word 'vertex' when learners refer to 'corners of a shape' as they make the transition to mathematically precise language. Throughout the unit it is important to encourage learners to seek clarification and confirmation of the mathematical language. This may involve prompting learners to call out and complete definitions, for example, following a demonstration of nets by saying: 'A net is… what a 3D shape would look like if it were opened out flat.'

Promoting Thinking and Working Mathematically

In Lesson 1, learners generalise (TWM.02) when they identify the similarity between the lower and upper sections of a compound shape.

In Lesson 2, learners critique (TWM.07) when they discuss their sketches of a compound 3D shape formed from two cuboids, suggesting how they can be improved.

Learners convince (TWM.04) in Lesson 3 when they answer the question: 'What kind of information can be obtained from a net of a prism about the solid it creates?'

Success criteria

Learners can:
- name the component shapes of a compound 3D shape and describe the compound shape
- identify and count the faces, vertices and edges on compound 3D shape
- sketch a compound 3D shape
- identify the net of a 3D shape including a cube, cuboid, sphere, cone, cylinder, pyramids and prisms
- sketch the net of a 3D shape.

Lesson 1: **Identifying and describing compound 3D shapes**

Learning objective

Code	Learning objective
6Gg.04	Identify, describe [and sketch] compound 3D shapes.

Resources

3D shapes: cube, cuboid, sphere, cone, cylinder, pyramids, prisms, hemispheres (per group)

Revise

Use the activity *Guess the 3D shape* from Unit 19: *3D shapes* in the Revise activities.

Teach [SB] 🖵 [TWM.02/05]

- Construct a compound 3D shape from a cuboid and a triangular prism, like this:
- Display the shape and name it a 'house'. **[T&T]** Ask: **What shape have I used to form the base of the house?** Take responses. Expect learners to identify a cuboid. Separate the 'base' from the 'roof' and hold up just the 'base'. Ask: **How do you know this shape is a cuboid?** Expect learners to describe a shape that has six faces and all angles that are right angles. Explain that all of its faces are rectangles (or 4 rectangles and 2 squares). Ask: **How many vertices does this shape have?** (8) **How many edges?** (12) **[TWM.05]** Say: **I think a cuboid is also a prism. Am I correct?** Take responses. Elicit that a cuboid is also a prism because it has the same cross-section along a length. In fact, it is a rectangular prism.
- Point to the 'roof' of the house. **[TWM.02]** Ask: **What does this 3D shape have in common with the 'base' of the house?** Take responses. Elicit that both shapes are prisms. Identify the 'roof' as a triangular prism. Separate the 'roof' from the 'base' and just hold up the 'roof'. Ask: **What are the properties of a triangular prism?** Remind learners that a triangular prism has two triangular ends connected by three rectangular faces. Ask: **How many vertices does this shape have?** (6) **How many edges?** (9)
- Hold up the 'house' shape and explain that it is an example of a compound shape – a shape made up of two or more component shapes. Ask: **How many faces does the shape have?** (7) **How many vertices?** (10) **How many edges?** (15). Ask: **What is the name of this shape?** (pentagonal prism)
- Organise learners into groups and provide them with a tray of 3D shapes. Ask them to work in pairs. Partners take turns to combine two or more 3D shapes to make a compound shape. They present it to their partner and ask them to identify the component shapes and describe their properties. Next, they describe the properties of the compound shape as a whole. Choose learners to share the compound shape they have made with the class.

- Display **Slide 1**. Point to each shape and choose learners to identify the component shapes. Ask them to describe the compound shape as a whole in terms of the number of faces, vertices and edges.
- 🖰 Assemble a compound shape from a pentagonal pyramid mounted on a cube. Ask learners to identify the two component shapes and describe the properties of the compound shape.
- Discuss the Let's learn section in the Student's Book.
- Introduce the paired activity in the Student's Book, asking learners to work out the number of cubes required. Choose learners to explain how they solved the problem.
- Discuss the Guided practice example.

Practise [WB] [TWM.05]

- Workbook

Title: Identifying and describing compound 3D shapes

Pages: 150–151

- Refer to Activity 1 from the Additional practice activities.

Apply 👥 🖵 [TWM.05]

- Display **Slide 2**.

Review 📊

- Use the **Shape set tool** set to 3D shapes or physical shapes to construct a compound shape, for example placing two triangular prisms (tetrahedrons) base to base to form an octahedron. Choose learners to describe the compound shape formed.

Assessment for learning

- What is a compound 3D shape?
- A square pyramid is placed on top of a cuboid. Describe the compound shape formed.

Same day intervention
Support

- Provide activities encouraging learners to name 3D shapes and their properties, for example, sorting games involving asking questions, such as: **What are the characteristics of this shape? Why did you sort the shapes?**

Lesson 2: **Sketching compound 3D shapes**

Learning objective

Code	Learning objective
6Gg.04	Identify, describe and sketch compound 3D shapes.

Resources

Resource sheet 19: Triangular dot paper (several sheets per learner); 3D shapes: cube, cuboid, sphere, cone, cylinder, pyramids, prisms, hemispheres (per class)

Revise

Use the activity *Sketch the shape* from Unit 19: *3D shapes* in the Revise activities.

Teach 💾 🖥 [TWM.05/07]

- On the board, remind learners how to sketch a cube and a cuboid. Explain that you use dashed lines to indicate the edges and faces that are hidden from view.
- Display **Slide 1**. Demonstrate how to sketch a compound 3D 'house' shape. Draw a cuboid. On top of the cuboid, draw a triangular prism. **[T&T] [TWM.05]** Begin by sketching an equilateral triangle with its base aligned with the front edge of the cuboid and then ask: **What shape would you draw next?** Take responses and establish that another equilateral triangle is needed to be the opposite base of the shape. Draw the shape with its base aligned with the back edge of the cuboid and ask: **How do you complete the prism?** Establish that you connect corresponding vertices of the two triangles using parallel lines. Demonstrate how to show edges invisible from view by joining vertices with dotted lines.
- Provide Resource sheet 19: Triangular dot paper (learners may require more than one sheet) and ask learners to sketch the 'house' shape. Explain that the triangular dot paper will help them to form the faces and edges of the shapes more easily.
- Repeat for a square pyramid on top of a cube. Learners attempt their own sketch of the compound shape.
- 🔁 Ask learners to sketch an 'L-shaped' object formed by two cuboids positioned at right angles to one another.
- Discuss the Let's learn section in the Student's Book.
- Introduce the paired activity in the Student's Book. Choose pairs of learners to share their sketches with the class. **[TWM.07]** Invite learners to comment on the drawings and make suggestions as to how they could be improved. For example, adding in edges hidden from view.
- Discuss the Guided practice example in the Student's Book.

Practise 📘

- Workbook

Title: Sketching compound 3D shapes

Pages: 152–153

- Refer to Activity 1 (Variation) from the Additional practice activities.

Apply 🖥

- Display **Slide 2**. Provide each learner with another copy of Resource sheet 19: Triangular dot paper. Give them time to complete their drawing and then choose learners to describe how they sketched the steps.

Review

- Construct a compound 3D shape from two physical component shapes, for example a pyramid and a cuboid. Ask learners to sketch the compound shape using another copy of Resource sheet 19: Triangular dot paper if necessary.

Assessment for learning

- How can you show edges hidden from view when sketching a compound 3D shape?
- A compound shape is formed from a cuboid attached to a cube. Sketch the shape on triangular dot paper.

Same day intervention
Support

- Some learners find it difficult to draw 3D shapes on triangular dot paper. Encourage the use of interlocking cubes or shapes to aid drawings.

Geometry and Measure – Geometrical reasoning, shapes and measurements

Lesson 3: Identifying nets

Learning objective

Code	Learning objective
6Gg.06	Identify [and sketch] different nets for cubes, cuboids, prisms and pyramids.

Resources

squared paper (several sheets per learner); scissors (per learner); cardboard box (per class); cereal box (per learner – they could be asked to bring in their own); sticky tape (per pair); Resource sheet 20: Nets of pyramids, enlarged to A3 if possible (per learner); Resource sheet 21: Nets of prisms, enlarged to A3 if possible (per learner)

Revise

Use the activity *Identify the net* from Unit 19: *3D shapes* in the Revise activities.

Teach [SB] [icons] [TWM.04/05]

- Display the **Nets tool**, showing a cube. Remind learners that a net is a pattern made when the surface of a 3D shape is laid out flat showing each face of the shape. Explain that a 3D shape may have more than one net.

- Demonstrate a net for a closed cube. Explain that the easiest solid shape to represent as a net is a cube, because all the faces are the same. Ask learners to draw a net for a closed cube on squared paper, then cut them out and fold them into a cube. **[TWM.04]** Choose a learner to ask: **Convince me that your pattern of squares is a net for a closed cube.**

- Display **Slide 1**. Ask learners to work in pairs. Provide them with squared paper. Ask them to copy the patterns on the slide and then try to fold each one into a cube. Ask: **Which patterns are successful?** Confirm that each successful pattern is a net of a cube (B, C, E and F).

- Demonstrate carefully unfolding a cardboard box by pulling back the tabs. Fold back the faces and flatten out the card to form a net.

- **[TWM.05]** Discuss the Let's learn section in the Student's Book and introduce the paired activity.

- Say: **We are going to make nets for shapes that have one or more triangular faces.** Hand out Resource sheet 20: Nets of pyramids to learners. Elicit descriptions for square, triangular- and hexagonal-pyramids. **[T&T]** Ask: **What do these nets have in common?** (triangles) **Which face will form the base of the shape when assembled?** (square pyramid: square; hexagonal pyramid: hexagon; tetrahedron: triangle)

- Ask each group to assemble a different solid shape from one of the nets on the resource sheet (so each group makes a different shape). Say: **Starting at a vertex, open up your shape and lay it flat on your table. What happens if you cut/take one of the side faces of your net and reattach it at some other place? Does it form a net that will make the same shape?**

- Repeat the activity for prisms using Resource sheet 21: Nets of prisms. **[TWM.04]** Ask: **What kind of**

information can be obtained from a net of a prism about the solid it creates? What is the relationship between the number of sides on the base and the number of faces on the prism? How does the net of a prism differ from the net of a pyramid?

- [icon] Ask: **What is the difference between a net for a cuboid and a net for a triangular pyramid?** (the net of a cuboid is all rectangles – or two squares and four rectangles; the net of a triangular pyramid is all triangles)

- Discuss the Guided practice example in the Student's Book.

Practise [WB] [TWM.04/06]

- Workbook

Title: Identifying nets

Pages: 154–155

- Refer to Activity 2 from the Additional practice activities.

Apply [icons]

- Display **Slide 2**. Ask learners to identify the 3D shape (hexagonal pyramid).

Review [icon]

- Display the **Nets tool**. Review the construction of nets for prisms and pyramids. Ask: **How are they different?**

Assessment for learning

- Ignoring any tabs, how many folds do I need to assemble a cube from its net? (5)

- To build a triangular prism using two equilateral triangles and six rectangles, what dimensions must be equal?

Same day intervention
Support

- Some learners find it difficult to represent a 3D object as a 2D drawing, particularly when trying to visualise the position, size and shape of the faces that form the shape. Provide activities that involve visualising 3D shapes from 2D shapes, for example recreating a 3D solid using multilink shapes.

Lesson 4: **Sketching nets**

Learning objective

Code	Learning objective
6Gg.06	Identify and sketch different nets for cubes, cuboids, prisms and pyramids.

Resources

squared paper or Resource sheet 17: Square dot paper (several sheets per learner); Resource sheet 19: Triangular dot paper (several sheets per learner)

Revise

Use the activity *Drawing a dice* from Unit 19: *3D shapes* in the Revise activities.

Teach [SB] 🖥

- Display **Slide 1**. Discuss the shapes in turn. Ask learners to identify each shape and name the polygonal or non-polygonal surfaces that comprise the shape.
- Display **Slide 2**. Discuss how to sketch a cuboid on squared paper. First look at the cuboid from all sides. Think about the shape and relative sizes of each of the faces. Begin by sketching all six faces. Then draw them as one complete pattern of shapes to form the net.
- Repeat the process for a second shape, for example a square pyramid drawing on the board the shapes of the faces of the square pyramid and then the net. Encourage learners to help in the process. Ask: **Which polygon forms the shape of the base?** (square) **[T&T] Does it matter what type of triangle forms the 'sides' of the shape?** (yes, they need to be equilateral or isosceles) **Can they be different sizes?** (they must all be identical in size)
- Return to **Slide 1**. Provide learners with squared paper and ask them to choose a shape to sketch. Once completed, choose learners to hold up their sketch and ask the class to name the 3D shape.
- 🖈 Ask learners to sketch the net for a triangular-based pyramid (tetrahedron).
- Discuss the Let's learn section in the Student's Book.
- Introduce the paired activity in the Student's Book, asking learners to sketch the net for an octagonal prism.
- Discuss the Guided practice example in the Student's Book.

Practise [WB] [TWM.07]

- Workbook

Title: Sketching nets

Pages: 156–157

- Refer to Activity 2 (Variation) from the Additional practice activities.

Apply 👥 🖥

- Display **Slide 3** and read the text to the class. Provide learners with square or triangular dot paper. Give them time to sketch a net for the shape. Choose learners to show their sketch to the class.

Review

- Ask learners to work in pairs. Provide square or triangular dot paper. They choose a 3D shape and take turns to draw one face at a time until it is complete.

Assessment for learning

- Sketch the net for a cuboid.
- What should you do before attempting to sketch a 3D shape?

Same day intervention
Support

- Some learners will get most benefit from cutting and folding pre-drawn nets, while others will enjoy drawing their own nets first, which can be done with the help of square or triangular dot paper.

Geometry and Measure – Geometrical reasoning, shapes and measurements

Additional practice activities

Activity 1

Learning objective
• Identify, describe and sketch compound 3D shapes.

Resources
3D shapes: cube, cuboid, sphere, cone, cylinder, pyramids, prisms, hemispheres (per group); Resource sheet 19: Triangular dot paper (per learner) (for variation)

What to do
• Provide learners with a selection of 3D shapes.
• Ask each learner to construct a compound shape from two of the shapes.
• Each learner writes a description of their shape that includes the number of faces, vertices and edges.

• Ask learners to get into groups of five. They display their compound shapes and descriptions on a table, but mixed-up.
• Groups swap tables and try to match the descriptions to the shape.

Variations
2 Learners sketch the 3D compound shape they have constructed on triangular dot paper. On the table, they mix up sketches and descriptions. Groups swap tables to match sketches to their descriptions.

3 Learners construct compound shapes from three or more of the shapes provided.

Activity 2

Learning objective
• Identify and sketch different nets for cubes, cuboids, prisms and pyramids.

Resources
interlocking triangles, squares and hexagons, or any other shape-building equipment, such as straws and sticky putty (per pair); Resource sheet 22: Pyramid properties (per pair); squared paper or Resource sheet 17: Square dot paper or Resource sheet 19: Triangular dot paper (per learner) (for variation)

What to do [TWM.02]
• Ask learners to investigate the relationship between the number of faces and vertices of any pyramid and complete the table on Resource sheet 22: Pyramid properties, by visualising or constructing each shape.
• They use the numbers in the table to find the relationship between the faces and the vertices and then share their results and conclusions. (The number of faces and the number of vertices is the same.)

Variation
2 Learners work in pairs. They each draw the net of a 3D shape on square or triangular dot paper but purposely make one mistake, for example drawing a face an incorrect shape or size, or a face in an incorrect position. Learners swap papers to identify the error. They discuss what the correct net should look like.

Unit 20: Angles

Collins International Primary Maths
Recommended Teaching and
Learning Sequence Term 1, Week 7

Learning objectives

Code	Learning objective
6Gg.09	Classify, estimate, measure and draw angles.
6Gg.10	Know that the sum of the angles in a triangle is 180°, and use this to calculate missing angles in a triangle.

Unit overview

This unit reviews the classification and estimation of angles as acute, right, obtuse or reflex.

Learners are introduced to the correct use of a full circle protractor and use the tool to measure and draw single angles.

By measuring or using paper folding, learners discover that the sum of interior angles in a triangle is 180°. They use this rule to solve missing angle problems in triangles.

Prerequisites for learning

Learners need to:

* know that an acute angle is less than a right angle and is between 0° and 90°
* know that an obtuse angle is less than a straight line, is greater than a right angle, and is between 90° and 180°
* know that a reflex angle is between 180° and 360°
* understand that two right angles equal a straight line
* know the properties of a triangle.

Vocabulary

angle, acute, right, obtuse, straight, reflex, protractor, vertex, benchmark angles, equilateral triangle, isosceles triangle

Common difficulties and remediation

Some learners may find a full-circle protractor difficult to use. Give these learners individual support and practice, confirming that they follow the key points in using a protractor as demonstrated. If available, half-circle protractors or measured angle templates may be preferable. Use of a circular protractor reinforces the relationship between the dynamic and static aspect of angle. Look out for learners who confuse the two scales of the protractor, inner and outer, for example: drawing a 135° angle for the 45° angle, or a 65° angle for the 115° angle.

Discuss acute and obtuse angles with the learners, asking them to consider angle types to determine if the angle drawn matches that required.

Supporting language awareness

Encourage frequent usage of mathematical language to help embed vocabulary. For example, use 'acute' and 'obtuse' alongside phrases 'less than a right angle/90 degrees' and 'greater than 90 degrees but less than 180 degrees/straight angle' as learners make the transition to mathematically precise language.

Encourage learners to seek clarification and confirmation of the mathematical language. This may involve prompting learners to call out and complete definitions, for example, saying 'add to 180 degrees' every time 'angles in a triangle' is mentioned, or praising learners when they ask for terms to be defined. A few times each lesson, ask a volunteer to summarise what has been learned so far so they get experience of explaining concepts in their own words.

How learners interpret mathematical text should also be addressed. Provide frequent opportunities for learners to read terms and phrases out loud and to explain, in their own words, what the text means.

Promoting Thinking and Working Mathematically

Opportunities to develop all four pairs of TWM characteristics are provided throughout the unit.

In Lesson 1, learners characterise (TWM.05) when they identify how 'benchmark' angles can be used to help estimate the size of unknown angles.

In Lesson 3, learners convince (TWM.04) when they determine how a missing angle can be found in a right-angled triangle when one angle (other than the right angle) is given.

Learners specialise (TWM.01) in Lesson 4 when they use their knowledge of the properties of an equilateral triangle and the sum of angles rule to determine the size of each angle in an equilateral triangle.

Success criteria

Learners can:

* use a protractor to measure and estimate angles to the nearest 10°
* draw acute and obtuse angles and use a protractor to measure to the nearest degree
* find unknown angles in triangles using the rule: *The sum of the interior angles of any triangle equals 180°.*

Lesson 1: **Measuring angles**

Learning objective

Code	Learning objective
6Gg.09	Classify, estimate, measure and draw angles.

Resources

ruler (per pair); Resource sheet 23: Angles (per learner); protractors (half-circle and full-circle) (per learner); books, wooden plank or length of stiff plastic, ruler, toy car (per group)

Revise

Use the activity *Book angles* from Unit 20: *Angles* in the Revise activities.

Teach 🔲 📊 🖥 [TWM.05]

- On the board, write: 'acute', 'right', 'obtuse', 'straight' and 'reflex'. Provide learners with rulers. Ask them to work in pairs to draw and label one example of each type of angle. Choose different pairs to share an example of each angle and confirm with the class that the angle size is correct. Ask each pair to define the angle in a sentence, for example, 'an acute angle is between 0° and 90°'.

- Explain the importance of being able to estimate the size of an angle before measuring it. Establish that a good way to estimate the size of angles is to compare them to 'benchmark angles', simple angles that you know the size of visually, such as 180°, 90° and 45°. **[TWM.05] [T&T]** Ask: **How can you use 'benchmark' angles such as these to estimate the size of other angles?** Take responses.

- Display the **Clock tool** showing two o'clock. Ask: **What is the size of the angle formed between the hands at two o'clock?** Take responses and accept a range of answers (acute angle, less than 90°, 60°) **At five o'clock?** (obtuse angle, between 90° and 180°, 150°) **At seven o'clock?** (reflex angle, between 180° and 360°, 210°)

- Display the **Geometry set tool** with the protractor showing. Say: **The scale starts at 0° and ends at 180°. The scale you use depends on the angle measured.** Explain that for obtuse angles, you use the scale with numbers greater than 90; for acute angles, you use the scale with numbers less than 90.

- Demonstrate measuring acute/obtuse internal angles of a shape. Choose a shape and place the centre point of the protractor on the vertex of the angle. Line up the zero line with one arm of the angle, make sure the protractor covers the other arm, and count round to the other arm.

- Ask learners to estimate the size of another angle to the nearest 10°. Discuss estimates, then measure them to the nearest degree with the protractor.

- Display **Slides 1, 2** and **3** in turn. Read the text aloud. Discuss the different parts of the two protractor types, their function and how to use them.

- Distribute a half-circle protractor, a full-circle protractor and Resource Sheet 23 to each learner.

Ask them to alternate between each type of protractor to measure and record the angles. Discuss and confirm the size of each angle as a class.

- 🖐 Ask learners to use a ruler to draw an angle of any size. They estimate the size of the angle and then use a protractor to measure it. How close was their estimate?

- Discuss Let's learn in the Student's Book.

- Introduce the paired activity and discuss and confirm the size of each angle as a class.

- Discuss the Guided practice example.

Practise 🔲

- Workbook

Title: Measuring angles

Pages: 158–159

- Refer to Activity 1 from the Additional practice activities.

Apply 👥 🖥 [TWM.04]

- Arrange learners in groups of three or four.

- Display **Slide 4** and discuss the investigation. Choose different groups to explain their findings and the conclusions they can draw from them. Investigate whether each group conducted the experiment in a similar way.

Review 📊

- Draw three angles on the board using a ruler. Ask learners to estimate the size of each angle. Choose learners to come to the board and measure the angle with a protractor.

Assessment for learning

- Which benchmarks do I use to estimate the size of unknown obtuse angles?

- Explain to a partner how you would use a protractor to measure an obtuse angle.

Same day intervention

Support

- Some learners assume that the size of an angle is proportional to the length of the arms of the angle. Have learners estimate multiple versions of the same angle but with a range of arm lengths.

Lesson 2: **Drawing angles**

Learning objective

Code	Learning objective
6Gg.09	Classify, estimate, measure and draw angles.

Resources

large protractor (per class); protractor (per learner); ruler (per learner); paper (per learner)

Revise

Use the activity *Estimating angles* from Unit 20: *Angles* in the Revise activities.

Teach ▣ ▢

- Display **Slide 1**. Discuss the text and image, talking through the instructions given for drawing an angle of 143°.
- Then say: **I will now show you how to draw an angle of 65°.**
  Step 1: Draw a horizontal line on the board. Label angle A.
  Step 2: Draw a dot at point A – the vertex.
  Line up the zero line of a large protractor ensuring the centre point is on vertex A.
  Step 3: Make a mark at 65°.
  Remove the protractor.
  Step 4: Draw a line with a ruler from A to the 65° mark. Label the angle 65°.
- Provide learners with protractors, rulers and paper.
- On the board, write: P: 114°; Q: 33°; R: 167°; S: 86°. **[T&T]** Ask: **What type of angle is angle P/Q/R/S?** (obtuse, acute, obtuse, acute) Ask learners to work in pairs to draw angles P, Q, R and S.
- Circulate, assessing their methods and accuracy and correcting any errors.
- ▣ Ask learners to use a protractor and ruler to draw an angle of 136°.
- Discuss the Let's learn section in the Student's Book to revise the four steps to drawing an angle.
- Introduce the paired activity in the Student's Book, asking learners to measure and draw the angle. Choose a pair of learners to explain how they worked out the size of the four equal angles (40°).
- Discuss the Guided practice example in the Student's Book.

Practise ▣

- Workbook
  Title: Drawing angles
  Pages: 160–161
- Refer to Activity 1 (Variation) from the Additional practice activities.

Apply ▣▣ ▢

- Display **Slide 2** and read the text to the class. Provide protractors and rulers and ask learners to draw the parallelogram with the angles given. Choose learners to share their diagrams and confirm they are correctly drawn.

Review

- Ask learners to work in pairs. They give each other three angles to draw, one acute, one obtuse and one reflex. Learners then use a protractor and ruler to draw the angles. Their partner measures the angles to confirm they are the correct size.

Assessment for learning

- Describe the four main steps to follow when drawing an angle.
- Draw an angle T that measures 87°.

Same day intervention

Support

- Provide reinforcement on how to use a protractor to draw angles of a given measure. Revise measuring angles. Once a learner is confident with measuring angles using a protractor, have them draw angles of a specified measure.

Geometry and Measure – Geometrical reasoning, shapes and measurements

Unit **20** Angles

Lesson 3: **Calculating missing angles in a triangle (1)**

Learning objective

Code	Learning objective
6Gg.10	Know that the sum of the angles in a triangle is 180°, and use this to calculate missing angles in a triangle.

Resources

protractor (per learner); ruler (per learner); paper (per learner)

Revise

Use the activity *Angles on a straight line* from Unit 20: *Angles* in the Revise activities.

Teach [SB] 🖵 [TWM.04]

- Display **Slide 1**. [T&T] Ask: **What do all of these triangles have in common?** Take responses. Discuss the properties of a triangle. Ask: **Look at the angles in each triangle. What do you notice about them?** Take responses but do not confirm at this stage the rule for angles in a triangle.

- Ask learners to draw three different triangles using a ruler: right-angled, isosceles, scalene. With a protractor, they measure and record the angles of each triangle.

- Ask learners to find the sum of each set of angles. Record ten results on the board. The totals should all be 180° (although there may be slight variation from rounding).

- Ask learners to mark the vertices A, B and C on one of the triangles. Then cut off the three angles of the triangle and arrange the vertices around a point along the edge of a ruler, forming a straight line. Ask: **What is the sum of angles on a straight line?** Elicit the sum of the three angles is 180°.

- On the board, draw a right-angled triangle and label the right angle A, angle B, 35° and the missing angle, C. **[TWM.04]** Ask: **How would you find the size of the missing angle, C?** Elicit that the sum of the angles A and B is 125° (90° + 35°), so angle C is 180° minus 125° = 55°. On the board, write: C = 180° − (A + B).

- Discuss how the expression can be used to calculate the size of any unknown angle in a triangle if two of the angles are known.

- 🗣 Say: **Two angles of a triangle are 48° and 67°. What is the size of the third angle?** (65°)

- Discuss the Let's learn section in the Student's Book.

- Introduce the paired activity in the Student's Book, asking learners to find the missing angle (43°).

- Discuss the Guided practice example in the Student's Book.

Practise [WB] [TWM.05]

- Workbook

Title: Calculating missing angles in a triangle (1)

Pages: 162–63

- Refer to Activity 2 from the Additional practice activities.

Apply 👥 🖵

- Display **Slide 2**. Give learners time to solve the problem and then choose pairs to explain their solution.

Review

- On the board, sketch a triangle with angles labelled A, B and C. Mark angle B with a small square. Ask: **If angle A is 46°, what is the size of angle C?** (44°)

Assessment for learning

- How would you prove that the angles of a triangle always add to 180°?

- In triangle ABC, angle A is 88° and angle B is 76°. What is the size of angle C? (16°)

Same day intervention
Enrichment

- Some learners may wish to write algebraic equations to help them solve missing angle problems. For example, p + 37° + 55° = 180. Provide more challenging problems where known angles are given as multiples of an unknown value, for example, 'The angles in a triangle are 75°, a and 2 × a. What is the value of a?' (35°)

Geometry and Measure – Geometrical reasoning, shapes and measurements

Lesson 4: **Calculating missing angles in a triangle (2)**

Learning objective

Code	Learning objective
6Gg.10	Know that the sum of the angles in a triangle is 180°, and use this to calculate missing angles in a triangle.

Resources

protractor (per pair); ruler (per pair); paper (per pair)

Revise

Use the activity *Guess the triangle* from Unit 20: *Angles* in the Revise activities.

Teach 🆂🅱 🖥 [TWM.01/03/04]

- Display **Slide 1**. Point to the equilateral triangle. **[T&T]** Ask: **What do you know about the angles in an equilateral triangle?** Take responses. Remind learners that the angles are all equal (a = b = c) **[TWM.01]** Ask: **How would you work out the size of each angle in an equilateral triangle?** Take responses. Elicit that the since the sum of the three angles (a + b + c) is 180°, each angle must be 60° (180° ÷ 3).

- Point to the isosceles triangle. Ask: **What do you know about the angles in an isosceles triangle?** Take responses. Remind learners that the two base angles in an isosceles triangle are equal (x = y). **[T&T]** Ask: **If angle z is 40°, what are angles x and y?** Give learners time to work out the answer. Choose a learner to explain their solution. Establish that since the sum of the angles x, y and z is 180° it follows that x + y = 180° − z = 180° − 40° = 140°. Since x = y, it follows that angles x and y are both 70° (140 ÷ 2).

- On the board, draw an isosceles angle ABC with base angles B and C. Label angle B: 37°. Ask learners to find angles A and C (C: 37°; A: 180° − (37° + 37°) = 106°).

- 🗣 Say: **Triangle PQR is an equilateral triangle. What size is angle Q?** (60°)

- Discuss the Let's learn section in the Student's Book.

- **[TWM.03/04]** Introduce the paired activity in the Student's Book, asking learners to work out the missing angle in the diagram. Ask questions to develop learners' understanding of the problem, for example: **What do you know about angles on a straight line? What do you know about the sum of angles in a triangle?**

- Discuss the Guided practice example in the Student's Book.

Practise 🆆🅱 [TWM.05]

- Workbook

Title: Calculating missing angles in a triangle (2)

Pages: 164–165

- Refer to Activity 2 (Variation) from the Additional practice activities.

Apply 👥 🖥

- Display **Slide 2** and read the text to the class. Give learners time to solve the problem and then choose pairs to explain how they found the missing angles.

Review

- Ask pairs of learners to design a flag problem similar to the one in **Slide 2**. The flag should be rectangular and include an isosceles triangle.

- Learners use a protractor to measure and mark the size of one angle and label missing angles with letters. They swap papers with another pair to calculate the missing angles. They return the papers for marking.

Assessment for learning

- What do you know about the angles in an equilateral triangle and an isosceles triangle?
- One base angle of an isosceles triangle is 22°. What are the other two angles?

Same day intervention

Enrichment

- Provide more challenging problems where known angles are given as multiples of an unknown value. For example, one of the base angles in an isosceles triangle has the value a and a non-base angle is 3 × a. What is the value of a? (36°)

Additional practice activities

Activity 1 👤 and 👥👥

Learning objective
• Classify, estimate, measure and draw angles.

Resources
protractor (per learner); ruler (per learner); paper (per learner); book (per learner)

What to do
• Learners use a ruler to draw a square and then several intersecting lines inside the square (see diagram for example).
• They label each of the angles with a letter.
• They construct a table with rows for each letter and columns headed 'estimated size'.
• They estimate and record the size of each angle. Then they measure the angles with a protractor and add the measurements to the table.
• Learners comment on how close their estimates are.

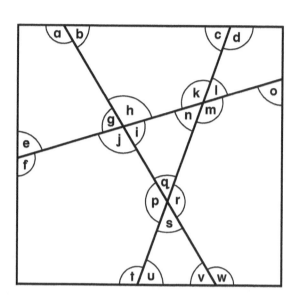

Variations

2 Learners work in teams of four. Learner A lists four angles A–D: two acute and two obtuse, and shares it with Learner B, who uses a ruler to draw and label estimates of angles A–D.

They share them with Learner C, who uses a protractor to measure the angles and record the measurements and then shares them with Learner D.

Learner D uses a protractor to draw the angles and passes the diagrams to Learner A, who uses a protractor to confirm the size of the angles.

The team scores one point for every angle within 10° of the original list and two points if the angle is within 5°.

The team with the highest score wins the game.

Activity 2 👤 and 👥👥 **2**

Learning objective
• Know that the sum of the angles in a triangle is 180°, and use this to calculate missing angles in a triangle.

Resources
Resource sheet 24: Missing angle problems (per pair); paper (per group); ruler (per group)

What to do
• Draw a triangle on the board with angles marked 40°, x° and y°.
• Ask learners to write a formula connecting x and y (x + y + 40 = 180).
• Write possible answers for x and y in the form, 'If x = 50°, then y = 90°'.

Variation

2 Give pairs of learners Resource sheet 24: Missing angle problems.

The aim of the activity is for the pair to turn the drawing into a missing angle problem to give to another pair to solve. They have to work out the minimum number of angles that must be labelled to make it possible to find all the remaining angles.

They label the angles and swap with another pair to solve.

Unit 21: Measurements, including time

Collins International Primary Maths
Recommended Teaching and
Learning Sequence Term 3, Week 8

Learning objectives

Code	Learning objective
6Gt.01	Convert between time intervals expressed as a decimal and in mixed units.
6Gg.05	Understand the difference between capacity and volume.

Unit overview

In this unit, learners convert units of time expressed as decimals and mixed units, focusing on hours and minutes, and minutes and seconds. They understand that time can be represented in decimal format (usually as the result of a calculation) and that it can be converted to hours and minutes, and even parts of a minute. Learners begin with examples that are recognisable and easy to calculate, for example, 3·25 hours is equivalent to $3\frac{1}{4}$ hours or 3 hours and 15 minutes, and then progress to more challenging conversions, for example, 3·7 hours is equivalent to 3 hours and 42 minutes (0·7 × 60 = 42). At this stage, learners do not convert from whole numbers to decimals, for example, 11 seconds into minutes.

They solve problems that involve conversion between units of time to calculate duration. The range of time intervals covered includes smaller units, such as seconds and minutes, minutes to hours.

In Lessons 3 and 4, learners understand the difference between capacity and volume. They read scales to determine the capacity of a container and the volume of liquid contained within.

Prerequisites for learning

Learners need to:
- know the conversion factors for converting between seconds and minutes, and minutes and hours
- know the units for capacity, millilitres and litres
- know how to read a scale.

Vocabulary

second, minute, hour, volume, capacity, millilitre, litre

Common difficulties and remediation

Look out for learners who apply whole number knowledge to time that uses a different base. For example, some learners will calculate the time interval between 10:45 and 11:25 as 80 minutes, not 40 minutes. It is important that learners apply their understanding that there are 60 seconds in one minute and 60 minutes in one hour. This also applies to calculations with larger units of time – days, weeks, months and years – where knowledge of time unit relationships is essential.

Many learners find the distinction between capacity and volume confusing. Help them to understand that

volume is how much space an object takes up and capacity is how much space an object has inside or how much it can hold. For example, a glass may have capacity to hold 350 ml, but the liquid volume of milk inside may only be 150 ml.

Supporting language awareness

Encourage frequent usage of mathematical language to help embed vocabulary. For example, model the use of the question, 'What is the time interval?' when learners ask: 'How much time goes by between…?' as they make the transition to mathematically precise language. At every stage, learners require a mathematical vocabulary to access questions and problem-solving exercises. You may find it useful to refer to the audio glossary on Collins Connect. If appropriate, when a new key word is introduced, ask learners to write a definition in their own words, for example: 'Capacity is the amount a container can hold when filled.' In addition, be aware of terms that have more than one definition, particularly meanings outside mathematics, for example: capacity, level and scale.

Promoting Thinking and Working Mathematically

Opportunities to develop all four pairs of TWM characteristics are provided throughout the unit.

In Lesson 3, learners convince (TWM.04) when they explain how it is possible for the capacity of a container and the volume of water inside to have the same value.

Learners classify (TWM.06) in Lesson 3 when they decide whether statements made about the volume and capacity of containers are sometimes true, always true or never true.

Success criteria

Learners can:
- convert between different units of time, for example: 3·75 minutes equals 3 minutes and 45 seconds
- explain the difference between capacity and volume
- convert between litres and millilitres using knowledge of place value, multiplication and division
- answer questions that involve comparing readings on different scales.

Geometry and Measure – Time; Geometrical reasoning, shapes and measurements

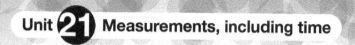
Lesson 1: Converting time intervals (1)

Geometry and Measure – Time: Geometrical reasoning, shapes and measurements

Learning objective

Code	Learning objective
6Gt.01	Convert between time intervals expressed as a decimal and in mixed units.

Resources

paper or mini whiteboard and pen (per learner)

Revise

Use the activity *Ordering units of time* from Unit 21: *Measurements, including time* in the Revise activities.

Teach 🔲 [TWM.06/07]

- On the board, write in a row: seconds, days, weeks, minutes, months, years, hours and write the numbers 60, 24, 7, 12, 10, below. Say: **Like other measurement systems, units of time have a particular relationship with each other. Unlike other measurement systems, time has a much wider range of units.** Point to the numbers on the board. **[TWM.06]** Ask: **Which of these numbers is the number of seconds in a minute?** Accept answers. Say: **Use the numbers on the board to write as many relationships between the time units as you can.** Take answers and record them in order of size, from seconds to years. Include abbreviations: s for seconds, min for minutes and h for hours.

- On the board, write: 1·5 minutes. Remind learners that like any measurement, you can write amounts of time in fractions or decimals. **[T&T]** Ask: **How many minutes and seconds are in 1·5 minutes?** Take responses. Confirm the answer 1 minute, 30 seconds. Choose a learner to explain how they worked out the conversion. Explain and record: $0 \cdot 5 \text{ min} = \frac{1}{2} \text{ min}$. Since there are 60 seconds in 1 min, $\frac{1}{2}$ min = 30 s. Therefore, 1·5 minutes is equal to 1 min 30 s.

- Discuss the Let's learn section in the Student's Book.

- On the board, write: 3·25 h = __ h __ min. Ask learners to convert 3·25 hours to hours and minutes. Take responses and confirm the answer: 3 h 15 min. Work through the problem with learners helping. Ask: **What is 0·25 as a fraction?** $(\frac{1}{4})$ **How many minutes are in 1 hour?** (60) **What is a quarter of 60?** (15) **How many hours and minutes is that?** (3 hours, 15 minutes)

- On the board, write: 5·75 h = __ h __ min and work through the problem with learners, establishing that $0 \cdot 75 \text{ h} = \frac{3}{4} \times 60 \text{ min} = 45 \text{ min}$. (5 h 45 min)

- Write some more conversion problems for learners to solve: 8·5 minutes to minutes and seconds, and 6·25 hours to hours and minutes. (8 min 30 s, 6 h 15 m)

- 📄 Say: **It takes Grace 12·75 minutes to walk to the bus stop. How much time is that in minutes and seconds?** (12 m 45 s)

- **[TWM.07]** Introduce the paired activity in the Student's Book. Ask a pair of learners to explain the error that Mandisa has made. Ask: **What would you say to Mandisa to correct her thinking?**

- Discuss the Guided practice example in the Student's Book.

Practise 🔳

- Workbook

Title: Converting time intervals (1)

Pages: 166–167

- Refer to Activity 1 from the Additional practice activities.

Apply 👥 🖥 [TWM.08]

- Display **Slide 1**. Give learners time to discuss the questions on the slide. Choose pairs to respond to the questions and ask other learners to say whether they agree or not.

Review

- On the board, write the following times in a row: 7 h 25 min, 7 h 45 m, 7 h 75 min, 7 h 15 min, 7 h 20 min, 7 h 30 min.

- In a row below, write 7·25 h, 7·5 h, 7·75 h.

- Ask learners to copy the time intervals and match numbers in the bottom row to those in the top row (there are some 'red herrings'). (7·25 h = 7 h 15 min; 7·5 h = 7 h 30 min; 7·75 h = 7h 45 min)

Assessment for learning

- Show me how you would convert 9·75 min to minutes and seconds. (9 min 45 s)

- How many minutes longer is 3·5 h than 2·75 h? (45 min)

Same day intervention

Enrichment

- Expand the range of questions to include conversions between days and hours, and years and months. For example:

  - **How many hours are there in 0·5/1·5/2·25 days?**

  - **How many months are there in 0·75/1·25/3·75 years?**

Lesson 2: **Converting time intervals (2)**

Learning objective

Code	Learning objective
6Gt.01	Convert between time intervals expressed as a decimal and in mixed units.

Resources

paper or mini whiteboard and pen (per learner)

Revise

Use the activity *Train arrivals* from Unit 21: *Measurements, including time* in the Revise activities.

Teach [SB] [TWM.07]

- On the board, write: 1·1 minutes. Ask: **How many minutes and seconds are in 1·1 minutes?** Take responses. Confirm the answer 1 minute, 6 seconds. Choose a learner to explain how they worked out the conversion. Explain and record: 0·1 min = $\frac{1}{10}$ min. Since there are 60 seconds in 1 min, $\frac{1}{10}$ min = 60 ÷ 10 = 6 s. Therefore, 1·1 minutes is equal to 1 min 6 s.
- Discuss the Let's learn section in the Student's Book.
- On the board, write: 4·3 h = __ h __ min. Ask learners to convert 4·3 hours to hours and minutes. Take responses and confirm the answer: 4 h 18 min. Work through the problem with learners helping. **[T&T]** Ask: **What is 0·3 as a fraction?** ($\frac{3}{10}$) **How many minutes are in 1 hour?** (60) **What is a tenth of 60?** (6) **What is three-tenths?** (3 × 6 = 18) **How many hours and minutes is that?** (3 hours, 18 minutes)
- On the board, write: 4·9 h = __ h __ min and work through the problem with the learners, establishing that 0·9 h = 0·9 × 60 min = 54 min. (4 h 54 min)
- Write some more conversion problems for learners to solve: 5·6 minutes to minutes and seconds, and 9·9 hours to hours and minutes. (5 min 36 s, 9 h 54 m)
- Write on the board: 3 min 4 s, 3 min 24 s, 3 min 40 s. Ask: **Which of these time intervals is equivalent to 3·4 hours?** Take responses. Confirm: 3 min 24 s. Ask: **Why is it not 3 min 40 s?**
- ⏸ Say: **On Tuesday, Logan's bus journey to school took 15·8 minutes. How long was his journey in minutes and seconds?** (15 min 48 s)
- **[TWM.07]** Introduce the paired activity in the Student's Book. Ask a pair of learners to explain the error that Danny has made. Ask: **What would you say to Danny to correct his thinking?**
- Discuss the Guided practice example in the Student's Book.

Practise [WB]

- Workbook

Title: Converting time intervals (2)

Pages: 168–169

- Refer to Activity 1 (Variation) from the Additional practice activities.

Apply 👥 🖥

- Display **Slide 1**. Discuss the problem. Ask a pair of learners to explain how they worked out the total amount of time Arthur spends on the two buses.

Review

- On the board, write the following times in a row: 5 h 24 min, 5 h 10 min, 5 h 20 min, 5 h 6 min, 5 h 40 min, 5 h 12 min.
- In a row below, write 5·2 h, 5·4 h, 5·1 h.
- Ask learners to copy the time intervals and match numbers in the bottom row to those in the top row (there are some 'red herrings'). (5·2 h = 5 h 12 min; 5·4 h = 5 h 24 min; 5·1 h = 5 h 6 min)

Assessment for learning

- Show me how you would convert 12·8 min to minutes and seconds. (12 min 48 s)
- How many seconds longer is 7·3 min than 6·4 min? (54 s)

Same day intervention
Enrichment

- Expand the range of questions to include conversions between days and hours, and years and months. For example:
  - **How many hours are there in 0·1/0·7/2·3 days?**
  - **How many months are there in 0·1/0·9/2·7 years?**

Unit **21** **Measurements, including time**

Lesson 3: **Capacity and volume (1)**

Learning objective

Code	Learning objective
6Gg.05	Understand the difference between capacity and volume.

Resources

measuring containers: cylinder, jug, etc. (per class/group); straight-sided plastic containers of various sizes (per class/group); paper (per learner)

Revise

Use the activity *Capacity conversions* from Unit 21: *Measurements, including time* in the Revise activities.

Teach 🔲 🖥 [TWM.04/05/06]

- Display and discuss **Slide 1**, then **Slide 2**.
- Hold up a plastic container and ask learners to estimate its capacity. Take responses. **[T&T] [TWM.04]** Ask: **How would you measure the capacity of this container?** Elicit that measuring the amount of water it takes to fill the container will give the capacity. Fill the container with water and then pour it in a measuring jug or cylinder.
- Choose a learner to read the scale. Say: **The container has a capacity of [x] millilitres**. Next, fill the container three quarters full. **[TWM.05]** Ask: **Have I filled the container to its capacity?** (no) **How do you know?** (the container is not full) **How would you describe the amount of water in the container?** Take responses. Pour the water into a measuring container and ask a learner to read the scale. Record the measurement ([y] millilitres) and then pour the water back into the plastic container. Ask: **Which is the correct sentence to use: 'The capacity of the water in the container is [y] millilitres' or 'The volume of water in the container is [y] millilitres'?** (volume of water) **Why?** Establish that you are now talking about an amount of water in the container and not the capacity of the container which is a measure of the maximum amount it *can* hold.
- **[TWM.04]** Ask: **Is it possible for the capacity of a container and the volume of water it holds to have the same value?** Take responses. Establish that this would be true for a container that is full.
- Organise learners into groups working at their table. Provide each table with plastic containers and measuring containers. Working in pairs, learners choose a plastic container and use the method previously described to find its capacity. Ask them to sketch the container and write a sentence below to describe the capacity of the container: 'The capacity of the container is [x] millilitres.' Ask learners to fill their containers with different volumes of water, for example, 300 ml, 600 ml. They draw a sketch of the container and mark the level of water. Below, they write a sentence to describe the volume of water.

- 🗣 Say: **I have a plastic cylinder that is labelled 1000 ml. I fill it halfway. Give me two sentences, one to describe the capacity of the container and another to describe the volume of water in the container.**
- Discuss the Let's learn section in the Student's Book.
- **[TWM.06]** Introduce the paired activity in the Student's Book.
- Discuss the Guided practice example.

Practise 🔲 [TWM.06]

- Workbook

Title: Capacity and volume (1)

Pages: 170–171

- Refer to Activity 2 from the Additional practice activities.

Apply 👥 🖥

- Display **Slide 3**.

Review

- Provide each group with a set of straight-sided plastic containers and measuring cylinders. With learners working in pairs, ask them to find the capacity of one container and then say the volume of water it would contain if it was half full.

Assessment for learning

- What is the difference between capacity and volume?
- It takes 1200 ml of water to fill a container. What volume of water should I pour into the container to fill it a quarter full? (300 ml)

Same day intervention
Support

- Look out for learners who make errors when reading capacity scales. Provide learners with opportunities to use and read a range of measuring scales, some at different orientations.
- Remind learners to keep containers on a flat surface and look at the base of the liquid meniscus at eye level.

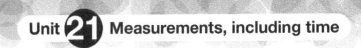

Lesson 4: **Capacity and volume (2)**

Learning objective

Code	Learning objective
6Gg.05	Understand the difference between capacity and volume.

Resources

paper, for the Workbook (per learner); measuring containers: cylinder, jug, etc. (per pair); straight-sided plastic containers of various sizes (per pair); paper (per pair)

Revise

Use the activity *Find the volume* from Unit 21: *Measurements, including time* in the Revise activities.

Teach SB

- Display the **Capacity tool** in two windows, both with the 2000 ml measuring bottle set to litres. Pour 1·7 litres in one bottle and 0·9 litres in the other. **[T&T]** Ask: **What is the combined volume of water in the bottles?** (2·6 ℓ) Ask: **How did you work this out?** Expect learners to describe a mental method, for example, partitioning, or place value and known facts.
- Now display 1000 ml measuring jugs, set to litres. Pour 0·825 ℓ in one bottle and 0·75 ℓ in the other. Say: **The volumes have different numbers of decimal places.** Ask: **What should we do before we add them?** (rewrite 0·75 as 0·750 by inserting a trailing zero) Ask learners to add the two volumes and explain their solution. Expect learners to solve using mental or written methods (1·575 ℓ).
- Introduce subtraction problems, for example: **What is the difference in volume between container A (3 ℓ 400 ml) and container B (0·625 ℓ)?** (2·775 ℓ)
- Introduce multiplication and division problems, for example: **I have a cup with a capacity of 0·35 ℓ. What would be the total capacity of 10 cups?** (3·5 ℓ) **How many cups could you fill with 7 litres of liquid?** (20)
- 🔁 Say: **Lucy pours 0·645 litres of orange juice into a container followed by 0·53 litres of apple juice. How much juice is in the container?** (1·175 ℓ)
- Discuss the Let's learn section in the Student's Book.
- Introduce the paired activity in the Student's Book. Ask a pair of learners to explain how they found the total volume of water in the two containers.
- Discuss the Guided practice example in the Student's Book.

Practise WB

- Workbook

Title: Capacity and volume (2)

Pages: 172–173

- Refer to Activity 2 (Variation) from the Additional practice activities.

Apply 👥

- Ask pairs of learners to find two pairs of containers that they estimate to have similar combined capacities. They measure them and calculate the difference between the combined capacities. Each pair of learners writes about their results. Ask: **How close were you?**

Review

- Say: **The combined capacity of seven identical plastic bottles is 12·6 ℓ.** Ask: **What is the capacity of one bottle?** (1·8 ℓ)

Assessment for learning

- Two buckets contain 5·47 ℓ and 2 ℓ 385 ml of water. I need to work out the total volume. What should I do first? Why? (7·855 ℓ)
- What is the difference between a capacity of 8·23 ℓ and a capacity of 3·567 ℓ? (4·663 ℓ)

Same day intervention

Support

- Some learners believe the volume of liquid has changed when a set amount is poured from one container to another of different size. They believe that the container with the higher level holds more. Provide learners with extended opportunities to transfer liquids from one container to another.

Additional practice activities

Activity 1

Learning objective
• Convert between time intervals expressed as a decimal and in mixed units.

Resources
1–10 number cards from Resource sheet 2: 0–100 number cards (per pair); Resource sheet 25: Fraction cards (per pair)

What to do [TWM.04]
• Learners shuffle the whole-number cards and place them in a pile face down. They do the same for the three fraction cards: $\frac{1}{2}$, $\frac{1}{4}$ and $\frac{3}{4}$.
• They take turns to turn over one card from each pile to make a mixed number of hours, for example, a '2' and a '$\frac{3}{4}$' would make $2\frac{3}{4}$ hours.

• Learners race against each other to be the first to call out the number converted to hours and minutes, for example, $2\frac{3}{4}$ hours would be converted to 2 hours 45 minutes. The first learner to do so scores a point.
• The winner is the player with the higher score after seven rounds.

Variations
2 Repeat the activity with learners using the tenths cards from the resource sheet as well as the $\frac{1}{2}$, $\frac{1}{4}$ and $\frac{3}{4}$.

3 Extend the range of fraction cards used to include thirds, fifths and tenths.

Activity 2

Learning objective
• Understand the difference between capacity and volume.

Resources
0–9 digit cards from Resource sheet 2: 0–100 number cards (per pair) (for variation); paper (per learner)

What to do
• Learners take turns to describe the volume of water in a jug to their partner, for example: 'When my jug is full, the volume of water in it is 600 ml. I fill the jug to half full.'
• Their partner responds by stating the capacity of the jug and the volume of water in the container, 'Your jug has a capacity of 600 ml. The volume of water in the jug is 300 ml.'
• Other descriptions could include: 'When my jug is full, the volume of water in it is 200 ml. I fill the jug to quarter full.' The required response would be: 'Your jug has a capacity of 200 ml. The volume of water in the jug is 50 ml.'

Variation
2 Learners take turns to use the 0–9 digit cards to represent the volume of water in each of two containers, A and B:

A: a number less than 10 with two decimal places (the number of litres in A, e.g. 2·78 ℓ)

B: a one-digit number that represents the number of litres and a three-digit number that represents the number of millilitres in B, e.g. 4 ℓ 156 ml.

Learners then calculate the total volume of water in both containers by finding the sum of A and B.

Geometry and Measure – Time: Geometrical reasoning, shapes and measurements

Unit 22: Area and surface area

Collins International Primary Maths
Recommended Teaching and
Learning Sequence Term 1, Week 9

Learning objectives

Code	Learning objective
6Gg.03	Use knowledge of area of rectangles to estimate and calculate the area of right-angled triangles.
6Gg.07	Understand the relationship between area of 2D shapes and surface area of 3D shapes.

Unit overview

In this unit, learners discover the area of a right-angled triangle as half the area of a rectangle. By bisecting a rectangle into two identical right-angled triangles, learners recognise this equivalence.

Using the formula for area of a rectangle, they derive the area of a right-angled triangle. The terms 'base' and 'perpendicular height' in relation to the triangle are introduced and learners practise finding the area of a right-angled triangle, given the base and height.

In Lessons 3 and 4, learners are introduced to the concept of surface area of 3D shapes. The surface area of a three-dimensional object is the measure of the total area of all its faces. This means that one way to find the surface area of a solid is to find the area of its net.

At this stage, learners are not expected to calculate surface area. This is introduced in Stage 7.

Prerequisites for learning

Learners need to:
- know the properties of triangles and rectangles
- understand how to find the area of a square or rectangle
- know how to draw a net for a cube and a cuboid.

Vocabulary

area, right-angled triangle, area, surface area

Common difficulties and remediation

Some learners may have difficulty with the concept of surface area. To support understanding of the term, provide learners with activities where they cover the surface area of a range of boxes representing cuboids. Learners use wrapping/brown paper to cover all the sides of the box. They should understand that the area of the wrapping paper used to cover the box is equivalent to the surface area of the box. This can be further supported by asking learners to take apart cuboidal packaging, such as cereal boxes. They reverse the process by taking a flat piece of cardboard and folding it up into a rectangular prism or box.

Supporting language awareness

At every stage, learners require a mathematical vocabulary to access questions and problem-solving exercises. You may find it useful to refer to the online glossary at Collins Connect. If appropriate, when a new key word is introduced, ask learners to write a definition in their own words, for example: 'Surface area is the combined area of all the 2D faces of a 3D shape' or 'It is the area you would need to paint in order to cover the solid.'

Promoting Thinking and Working Mathematically

Opportunities to develop all four pairs of TWM characteristics are provided throughout the unit.

In Lesson 1, learners conjecture (TWM.03) when they suggest how a rectangle can be turned into two identical right-angled triangles.

In Lesson 2, learners classify (TWM.06) when they investigate whether it is possible to switch the values for base and height when calculating the area of a right-angled triangle and still get the same answer.

In Lesson 4, learners are asked to make conjectures (TWM.03) about finding the surface areas of a cube and cuboid.

Success criteria

Learners can:
- recognise the area of a triangle as half the area of a related parallelogram by bisecting a rectangle into two congruent right-angled triangles
- identify the height of a triangle as the perpendicular distance from the base to the apex
- find the area of a right-angled triangle given its base and height
- solve simple contextual problems that involve finding the area of a triangle
- explain where the surface is on a 3D object
- use nets to understand the relationship between area of 2D shapes and surface area of 3D shapes.

Geometry and Measure – Geometrical reasoning, shapes and measurements

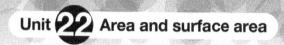
Lesson 1: **Calculating the area of a triangle (1)**

Geometry and Measure – Geometrical reasoning, shapes and measurements

Learning objective

Code	Learning objective
6Gg.03	Use knowledge of area of rectangles to estimate and calculate the area of right-angled triangles.

Resources

squared paper or Resource sheet 17: Square dot paper (several sheets per learner); coloured pencil (per learner); scissors (per learner); ruler (per learner)

Revise

Use the activity *Double the size, half the size* from Unit 22: *Area and surface area* in the Revise activities.

Teach SB [TWM.02/03]

- Provide each learner with squared paper, a ruler and a coloured pencil. Ask them to draw a rectangle and calculate and write down the area of the shape. Have them pick any three corners of their rectangle, draw lines between them and colour in the triangle formed. Ask learners to estimate and record the area of their triangle by counting the squares inside.
- [T&T] [TWM.02] Ask learners to discuss and compare their answers. Ask: **What do you notice about the rectangles and triangles you have drawn and the areas you recorded?** Expect learners to make comments similar to the following:
  - all the triangles are right-angled triangles
  - the triangles cut the rectangles in half
  - some areas are whole numbers, but some are not
  - the area of the triangles is always smaller than the area of the rectangles
  - the area of the triangles is (approximately) half of the area of the rectangles.
- Ask learners to work in pairs. Provide each pair with another sheet of squared paper and scissors. Ask them to draw a rectangle with its base on a horizontal gridline. [TWM.03] Set this challenge to learners: **How do you turn a rectangle into two identical right-angled triangles?** Give learners time to solve the problem and then ask: **Who has found a solution to the problem?** Choose a pair to demonstrate their method. If learners have not found an answer, hint that they should cut out the rectangle, fold it along the diagonal and then cut along the fold. Establish that cutting along one of the diagonals of the rectangle splits the shape into two identical right-angled triangles. Ask them to confirm this by turning one of the triangles and overlaying it on the other triangle.
- Ask: **What does this tell you about the area of each triangle?** Take responses. Confirm that it is half the area of the related rectangle.
- 🗣 Say: **You are presented with a drawing of a right-angled triangle. How would you change the shape so that it has double the area?**

- Discuss the Let's learn section in the Student's Book.
- Introduce the paired activity in the Student's Book, asking learners to work out the area of the yellow triangle.
- Discuss the Guided practice example in the Student's Book.

Practise WB

- Workbook

Title: Calculating the area of a triangle (1)

Pages: 174–175

- Refer to Activity 1 from the Additional practice activities.

Apply 👥 🖥

- Display **Slide 1**. Learners use the rule that the area of a right-angled triangle is half the area of a rectangle to work out the area of the triangular lawn.

Review

- Provide learners with another sheet of squared paper and ask them to draw a rectangle 8 units by 10 units. Next, have them draw a diagonal to split the rectangle into two right-angled triangles. Ask: **What is the area of the rectangle?** (80 square units) **What is the area of each triangle?** (40 square units) Choose a learner to explain how they found the area of each triangle.

Assessment for learning

- Show me that the area of a right-angled triangle is half the area of a related rectangle.
- Draw a right-angled triangle 6 units by 8 units on squared paper. What is its area? (24 square units)

Same day intervention
Support

- Some learners may find it difficult to count the squares inside the border of a shape drawn on a grid. Show them a counting technique where fractions of squares are rounded to 0, 0·5 or 1 depending on the portion of the square covered by the interior of the shape.

Lesson 2: **Calculating the area of a triangle (2)**

Learning objective

Code	Learning objective
6Gg.03	Use knowledge of area of rectangles to estimate and calculate the area of right-angled triangles.

Resources

set square (per class) paper (optional), for the Workbook (per learner)

Revise

Use the activity *Field areas* from Unit 22: *Area and surface area* in the Revise activities.

Teach [SB] 🖵 [TWM.06]

- Display **Slide 1**. Point to the first diagram. Discuss the base and height of a right-angled triangle. Explain that the base and the height of a right-angled triangle are the two sides that form the right angle.
- Point to the second diagram and explain that as both right-angled triangles share the same base (b) and height (h) as the parent rectangle, the area of each triangle can be found by halving the area of the rectangle, that is half the product of the base and height.
- Ask: **What is the length of the base of the triangle?** (5 units) **What is the height?** (3 units) **How do you know?** (the base and height are the sides that form the right angle) **What is the area of the triangle?** Take responses. Choose a learner to come to the board to demonstrate the calculation that will give the area. Expect that they find the area of the related rectangle and halve this value (half of $5 \times 3 = 7 \cdot 5$ square units).
- **[T&T] [TWM.06]** Ask: **Does it matter if you decide to switch base and height? Why?** Take responses. Elicit that since multiplication is a commutative operation, the area will be the same whether you multiply base × height or height × base. This means that for a right-angled triangle, base and height can be switched around.
- Draw a right-angled triangle on the board. Label the base 6 cm and the height 10 cm. Ask: **What is the area of the triangle?** (30 cm^2) **How do you know?** (I found the area of the rectangle 60 cm^2 and then divided it by 2 to give 30 cm^2)
- 🗩 Say: **I draw a rectangle 9 cm by 6 cm and split it into two triangles by drawing the diagonal of the rectangle.** Ask: **What is the area of each triangle?** (27 cm^2)
- Discuss the Let's learn section in the Student's Book.
- Introduce the paired activity in the Student's Book. Choose pairs of learners to share how they calculated the area of each triangle.

- Discuss the Guided practice example in the Student's Book.

Practise [WB]

- Workbook

Title: Calculating the area of a triangle (2)

Pages: 176–177

- Refer to Activity 1 (Variation) from the Additional practice activities.

Apply

- Display **Slide 2**. Ensure learners understand the problem. Remind them that the area of a right-angled triangle can be found by finding the area of a related rectangle and halving the value. Choose a pair of learners to explain how they solved the problem.

Review 📊

- Display the **Geoboard tool**. Draw a right-angled triangle 4 units by 5 units. Ask learners to work out the area of the triangle by visualising a related rectangle and finding half its area (10 square units).
- Repeat for two more right-angled triangles with different dimensions.

Assessment for learning

- What is the area of a rectangle 12 cm by 6 cm? (72 cm^2)
- A rectangle 8 cm by 4 cm is split into two identical right-angled triangles. What is the area of each triangle? (16 cm^2)

Same day intervention

Enrichment

- Ask learners to write a formula for the area of a triangle. Have them use it to calculate the area of a triangle with base 16 cm and height 18 cm (144 cm^2).

Unit **22** **Area and surface area**

Lesson 3: **Surface area (1)**

Learning objective

Code	Learning objective
6Gg.07	Understand the relationship between area of 2D shapes and surface area of 3D shapes.

Resources

large cube (per class); squared paper or Resource sheet 17: Square dot paper (per learner); ruler (per learner); scissors (per learner); empty cereal box (per learner) (learners could be asked to bring in their own); coloured pencils, for the Workbook (per learner)

Revise

Use the activity *Guess the net* from Unit 22: *Area and surface area* in the Revise activities.

Teach SB

- Hold up a large cube and ask: **Are all the faces of a cube the same or different?** (the same) **What shape are all the faces of a cube?** (square) **How many faces does a cube have?** (6)
- Introduce the term 'surface area' as the sum of all the areas of all the shapes that cover the surface of a 3D shape.
- Using the **Nets tool** set to the 'cube' video and pause play at the point where the net is displayed.
- Discuss the faces of a cube and its net, and how the area of each face (2D shape) is the same.
- Provide each learner with squared paper and ask them to sketch the net of a cuboid. If they need help, then display the **Nets tool** set to the 'cuboid' video and pause play at the point where the net is displayed. Ask learners to cut out their net and fold it into the shape. Ask: **How would you find the surface area of the cuboid?** Take responses.
- Display **Slide 1**. Establish that for a cube and a cuboid, the surface area is the sum of the areas of the six faces.
- If not already discussed, remind learners that all the faces of a cube are identical. Ask: **What can you tell me about the faces of a cuboid?** Praise any learner who identifies that opposite faces of a cuboid are identical.
- Discuss the Let's learn section in the Student's Book.
- Introduce the paired activity in the Student's Book.
- Discuss the Guided practice example.

Practise WB

- Workbook

Title: Surface area (1)

Pages: 178–179

- Refer to Activity 2 from the Additional practice activities.

Apply

- Display **Slide 2**. Learners talk with their partner about the different 2D shapes that cover the surface of this cereal box and discuss how they would find the surface area of the box.

Review

- Ask: **What have you learned about finding the surface area of a cube and of a cuboid?** Ask pairs to discuss how the surface area of a cube will relate to the surface area of its net and to share this with the class. Take feedback and elicit that both will have the same area.

Assessment for learning

- How many faces does this 3D shape have?
- What are the different 2D shapes that cover the surface of this 3D shape?
- How many of each 2D shape are there?
- What does the term 'surface area' mean?

Same day intervention

Support

- Provide lots of opportunity for learners to open the nets of cubes and different cuboids in order to identify the different 2D shapes.

240

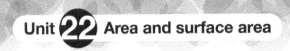
Lesson 4: **Surface area (2)**

Learning objective

Code	Learning objective
6Gg.07	Understand the relationship between area of 2D shapes and surface area of 3D shapes.

Resources

squared paper or Resource sheet 17: Square dot paper (several sheets per learner); ruler (per learner); coloured pencils, for the Workbook (per learner)

Revise

Use the activity *Guess the net* from Unit 22: *Area and surface area* in the Revise activities.

Teach 📖 🖥

- Display **Slide 1**. Explain that the box is a cube. **[T&T] How would you work out the surface area of the box?** Take responses. Confirm that since each face is a square and a cube has six identical faces, then the surface area can be found by multiplying the surface area of one face by six.
- Display Slide 2. Discuss the cuboid and its net.
- Point to a face on the box and ask them to identify the corresponding face on their net. Confirm they have located the correct rectangle and shade the face blue. Ask: **Is this the only rectangle that has these dimensions?** Take responses. Elicit that since the opposite faces of a cuboid are identical, you also know the measurements of a corresponding rectangle in the net. Ask learners to identify the corresponding rectangle on the net and also shade this face blue.
- Repeat for the remaining two pairs of opposite faces, colouring each pair of identical faces the same colour.
- Discuss the Let's learn section in the Student's Book.
- Introduce the paired activity in the Student's Book.
- Discuss the Guided practice example in the Student's Book.

Practise 📒 [TWM.04]

- Workbook

Title: Surface area (2)

Pages: 180–181

- Refer to Activity 2 (Variation) from the Additional practice activities.

Apply 👥 🖥

- Display **Slide 3** and read the text to the class. Ask pairs to show their sketches of other shapes with the same surface area.

Review

- Conclude by reviewing the concept of surface with the class. Ask questions such as: **What is the same about a cube and a cuboid? What is different? What does the term 'surface area' refer to? What is the relationship between area of 2D shapes and surface area of 3D shapes?**

Assessment for learning

- What can you tell me about the faces of a cube/cuboid?
- How does knowing about the faces on a cube/cuboid help you to understand surface area?

Same day intervention
Support

- Provide lots of opportunity for learners to open the nets of cubes and different cuboids in order to identify the different 2D shapes and also develop an understanding that opposite faces of a cuboid are identical.

Geometry and Measure – Geometrical reasoning, shapes and measurements

Additional practice activities

Activity 1 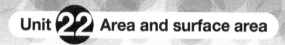 and 2

Learning objective
• Use knowledge of area of rectangles to estimate and calculate the area of right-angled triangles.

Resources
squared paper or Resource sheet 17: Square dot paper (per learner); ruler (per learner)

What to do
• Each learner draws three different rectangles on squared paper. They divide the rectangle into two right-angled triangles by marking a diagonal.
• They work out the area of the rectangle and then halve this value to find the area.
• Learners then check to confirm the triangle is half the area of the related rectangle by counting squares in one of the triangles.

Variation
2 Each learner draws four rectangles. They mark one diagonal in each shape to divide the rectangle into two identical right-angled triangles. They label the sides of each rectangle with the length and the width (equivalent to the base and height of the triangles). Learners swap papers with their partner to calculate the area of each triangle. They return their papers for marking.

Activity 2 2

Learning objective
• Understand the relationship between area of 2D shapes and surface area of 3D shapes.

Resources
Resource sheet 19: Triangular dot paper (per learner); Resource sheet 17: Square dot paper (per learner); ruler (per learner); coloured pencils (per pair) (for the variation)

What to do
• Each learner sketches a cuboid on triangular dot paper.
• Learners then swap papers with their partner to draw a net of the cuboid on square dot paper.
• When both learners have finished they compare and discuss each sketch and its net.

Variation
2 Working together, learners look at each sketch and it's corresponding net drawn in 'What to do', and use different coloured pencils to shade on the net faces with the same area.

Unit 23: Coordinates

Learning objectives

Code	Learning objective
6Gp.01	Read and plot coordinates including integers, fractions and decimals, in all four quadrants (with the aid of a grid).
6Gp.02	Use knowledge of 2D shapes and coordinates to plot points to form lines and shapes in all four quadrants.

Unit overview

In this unit, learners extend their knowledge of the coordinate system to reading and plotting coordinates in all four quadrants. They are introduced to coordinates that include integers, fractions and decimals, and examine the signs (+, –) of x- and y-coordinates in all four quadrants.

In Lessons 3 and 4, they use knowledge of 2D shapes and coordinates to plot points to form lines and shapes in all four quadrants. Activities include finding the missing vertices of squares, rectangles and triangles.

Prerequisites for learning

Learners need to:
- be able to apply their knowledge of coordinates to read and plot points in the first quadrant of the coordinate plane
- be able to apply their knowledge of polygons to locate the position of a missing vertex in the first quadrant of the coordinate plane and complete the shape.

Vocabulary

x-axis, y-axis, coordinates, quadrant, vertex (vertices), line, coordinate plane, origin

Common difficulties and remediation

Learners may make the assumption that coordinates only appear in the first quadrant, so it is important to give them opportunities to work with coordinate points in all four quadrants. For this, learners will need to be secure in their understanding of negative integers so that they can find points such as (–4, –3).

As the x- and y-axes are extended to include negative integers, help learners identify the point where the axes intersect as the origin. In addition, help them to recognise the four quadrants and how to identify which quadrant an ordered pair should be plotted in, based on the signs of the coordinates: Quadrant I (+, +), Quadrant II (–, +), Quadrant III (–, –), and Quadrant IV (+, –).

Look out for learners who have difficulty with the zero coordinates of points on the axes, especially when plotting. For example, some learners may read the point (–6, 0) as (–6, 1) or leave the y-coordinate blank because there is no movement involved along the y-axis. Learners making these errors require extended practice at writing ordered pairs of coordinates for points plotted on both axes.

Supporting language awareness

'Axis' and 'axes' are two words that are often confused. As homophones, they are pronounced in the same way but are spelt differently and have different meanings in the context of geometry: 'axis' as in a fixed line on a graph used to show the position of a point and 'axis' as in a real or imaginary straight line going through the centre of an object that is spinning (such as the Earth).

The word 'axes' has two different pronunciations dependent on its meaning. When pronounced with a 'short' e, axes is the plural form of the tool 'axe'. When pronounced with a 'long' e, axes is the plural form of the geometry term 'axis'.

Promoting Thinking and Working Mathematically

Opportunities to develop all four pairs of TWM characteristics are provided throughout the unit.

In Lesson 1, learners generalise (TWM.02) when they identify what points plotted in the same quadrant of the coordinate plane have in common.

In Lesson 3, learners critique and improve (TWM.07) when they suggest methods for finding the missing vertex of a shape drawn on the coordinate plane.

Learners conjecture (TWM.03) in Lesson 4 when they find ways to describe points on a line on the coordinate plane.

Success criteria

Learners can:
- apply their knowledge of coordinates to read and plot points in all four quadrants of the coordinate plane
- read and plot points in all four quadrants described by ordered pairs of coordinates that include negative integers, fractions and decimals
- apply their knowledge of polygons to locate the position of a missing vertex in any of the four quadrants of the coordinate plane and complete the shape.

Geometry and Measure – Position and transformation

Lesson 1: **Reading and plotting coordinates (1)**

Geometry and Measure – Position and transformation

Learning objective

Code	Learning objective
6Gp.01	Read and plot coordinates including integers, fractions and decimals, in all four quadrants (with the aid of a grid).

Resources

Resource sheet 26: Four quadrant grid (per learner)

Revise

Use the activity *Kick the ball* from Unit 23: *Coordinates* in the Revise activities.

Teach 📖 💻 📊 [TWM.02]

- Display **Slide 1**. Explain that the axes of the coordinate plane extend beyond zero to include negative numbers. This divides the plane into quarters, called quadrants. We number them from one to four, anticlockwise from the upper right quadrant. Explain that the quadrants are usually numbered using Roman numerals.
- Display the **Coordinates tool** set to one quadrant. On the board, write: (3, 4). Ask: **What do the numbers in brackets tell you?** Remind learners that the first number, 3, the x-coordinate, tells you how many units to take to the right and the second number, 4, the y-coordinate, tells you how many units to move up.
- Reset the **Coordinates tool** to four quadrants. Plot and label point A (–3, 4) Ask: **Which quadrant is this point in?** (II) **What are the coordinates of this point?** Take responses. Explain that for a negative coordinate, the process is similar to locating a positive coordinate. You read to the left instead of right for an x-coordinate and down instead of up for the y-coordinate.
- Plot and label the following points B (–2, –5), C (6, –7), D (–8, –8), E (–9, 2), F (0, –1). Ask: **Which of these points are in Quadrant II?** (E) **Quadrant III?** (B, D) **Quadrant IV?** (C) **What about point F? Where does this lie?** Take responses. Explain that points with a zero x- and/or y-coordinate, which will fall on the axes, are not considered to be in any quadrant.
- Ask learners to write the coordinates of each point, B to F. Choose one point at a time and ask a learner to say the coordinates. Confirm with the class.
- **[T&T] [TWM.02]** Ask: **What do the points in each quadrant have in common?** Have learners discuss with their partner what happens to the x- and y-coordinates in each quadrant. Take responses and establish that for a point:
  - in the first quadrant, both coordinates are positive
  - in the second quadrant, the x-coordinate is negative and the y-coordinate is positive
  - in the third quadrant, both coordinates are negative
  - in the fourth quadrant, the x-coordinate is positive and the y-coordinate is negative.
- Display **Slide 2**. Plot points G (2, 4·5), H (–3·5, 4), I (–5, –1·5), J (–2·5, 0). Explain that not only can

coordinates be positive or negative, they can also be fractions or decimals. When you use decimals or fractions, you might have to estimate the position of the point a little bit. Ask learners to identify the positions of points G and H as decimals, and I and J as fractions (–5, –1$\frac{1}{2}$), (–2$\frac{1}{2}$, 0).

- 📝 Plot the points P (–2, 6), S (–7, –8) and T (9·5, –4). Say: **Give me the coordinates of each point.**
- Discuss Let's learn in the Student's Book.
- Introduce the paired activity in the Student's Book, asking learners to work out possible coordinate positions of two 'treasure' items. Choose learners to explain how they found coordinates.
- Discuss the Guided practice example.

Practice 📒

- Workbook

Title: Reading and plotting coordinates (1)

Pages: 182–183

- Refer to Activity 1 from the Additional practice activities.

Apply 👥 💻

- Display **Slide 3** and read the text to the class. Choose a pair of learners to explain how they found the coordinates of the five points following the movement described.

Review

- Ask learners to work in pairs. Provide each learner with a copy of Resource sheet 26: Four quadrant grid. Each learner marks five points on their grid, A to E. They swap papers and ask their partner to identify and write the coordinates of each point. Learners return their papers for marking.

Assessment for learning

- Which is further away from the y-axis, point J (–7, 8) or point K (–6, –9)? How do you know?
- Give me the coordinates of two points that have negative x-coordinates, are 5 units away from the y-axis and 7 units away from the x-axis.

Same day intervention
Support

- Check that learners use the term 'negative' and not 'minus' when they read coordinates.

Lesson 2: **Reading and plotting coordinates (2)**

Learning objective

Code	Learning objective
6Gp.01	Read and plot coordinates including integers, fractions and decimals, in all four quadrants (with the aid of a grid).

Resources

Resource sheet 26: Four quadrant grid (2 copies per learner and 1 copy per pair)

Revise

Use the activity *Find the treasure* from Unit 23: *Coordinates* in the Revise activities.

Teach [SB] [📊] [TWM.01/06]

- Display the **Coordinates tool** set to four quadrants. Ask learners to work in pairs. Provide Resource sheet 26: Four quadrant grid.
- On the board, write: A (–4, 6) and B (–8, –3) **[T&T]** Say: **Take turns to describe to your partner how you would plot these points on the coordinate grid.** Ask: **In which quadrant does point A lie? How do you know?** Take responses. Confirm: Quadrant II. Remind learners that all points in the second quadrant have an *x*-coordinate that is negative and a *y*-coordinate that is positive. Choose a learner to plot point A using the **Coordinates tool**. Confirm the distance along the *x*-axis (left 4 units from the *y*-axis) and *y*-axis (up 6 units from the *x*-axis). Ask learners to plot point A on their own grid.
- Ask: **In which quadrant does point B lie? How do you know?** Take responses. Confirm: Quadrant III. Remind learners that in the third quadrant, both coordinates are negative. Choose a learner to plot point B using the **Coordinates tool**. Ask: **How many units to the left of the *y*-axis is point B?** (8) **How do you know?** (the size of the *x*-coordinate is 8) **How many units down from the *x*-axis is point B?** (3) **How do you know?** (the size of the *y*-coordinate is 3) Ask learners to plot point B on their own grid.
- On the board, write: C (0, –6), D (8, 7), E (5, –9), F (–6, 3). **[TWM.01]** Ask: **Which of these points lies in Quadrant IV?** (E) **How do you know?** (points in the fourth quadrant have an *x*-coordinate that is positive and a *y*-coordinate that is negative)
- On the board, write: G (1, 1·5), H (–$2\frac{3}{4}$, 3), I ($5\frac{1}{2}$, –2), J (–3, 4·25). Ask learners to plot these points, reminding them to estimate the positions where the value of the *x*- or *y*-coordinate is a decimal or mixed number. Choose learners to show their plots and confirm positions on the grid.
- 👂 Ask learners to plot the point K (–5, –6·5) and a point L, 8 units to the right and 8 units up from point K. Ask: **What are the coordinates of point L?** (3, 1·5)

- Discuss the Let's learn section in the Student's Book.
- **[TWM.06]** Introduce the paired activity in the Student's Book. Choose learners to explain how they identify in which quadrant each set of coordinates should be plotted.
- Discuss the Guided practice example in the Student's Book.

Practice [WB]

- Workbook

Title: Reading and plotting coordinates (2)

Pages: 184–185

- Refer to Activity 1 (Variation) from the Additional practice activities.

Apply 👥 🖥

- Display **Slide 1** and distribute Resource sheet 26: Four quadrant grid, for pairs of learners to answer the question.

Review

- Ask learners to work in pairs. Provide each learner with Resource sheet 26: Four quadrant grid. Each learner writes the coordinates of points A to H in the margin of the grid. The points should include at least two sets of coordinates with an *x*- or *y*-coordinate that is a decimal or fraction. They swap papers and ask their partner to plot the points. Learners return their papers for marking.

Assessment for learning

- Show me how to plot point A (–7, –8).
- Plot a point C halfway between point B (–6, 4) and D (4, 4). Answer: C (–1, 4).

Same day intervention

Enrichment

- Challenge learners to write real-world problems that require graphing points in the coordinate plane, for example, adapting the problem given in the Apply section.

Unit 23 Coordinates

Lesson 3: **Plotting lines and shapes across all four quadrants (1)**

Learning objective

Code	Learning objective
6Gp.02	Use knowledge of 2D shapes and coordinates to plot points to form lines and shapes in all four quadrants.

Resources

Resource Sheet 26: Four quadrant grid (2 copies per learner and 1 copy per pair); ruler (per learner))

Revise

Use the activity *Around the shape* from Unit 23: *Coordinates* in the Revise activities.

Teach [SB] 🖵 📊 [TWM.03/04/05/07]

- Display **Slide 1**. Ask: **What are the coordinates of the four vertices of the square?** (A: (–4, 8), B: (2, 8), C: (2, 2), D: (–4, 2)) **What is the length of each side?** (6 units) Cover one vertex with a small piece of paper. Ask: **How does knowing the length of the side of the square help you find a missing vertex?** Take responses.

- **[T&T] [TWM.07]** Say: **You are given the coordinates for three vertices of a rectangle marked on a coordinate grid. How would you find the coordinates of the fourth vertex?** Take responses. Ask learners to comment on how effective they think each strategy is.

- Display the **Coordinates tool** set to four quadrants. Plot three vertices of a parallelogram ABCD, A: (–4, –4), B: (–1, 2), C: (4, 2). **[TWM.05]** Ask: **I think the missing vertex D is at (2, –4). Am I right?** Expect learners to correct the mistake (1, –4) Ask: **How did you know the missing vertex was at (1, –4)?** Establish that sides AB and DC are parallel. Vertex B to vertex A is a movement 3 squares left and 6 squares down. Repeating the movement from vertex C will give the coordinates of vertex D (3 squares left, 6 squares down).

- Plot one side of an isosceles triangle ABC: A (–4, –3), B (–1, 5). Explain that the base of the triangle AC is parallel to the *x*-axis. Provide Resource sheet 26: Four quadrant grid and ask learners to plot and join points AB. Ask: **What are the coordinates of the missing vertex C?** Give learners time to solve the problem and then take responses. Expect: (2, –3). Ask: **How did you work out the position?** Revise the properties of an isosceles triangle. Establish that the *x*-coordinate of vertex B must lie halfway along the base AB. Since the distance of A to this midpoint is 3 units, the distance from the midpoint to vertex C must be 3 units. This means the base is 6 units and vertex C must sit 6 units to the right of vertex A (2, –3). **[TWM.07]** Ask: **Did anyone use a different method?** Take responses and discuss the efficiency of the different strategies suggested.

- Discuss the Let's learn section in the Student's Book.
- **[TWM.03/04]** Introduce the paired activity in the Student's Book, ensuring that learners know the properties of an isosceles triangle. Ask: **How did you work out the position of vertex C?**
- Discuss the Guided practice example.

Practice [WB]

- Workbook

Title: Plotting lines and shapes across all four quadrants (1)

Pages: 186–187

- Refer to Activity 2 from the Additional practice activities.

Apply 👥 🖵

- Display **Slide 2** and distribute Resource sheet 26. Learners locate the position of the sixth item and explain how they identified the coordinates.

Review

- Ask learners to work in pairs. Provide each learner with Resource sheet 26. Have them draw in pencil a shape on the grid. The shape should be a square, rectangle or an isosceles triangle. They plot and label all but one of the vertices of the shape using the notation ABC… and then rub out their pencil lines. Learners swap papers to identify the missing vertex. They return papers for marking.

Assessment for learning 📊

- Use the **Coordinates tool** to draw three vertices of a square. At least two of the vertices must have a negative *x*-coordinate. What are the coordinates of the missing vertex?
- Three vertices of rectangle ABCD are at positions A (–7, 5), B (–3, 5), C (–3, –3). What is the position of vertex D? (–7, –3)

Same day intervention

Support

- Ask learners to use what they know about the properties of the shape to help them to predict the coordinates of the missing vertex.

Lesson 4: **Plotting lines and shapes across all four quadrants (2)**

Learning objective

Code	Learning objective
6Gp.02	Use knowledge of 2D shapes and coordinates to plot points to form lines and shapes in all four quadrants.

Resources

Resource Sheet 26: Four quadrant grid (1 copy per learner and 2 copies per pair); ruler (per learner)

Revise

Use the activity *Completing quadrilaterals* from Unit 23: *Coordinates* in the Revise activities.

Teach 📖 📊 🖥 [TWM.03/04]

- Display the **Coordinates tool**. Plot the points A (–3, 8) and B (6, –4). **[T&T]** Ask: **I have plotted two points on the coordinate grid. What is the shortest distance between these points?** Take responses and confirm that it is a straight line. Join points A and B. Ask: **What other points are on this line?** Take responses and confirm that the sets of coordinates given by the learners are on the line. Ask: **Which point on the line has a y-coordinate of zero?** Confirm and plot point C (3, 0). **Which point on the line has an x-coordinate of zero?** Confirm and plot point D (0, 4). **[TWM.03]**

- Display **Slide 1**. Ask learners to describe the line AB. Take responses. Expect learners to comment on the points where the line crosses the axes (C and D) and the positions of other points on the line.

- Ask learners to work in pairs. Provide each learner with a copy of Resource sheet 26: Four quadrant grid. Each learner draws two points A and B on their grid and uses a pencil and ruler to join the points. Learners swap papers and write three comments to describe the characteristics of the line. Choose learners to share their line drawings and comments.

- 🖉 Mark the points A (–7, –2) and B (4, 9). Ask: **What are the coordinates of point C where the line crosses the x-axis and point D where the line crosses the y-axis?** C (–5, 0), D (0, 5)

- Discuss the Let's learn section in the Student's Book.

- **[TWM.03/04]** Introduce the paired activity in the Student's Book, ensuring that learners understand the problem. Choose a learner to explain how they found the positions of points H and K.

- Discuss the Guided practice example in the Student's Book.

Practice 🆆🅱 [TWM.05/06]

- Workbook

Title: Plotting lines and shapes across all four quadrants (2)

Pages: 188–189

- Refer to Activity 2 (Variation) from the Additional practice activities.

Apply 👥 🖥

- Display **Slide 2** and read the text to the class. Give learners time to complete the task and then ask them to explain how they found the position and direction of the other side of the path.

Review

- Working in pairs, learners take turns to draw a straight line that passes across both axes. They ask questions in the form: 'Point A sits on the line. If the x-coordinate of point A is [value] then what is the y-coordinate?' and 'Point B sits on the line. If the y-coordinate of point B is [value] then what is the x-coordinate?'

Assessment for learning

- Write the coordinates of two other points that sit on the line that connects point A (–3, 6) to point B (2, –9).

- A line is drawn through the points C (6, 9) and D (–3, –9). Point E sits on this line. Its x-coordinate is 0. What is the value of its y-coordinate?

Same day intervention

Enrichment

- Challenge learners to write real-world problems that require graphing points in the coordinate plane, for example, adapting the problem given in the Apply section.

Additional practice activities

Activity 1

> ### Learning objective
> • Read and plot coordinates including integers, fractions and decimals, in all four quadrants (with the aid of a grid).

Resources
• barrier, such as a book (per pair); Resource sheet 26: Four quadrant grid (per learner); two different -coloured pencils (per pair); Resource sheet 27: Spinners (3) (per pair); pencil and paper clip, for the spinner (per pair)

What to do
• Pairs sit opposite each other and place a barrier, such as a book, between them.
• Provide each learner with Resource sheet 26: Four quadrant grid. Ask them to mark a square with sides of 10 units that has its centre at the origin. The area inside the square represents the land occupied by a golf course.
• Learners plot the positions of five holes on the golf course, labelled A to E, and label the points with the coordinates.
• Learners take turns to 'take a swing of a golf club' and state the coordinates of the position of a hole. If

the hole is not found, their partner provides a clue, for example, 'closer to the x-axis' or 'further away from the y-axis'. The learner gets a second attempt to find the hole.
• The winner is the first person to land a golf ball in every hole.

Variation
Learners use different-coloured pencils to plot points on the same Resource sheet 26: Four quadrant grid.

They spin a the 1 to 10 spinner twice. The first and second spin give the coordinates. After each spin, they spin the +/– spinner to give the sign of the number.

Learners plot a point at these coordinates.

The winner is the first player to plot four points in the same quadrant.

Activity 2

> ### Learning objective
> • Use knowledge of 2D shapes and coordinates to plot points to form lines and shapes in all four quadrants.

Resources
• barrier, such as a book (per pair); Resource sheet 26: Four quadrant grid (two copies per learner); ruler (per learner)

What to do
• Pairs sit opposite each other and place a barrier, such as a book, between them.
• On one grid, players draw two isosceles triangles, two squares and two rectangles on or across any of the four quadrants. The shapes should not overlap or touch at any point.
• At the vertices of each shape, they draw an item of 'treasure'.
• Player 1 calls out a coordinate on their opponent's map. If they locate a treasure item, Player 2 must

inform Player 1, who continues to call out coordinates until they fail to find an item. They record all hits and misses on their second grid as they go along.
• Players swap turns. The game ends when all of a player's treasure items have been found.

Variation
One learner writes the position of a point on the x-axis.

The other learner draws a line that passes through this point and then describes three characteristics of the line.

Learners switch roles and repeat the activity.

Unit 24: Translation, reflection and rotation

Collins International Primary Maths Recommended Teaching and Learning Sequence Term 2, Week 9

Learning objectives

Code	Learning objective
6Gp.03	Translate 2D shapes, identifying the corresponding points between the original and the translated image, on coordinate grids.
6Gp.04	Reflect 2D shapes in a given mirror line (vertical, horizontal and diagonal), on square grids.
6Gp.05	Rotate shapes 90° around a vertex (clockwise or anticlockwise).

Unit overview

In this unit, learners extend their skills in translating 2D shapes to working with translations on a coordinate grid. They use all four quadrants to translate shapes and identify the corresponding points between the original and the image.

Reflection in horizontal and vertical mirror lines on square grids is revised and learners are introduced to reflection in diagonal mirror lines. Tasks include translating shapes where the mirror line is parallel to one of the sides of the shape, or oblique.

The unit introduces learners to a third form of transformation, rotation. They practise rotating shapes 90° around a vertex (clockwise or anticlockwise) and contrast this type of transformation to reflection and translation. They join the corresponding corners of a pair of rotated shapes together and learn that, unlike reflection and translation, the lines formed are not parallel.

Prerequisites for learning

Learners need to:
- know how to translate a 2D shape using a combination of movements in two directions, up/down, left/right, on a square grid
- know how to reflect a 2D shape in vertical or horizontal mirror lines
- know how to make clockwise and anticlockwise turns of 90°.

Vocabulary

transformation, image, translate, quadrant, corresponding vertices, reflect, vertex (vertices), perpendicular distance, diagonal, rotate, centre or point of rotation, clockwise, anticlockwise

Common difficulties and remediation

Look out for learners who inaccurately reflect a point in a mirror line. The learner may have confused reflection with translation or may have a fundamental misunderstanding of how a shape changes orientation through reflection. It may also indicate that they do not understand how the distance between points is preserved in a reflection. Help learners to connect work on reflection to that of folding and the properties of reflection discovered through investigative work. Using computer software

allows learners to investigate the properties of a reflection that remain unchanged as the object, image and mirror line are changed in various ways.

Learners are usually introduced to rotation by finding the centre of rotation of familiar shapes. When they progress to rotating a shape around a vertex, they may confuse the transformation with that around the centre of rotation. Some learners may not understand the principle that rotation, like reflection and translation, is a rigid transformation. Use examples of rigid motions to determine if two figures are congruent. Ask learners to compare a given shape and its image to verify that corresponding sides and corresponding angles are congruent.

Supporting language awareness

If appropriate, when a new key word is introduced, ask learners to write a definition in their own words, for example: 'A rotation turns a figure about a fixed point.' In addition, be aware of terms that have more than one definition, particularly meanings outside mathematics, for example: coordinate, plot, reflect, translate.

Promoting Thinking and Working Mathematically

In Lesson 1, learners specialise (TWM.01) when they identify the relationship between the coordinates of corresponding vertices after a shape is translated.

In Lesson 2, classify (TWM.06) when they determine that for a reflection, all pairs of corresponding points in the object and image are the same distance from the mirror line.

Learners improve (TWM.08) in Lesson 3 when they compare two methods for determining the position of the vertices of a shape after reflection in a diagonal mirror line.

Success criteria

Learners can:
- translate shapes into all four quadrants on a coordinate grid
- reflect a 2D shape in a diagonal mirror line where the line is parallel to one or more of the sides of the shape, or oblique to the shape
- identify, describe and represent the position of a shape following a rotation of 90° around a vertex.

Geometry and Measure – Position and transformation

Lesson 1: **Translating 2D shapes on coordinate grids**

Geometry and Measure – Position and transformation

Learning objective

Code	Learning objective
6Gp.03	Translate 2D shapes, identifying the corresponding points between the original and the translated image, on coordinate grids.

Resources

Resource sheet 26: Four quadrant grid (three copies per pair); ruler, for the Workbook (per learner)

Revise

Use the activity *Translating shapes* from Unit 24: *Translation, reflection and rotation* in the Revise activities.

Teach 🖥 🖳 📊 [TWM.01]

- Explain how this unit looks at three types of 'rigid' transformations that change the position and orientation of a shape, but not a shape's size. Display **Slide 1**. Remind learners that a translation is movement along a straight line. Every point of the object is moved in the same direction and distance.

- Display the **Coordinates tool** set to four quadrants. Plot two squares: A (0, 8), (2, 8), (2, 6), (0, 6) and B (4, 5), (6, 5), (6, 3), (4, 3). Say: **Square A has been translated to square B. Square A has been translated four units to the right, three units down, to make square B.** Explain that you only need to find the images of the vertices, then join them, to reform the shape in its new position.

- Distribute Resource sheet 26. Learners work in pairs to copy the translation A to B, then make a second translation to square C, using the same rules.

- Ask learners to circle a vertex of square A and the corresponding vertices on the images, then list their coordinates. **[T&T] [TWM.01]** Ask: **What do you notice about the three vertices?** Elicit that the numbers in the x-coordinate column increase by four and the numbers in the y-coordinate column decrease by three each time.

- Plot two triangles using the **Coordinates tool**: A (–6, 8), (–6, 6), (–3, 6) and B (–1, 4), (–1, 2), (2, 2). Say: **Triangle A has been translated five units to the right, four units down to make triangle B.** Ask: **Knowing the translation, how could you have calculated the position of triangle B, the image of triangle A?** Establish that you find the horizontal and vertical distance between the vertices of the object and the image. The x-coordinate has increased by five units and the y-coordinate has decreased by four units. Say: **When you translate to the right and upwards, the x- and y-coordinates increase; when you translate to the left and downwards, the coordinates decrease.**

- Distribute another copy of Resource sheet 26. Learners work in pairs using the same translation as in A to B to translate triangle B to a new position, triangle C, by first calculating the coordinates of triangle C, then using the grid to confirm. Ask: **What**

are the coordinates of triangle C? ((4, 0), (4, –2), (7, –2)). Invite a volunteer to plot triangle C and explain how they calculated the positions of the corresponding vertices, with and without a grid.

- 🖳 Ask learners to plot the rectangle PQRS: P (2, –2), Q (7, –2), R (7, –5), S (2, –5). Ask: **The rectangle is translated left ten squares and up nine squares. What are the coordinates of the vertices in the translated image?** Complete the translation using the **Coordinates tool**. Label the corresponding vertices of the image P' (P' (–8, 7), Q' (–3, 7), R' (–3, 4), S' (–8, 4)). Explain that you usually label the image of the vertex of a shape using the prime symbol (').

- Discuss the Let's learn section in the Student's Book.

- Introduce the paired activity in the Student's Book.

- Discuss the Guided practice example.

Practice 📒

- Workbook

Title: Translating 2D shapes on coordinate grids

Pages: 190–191

- Refer to Activity 1 from the Additional practice activities.

Apply 👥 🖥

- Display **Slide 2** and distribute Resource sheet 26. Pairs solve the problem, then share their solutions.

Review 📊

- Display the **Coordinates tool**. Plot a triangle (–5, –1), (–5, –4), (–1, –4). Learners translate the shape four units to the right, two units up. Ask: **What are the coordinates of the translation?**

Assessment for learning

- How would you translate a point from (–8, –9) to (7, 6)?

- A triangle ABC (A: (4, –2), B (1, –6), C (7, –6)) is translated nine units left and three units down. What are the coordinates are the positions of the vertices A', B' and C'?

Same day intervention

Enrichment

- Provide grids that have more rows and columns. Ask learners to translate shapes over greater distances.

Lesson 2: **Reflecting 2D shapes in a mirror line (1)**

Learning objective

Code	Learning objective
6Gp.04	Reflect 2D shapes in a given mirror line (vertical, horizontal and diagonal), on square grids.

Resources

squared paper (per pair and per learner); ruler (per learner)

Revise

Use the activity *Reflecting shapes* from Unit 24: *Translation, reflection and rotation* in the Revise activities.

Teach [SB] 🖥 📊 [TWM.06/07]

- Display **Slide 1**. Ask: **What type of transformation does the diagram show?** Take responses. Confirm that it shows a reflection. **[T&T] [TWM.06]** Ask: **What can you say about the distances a to f?** Confirm that a = b, c = d and e = f. Ask: **How do you know?** Elicit that for a reflection, all pairs of corresponding points in the object and image are the same distance from the mirror line. Ask: **What else do you know about a reflection?** Take responses. Remind learners that when an object undergoes a reflection, the shape and size of the object remains unchanged.

- Display the **Rotate and reflect tool**. Click the vertical mirror line and position the red rectangle 1·5 grid squares to the right of it. Ask a volunteer to show where the shape is after reflection in the mirror line. Expect them to count squares to check that the vertices and their images are the same perpendicular distance from the mirror line.

- Position the orange triangle 2·5 grid squares above a horizontal mirror line. Ask a volunteer to show where the image of the shape is after reflection. Join the corresponding corners of the object and the image. Ask: **What do you notice? How does this differ from translation?** (the lines are parallel but different lengths, unlike translation where the lines are the same length)

- Refer to **Slide 2** that shows four shapes on grids, each positioned so that the sides of the shape are not parallel or perpendicular to the mirror line. Set up each shape and position on the **Rotate and reflect tool** one at a time. For each, choose a learner to show the position of the image of the shape after reflection in the mirror line. Ensure that learners are counting or estimating the perpendicular distance from the vertex to the mirror line.

- ⌨ Ask learners to draw a vertical mirror line on squared paper. Then ask them to draw an isosceles triangle with a height of five squares so that the base of the triangle is parallel with the line and 4·5 grid squares away. Say: **Reflect the shape in the mirror line.**

- Discuss the Let's learn section in the Student's Book.
- **[TWM.07]** Introduce the paired activity in the Student's Book. Choose learners to explain the mistakes that Frankie has made when reflecting the triangle across the mirror line.
- Discuss the Guided practice example in the Student's Book.

Practice [WB]

- Workbook

Title: Reflecting 2D shapes in a mirror line (1)

Pages: 192–193

- Refer to Activity 1 (Variation) from the Additional practice activities.

Apply 👥 🖥

- Display **Slide 3** and distribute squared paper. Ensure the learners understand the task. Choose learners to demonstrate the reflection patterns they have designed.

Review

- Provide squared paper to each learner. Learners work in pairs. They each draw a horizontal mirror line and a shape above it. In another part of the grid, they draw a vertical mirror line. The idea is to reflect the shape in the horizontal mirror line and then reflect the image in the vertical mirror line. Learners swap papers to complete the reflections.

Assessment for learning

- How does the image of a shape after reflection differ from the object? How is it the same?
- How can you use the vertices of a shape, and the corresponding vertices, to check if the shape has been reflected correctly?

Same day intervention

Support

- Look out for learners who fail to reflect shapes correctly. Provide them with practical opportunities to paint dots, lines and shapes on grids and then fold them to locate the position of the image.

Geometry and Measure – Position and transformation

Lesson 3: **Reflecting 2D shapes in a mirror line (2)**

Geometry and Measure – Position and transformation

Learning objective

Code	Learning objective
6Gp.04	Reflect 2D shapes in a given mirror line (vertical, horizontal and diagonal), on square grids.

Resources

squared paper (2 copies per learner and 1 copy per pair); ruler (per learner)

Revise

Use the activity *What's the shape?* from Unit 24: *Translation, reflection and rotation* in the Revise activities.

Teach ⬛ 🖥 📊 [TWM.04/08]

- Display **Slide 1**. Point out that the shape is reflected in a diagonal mirror line. Explain that not all mirror lines will be parallel or perpendicular to one of the sides of a shape; sometimes they will be diagonal. **[T&T]** Say: **Look at the diagram. How does reflection in a diagonal mirror line compare with a horizontal or vertical line? What are the similarities and differences?** Take responses. Establish that you use the same method to reflect the shape, no matter the orientation of the mirror line. Indicate on the diagram the arrowed lines perpendicular to the mirror line that can be drawn from the vertices of the shape, and extended the same distance on the other side of the mirror line as before. Reflecting all the vertices in the same way locates the position of the reflected vertices. Joining the vertices with straight lines makes the reflected shape.

- Display the **Rotate and reflect tool**. Click the diagonal mirror line and position the blue square below the line, with its base parallel to the bottom edge of the grid and its top right vertex at a distance of 0·5 of the diagonal of a grid square from the mirror line. Ask a volunteer to show where the shape is after reflection in the mirror line. Expect them to count squares along imaginary lines drawn from the vertices perpendicular to the mirror line. They confirm that the vertices and their images are the same perpendicular distance from the mirror line. Repeat for other shapes and the opposite diagonal mirror line.

- Refer to **Slide 2** that shows two shapes on grids, each positioned so that the sides of the shape are not parallel or perpendicular to the diagonal mirror line. Set up each shape and position on the **Rotate and reflect tool**, one at a time. For each, choose a learner to show the position of the image of the shape after reflection in the mirror line. Ensure that learners are counting or estimating the perpendicular distance from the vertex to the mirror line.

- 🖾 Ask learners to draw a diagonal mirror line on squared paper. Then ask them to draw a rectangle that has its base parallel with the mirror line and three grid squares away. The length of the rectangle

should be five (grid square) diagonals long and two diagonals wide. Say: **Reflect the shape in the mirror line.**

- Discuss the Let's learn section in the Student's Book.

- **[TWM.04]** Introduce the paired activity in the Student's Book. Choose a pair of learners to explain how Alice's method works and to suggest how it could be improved. **[TWM.08]** Ask: **Which method do you think is more efficient and why?**

- Discuss the Guided practice example in the Student's Book.

Practice 📒

- Workbook

Title: Reflecting 2D shapes in a mirror line (2)

Pages: 194–195

- Refer to Activity 2 from the Additional practice activities.

Apply 👥 🖥

- Display **Slide 3** and distribute squared paper. Ensure learners understand the task. Choose learners to demonstrate the reflection patterns they have designed.

Review

- Provide another sheet of squared paper to each learner. Learners draw two parallel diagonal mirror lines a distance of ten diagonal units apart. They draw a rectangle above the first diagonal mirror line with a pair of sides parallel to the mirror line. Learners then reflect the shape across the first mirror line and then its image across the second diagonal line. Choose learners to share their diagrams to confirm that the shape has been reflected correctly.

Assessment for learning

- How does reflecting a shape in a diagonal mirror line differ from reflecting a shape in horizontal and vertical mirror lines?

Same day intervention
Enrichment

- Provide learners with grids that have more rows and columns and ask them to reflect shapes that are a greater distance away from the mirror line.

Lesson 4: **Rotating shapes 90° around a vertex**

Geometry and Measure – Position and transformation

Learning objective

Code	Learning objective
6Gp.05	Rotate shapes 90° around a vertex (clockwise or anticlockwise).

Resources

ruler (per learner); tracing paper (per pair and per learner); squared paper (2 copies per pair); scissors (per pair)

Revise

Use the activity *Quarter turns* from Unit 24: *Translation, reflection and rotation* in the Revise activities.

Teach 🔲 💻 📊 [TWM.04]

- Ask learners to stand up with an arm outstretched and move through clockwise and anticlockwise movements of 90° to 360° (quarter to full turns). Say: **In all of these movements, you rotated about a given point. Where is this point?** Establish it is the position where each learner stands. Explain that you can apply the same rotational movements to shapes.
- Display **Slide 1**. Read the text and discuss the rotation of the triangle. Ask: **Compare the triangle and its image after rotation. What has remained the same? What has changed?**
- Display the **Rotate and reflect tool**, showing the grid with a horizontal mirror line and the right-angled triangle on the line. Ask: **What information do you need to rotate a shape?** (the amount and direction of rotation, and the centre of rotation) Demonstrate rotating the triangle 90°clockwise about one of its vertices. Ask: **What do you notice about the position of the image?** Take responses. Establish that, while the vertex has remained in the same position, the other vertices have all moved a quarter turn (90°) in the same direction.
- Point out that the object and its resulting image are identical in shape and size, but the orientation of the shape and image are different.
- Distribute tracing paper, rulers and squared paper. Ask pairs of learners to draw a right-angled triangle and use tracing paper to practise rotating the shape through 90°. Have them join the corresponding corners of the object and the image. Ask: **What do you notice about these lines? How do they compare with those drawn for a reflection or translation?** Establish that unlike reflection and translation, the lines formed are not parallel. Learners complete four 90°anticlockwise rotations about one vertex. Ask: **What is the position of the final image?** (the original position)
- Demonstrate four 90° anticlockwise rotations on the tool.
- Draw a rectangle on the board. Label one vertex as a point of rotation. [TWM.04] Ask: **In a full rotation of 360°, stopping every 90°, how many times would**

the image look like the object? How do you know? Ask learners to work in pairs to find the answer. (once)

- 🔲 Give the following problem to learners: **Freya draws an arrow pointing east. She rotates it anticlockwise 90°around a vertex at the opposite end to the arrowhead. In which direction does the arrow point now?** (north)
- Discuss the Let's learn section in the Student's Book.
- [TWM.04] Introduce the paired activity in the Student's Book, ensuring that learners understand the problem. Choose a learner to explain how they would use the parallelogram to create the pattern shown.
- Discuss the Guided practice example.

Practice 🔲

- Workbook

Title: Rotating shapes 90° around a vertex

Pages: 196–197

- Refer to Activity 2 (Variation) from the Additional practice activities.

Apply 👥 💻

- Display **Slide 2**. Learners make patterns from their triangles, then share their patterns and describe how they constructed them.

Review 💻

- Use the **Rotate and Reflect tool** to rotate a trapezium around one vertex, first a clockwise rotation and then an anticlockwise rotation of 90°. Ask learners to describe the differences between the shape and its two images.

Assessment for learning

- What information do you need to rotate a shape about a vertex?
- Tell me the procedure for rotating a triangle ABC 90° clockwise about its vertex B.

Same day intervention

Support

- Use tracing paper to rotate objects to show that the image is in a different position but the size and shape remain unaltered.

Geometry and Measure – Position and transformation

Additional practice activities

Activity 1

Learning objectives

- Translate 2D shapes, identifying the corresponding points between the original and the translated image, on coordinate grids.
- Reflect 2D shapes in a given mirror line (vertical, horizontal and diagonal), on square grids.

Resources

Resource sheet 26: Four quadrant grid (per pair); ruler (per learner); coloured pencils (per pair); squared paper (per learner) (for variation)

What to do

- Pairs of learners face the four-quadrant grid from the same direction. They each draw an L-shaped object formed from five squares in opposite corners of the grid and shade it with a chosen colour.
- Players take turns to call out a translation that will move their shape to a part of the grid that is completely empty. They shade the image formed.
- If a player finds that one or more of the squares are filled, they miss the next turn.
- The game continues until a chosen time limit has elapsed, or the grid is full.

Variations

1 Learners are allowed to mark in the image of their translated shape before naming the translation to ensure that the translation into a clear grid is possible.

2 Learners nominate a shape to draw, for example, a triangle, rectangle or square.

They each draw a horizontal or vertical mirror line on squared paper lightly in pencil and then draw the shape on one side of the line. Without their partner seeing, they reflect the shape over the mirror line and then erase the mirror line.

They swap papers to try to locate and mark the position of the mirror line. They return papers for marking.

Activity 2

Learning objectives

- Reflect 2D shapes in a given mirror line (vertical, horizontal and diagonal), on square grids.
- Rotate shapes 90° around a vertex (clockwise or anticlockwise).

Resources

squared paper (per learner); ruler (per learner)

What to do

- Each learner draws a diagonal mirror line on squared paper.
- They draw a triangle on one side of the line. One side of the triangle should be parallel with the mirror line.
- Learners swap papers to reflect the shape.
- They return their papers for marking.

Variation

2 Learners draw twenty 10 by 1 rectangles on squared paper.

They take turns to nominate one of the rectangles and a vertex and rotate the rectangles 90° clockwise or anticlockwise about the chosen vertex. The aim is for the rotating shape to contact as many other rectangles as possible. The player scores a point for every shape contacted.

The rectangles rotated are removed from play.

The game ends when it is no longer possible to 'knock out' any other rectangles. The winner is the player with more points.

Unit 25: Statistics (A)

Learning objectives

Code	Learning objective
6Ss.01	Plan and conduct an investigation and make predictions for a set of related statistical questions, considering what data to collect (categorical, discrete and continuous data).
6Ss.02	Record, organise and represent categorical, discrete and continuous data. Choose and explain which representation to use in a given situation: - Venn and Carroll diagrams - tally charts and frequency tables - bar charts - waffle diagrams and pie charts [- frequency diagrams for continuous data - line graphs - scatter graphs - dot plots].
6Ss.03	Understand that the mode, median, mean and range are ways to describe and summarise data sets. Find and interpret the mode (including bimodal data), median, mean and range, and consider their appropriateness for the context.
6Ss.04	Interpret data, identifying patterns, within and between data sets, to answer statistical questions. Discuss conclusions, considering the sources of variation and check predictions.

Unit overview

In this unit, learners plan and conduct investigations to answer statistical questions. They revise the definition of a statistical question and devise their own examples to investigate. The work completed in Stage 5 is extended to include making predictions. Learners are asked to make predictions before conducting their investigation, for example, 'Do children that watch more TV read fewer books?'

Learners decide what data they need to collect and the best format for recording it. The data may be categorical, discrete or continuous but, at this stage, learners do not need to know the formal definitions. Once they have collected the data, learners explore different ways to represent it: Venn and Carroll diagrams, tally charts and frequency tables, bar charts, waffle diagrams and pie charts.

Each lesson in Units 25 and 26 focuses on a different way of representing data. Within each lesson, the statistical question investigated and the type of data collected match this representation.

In Lesson 4, learners describe and interpret the modes, medians, means and ranges of various data sets. They find modal values and continue to investigate the use of 'averages' in a range of contexts, considering their appropriateness.

Prerequisites for learning

Learners need to:
- be able to suggest appropriate statistical questions to investigate
- be able to decide on a suitable way to collect data and how to represent it
- know how to interpret data represented in various charts and graphs and draw conclusions related to the statistical question that founded the investigation.

Vocabulary

statistical question, prediction, data, Venn diagram, Carroll diagram, tally chart, frequency table, bar chart, waffle diagram, pie chart, mode, median, mean, range

Common difficulties and remediation

Look out for learners who assume that mean and median values of a data set are always close in number. Provide examples where data is clustered, to show that the mean is pulled towards the extreme data cluster. Some learners assume that the mode of a data set is always numerical. Provide examples where the data set is non-numerical and point out that the mode is the datum that appears most often.

Supporting language awareness

If appropriate, when a new key word is introduced, ask learners to write a definition in their books, drawing a box around it for emphasis.

Promoting Thinking and Working Mathematically

In Lesson 1, learners specialise (TWM.01) when they suggest the type of statistical question that would give data suitable for representing in a Carroll diagram.

Learners critique (TWM.07) in Lesson 3 when they compare and contrast how data is represented in a waffle diagram and a pie chart. They point out the strengths of each type of graph and which of the two they think displays data in the clearest way.

Success criteria

Learners can:
- suggest appropriate statistical questions to investigate and make predictions about the data
- say what data is best represented in a pie chart and construct a pie chart with appropriate proportions
- find the mode median, mean and range.

Unit 25 Statistics (A)

Statistics and Probability – Statistics

Lesson 1: Venn and Carroll diagrams

Learning objectives

Code	Learning objective
6Ss.01	Plan and conduct an investigation and make predictions for a set of related statistical questions, considering what data to collect (categorical, discrete and continuous data).
6Ss.02	Record, organise and represent categorical, discrete and continuous data. Choose and explain which representation to use in a given situation: - Venn and Carroll diagrams - tally charts and frequency tables
6Ss.04	Interpret data, identifying patterns, within and between data sets, to answer statistical questions. Discuss conclusions, considering the sources of variation and check predictions.

Revise

Use the activity *Cycling or scootering?* from Unit 25: *Statistics (A)* in the Revise activities.

Teach [SB] 🖥 [TWM.01/07]

- Display **Slide 1**. Read and discuss the text with the learners.

- Ask learners to work in pairs to devise a recording method. **[TWM.07]** Choose pairs to explain their method and invite the class to comment on its strengths and weaknesses, suggesting possible improvements.

- Go around the class asking learners to state one conclusion that they can draw from the results and to comment on how well they did with their predictions.

- Display **Slide 2**. Explain how data can be represented in a Carroll diagram. Ask learners to say what the data in each part of the diagram represents, for example: **What do the crosses in the first row, second column tell us?** (learners that have a brother but no sister) Choose learners to state one conclusion that they can draw from the data.

- **[TWM.01] [T&T]** Ask: **What type of data should be represented in a Carroll diagram?** Take responses. Remind learners that a Carroll diagram is an easy way to sort objects, numbers or concepts into two different categories, using yes/no situations. Establish that the question must consist of two parts, for example: 'Do you read a book every day? (yes/no) Do you watch the TV every day? (yes/no)' and is particularly useful for establishing if there is a connection between the parts. Ask: **What connection could a statistical question look to answer?** (Do children that watch more TV read fewer books?) Ask learners to predict: **What do you think the data would show?**

- Learners devise a suitable statistical question and gather data. They represent the data in a Carroll diagram and use it to draw conclusions. They compare these with their predictions. Choose learners to comment on what they found out and how it compared with their predictions. Ask: **What answer to the statistical question does the data provide?** Take responses and discuss.

- 📖 Ask: **How could you extend your investigation? What statistical question could you ask next?**

- Discuss the Let's learn section in the Student's Book and address any misunderstandings.

- Introduce the paired activity in the Student's Book, asking learners to decide how they would investigate Maha's prediction that more boys own bikes than girls.

- Discuss the Guided practice example in the Student's Book.

Practice [WB] [TWM.03/04]

- Workbook

Title: Venn and Carroll diagrams

Pages: 198–200

- Refer to Activity 1 from the Additional practice activities.

Apply 👥 🖥

- Display **Slide 3**. Read the problem to the class. Choose pairs of learners to discuss how they would complete a class survey to test Florence's prediction. Have them conduct the survey and then choose learners to explain their findings.

Review

- Working in pairs, ask learners to make a prediction based on a preference, for example: 'Are learners who have cereal for breakfast more likely to also have a glass of fruit juice?' Go around the class and ask learners their question. Discuss how easy or difficult it would be to answer the question.

Assessment for learning

- Give me an example of a statistical question that you could test at home or in class.
- What would be your prediction of the outcome?

Same day intervention
Support

- Some learners may have difficulty identifying the correct region of a Venn diagram when there are two events occurring, for example, when given a choice of two ice-cream flavours, a learner states that they like both. Provide two tissue paper circles to model the sets of learners that choose each flavour, for example, yellow for lemon and red for strawberry. Overlap the circles slightly and comment on the change of colour in the area of overlap. Explain that this is the set of learners who like both flavours.

Lesson 2: **Tally charts, frequency tables and bar charts**

Learning objectives

Code	Learning objective
6Ss.01	Plan and conduct an investigation and make predictions for a set of related statistical questions, considering what data to collect (categorical, discrete and continuous data).
6Ss.02	Record, organise and represent categorical, discrete and continuous data. Choose and explain which representation to use in a given situation: - tally charts and frequency tables - bar charts
6Ss.04	Interpret data, identifying patterns, within and between data sets, to answer statistical questions. Discuss conclusions, considering the sources of variation and check predictions.

Resources

squared paper (per pair); ruler (per pair); coloured pencils (per pair); spinner from Resource sheet 28: Spinner (4), for the Workbook (per pair); pencil and paper clip, for the spinner used in the Workbook (per pair)

Revise

Use the activity *Finding information* from Unit 25: *Statistics (A)* in the Revise activities.

Teach 🆂🅱 🖥 [TWM.03/05]

- Display **Slide 1**. Explain that the frequency table is used to gather data to answer the statistical question: 'How many pets do learners own?' **[T&T] [TWM.05]** Ask: **Why do you think the data has been grouped in this way?** Take responses. Elicit that when a data set has a wide range of values, it is easier to organise and record the values as groups arranged in intervals. This makes the data more readable and easier to interpret.
- Copy the table on the board. Ask learners to make predictions about which intervals will be the most frequent and why. Write some of these predictions on the board.
- Begin the survey. Ask learners to respond to the question by making tally marks in the table. Count these to complete the frequencies.
- On the board, write the statistical question: 'Giving the time in seconds, how long does it take learners to run around the school field?' **[TWM.05]** Ask: **If you were to draw a frequency table to record the results for this run around the school field, how would you group the times taken for learners to complete the run? For example, would intervals of two seconds be a good way of organising the data? If not, why?** Take responses. Establish that because the entire range of data could take on many values, it would be best to group the data in larger intervals, such as ten seconds.
- Organise a class run around the field or playground and collect the results. Before completing the run, ask learners to make predictions about the data. For example, 'I think older children will complete the distance in a faster time.' Write some of these predictions on the board.
- Display and discuss **Slide 2**. **[TWM.05]**
- ▣ Ask learners to consider the following: **Fred is going to investigate his prediction that the further you live away from school, the more likely you are to travel** to school by bus or car. **What question should he ask the learners in his class? How should Fred collect and represent his data? What should he do to find out if his prediction is correct or not?**
- Discuss the Let's learn section in the Student's Book.
- **[TWM.03]** Introduce the paired activity.
- Discuss the Guided practice example.

Practice 🆆🅱 [TWM.04]

- Workbook

Title: Tally charts, frequency tables and bar charts

Pages: 201–203

- Refer to Activity 1 (Variation) from the Additional practice activities.

Apply 👥

- Display **Slide 3**. Ask learners how they plan to test Logan's and Sam's predictions. Who made the correct prediction? How do you know?

Review

- Referring to the appropriate questions in the Workbook, choose individual learners to present their conclusions to the class and to explain if their predictions were correct.

Assessment for learning

- Give me an example of a statistical question where data could be collected as measurements organised in grouped intervals.

Same day intervention
Support

- Learners may fail to understand why data is sometimes 'grouped' into ranges of values. Explain, through examples, that grouping data enables graphs and frequency diagrams to be drawn showing class intervals rather than individual values. Demonstrate that graphing data with multiple values be unmanageable and time consuming.

Unit **25** **Statistics (A)**

Statistics and Probability – Statistics

Lesson 3: **Waffle diagrams and pie charts**

Learning objectives

Code	Learning objective
6Ss.01	Plan and conduct an investigation and make predictions for a set of related statistical questions, considering what data to collect (categorical, discrete and continuous data).
6Ss.02	Record, organise and represent categorical, discrete and continuous data. Choose and explain which representation to use in a given situation: - tally charts and frequency tables - waffle diagrams and pie charts
6Ss.04	Interpret data, identifying patterns, within and between data sets, to answer statistical questions. Discuss conclusions, considering the sources of variation and check predictions.

Resources

Resource 29: Pie charts (10 divisions) (per learner); Resource 29: Pie charts (20 divisions) (per pair); ruler (per learner); waffle diagram from Resource sheet 14: Waffle diagrams (1 copy per learner and 2 copies for pair); coloured pencils (per learner)

Revise

Use the activity *Predicting the weather* from Unit 25: *Statistics (A)* in the Revise activities.

Teach [SB] 🖵 ☐ **[TWM.04/05/07]**

- Display **Slide 1**. Discuss the data that represents the results of a class survey that asked learners their favourite colour. Explain that Alfie, a learner in the class, drew a waffle diagram and pie chart to represent the data. Remind learners that a waffle diagram is like a 100 square that uses squares to represent percentages. This waffle diagram shows the percentages when 20 learners voted for their favourite colour. Each cell in the 10 by 10 grid represents 1%. **[T&T] [TWM.05]** Point to the pie chart and highlight that the circumference is divided into ten equal divisions. Ask: **How did Alfie decide on the size of each slice (sector)?** Take responses. Establish that Alfie knows that each mark on the circle divides the circle into ten equal divisions and therefore, each division must represent 10% of the circle.

- On the board, write the statistical question: 'What is your favourite smell: lemon, fresh bread, mown grass or roast dinner?' Choose ten learners to take part in the survey and ask them to vote for their favourite smell. Record the results in a frequency table with an extra column for 'Percentage'. With learners helping to calculate, complete the percentage column of the table.

- Provide Resource 29: Pie charts (10 divisions). Ask learners to calculate the size of each sector of the pie chart for the four favourite smells. Ask them to draw a rough sketch of what they think the pie chart would look like. Have them label each sector with the favourite smell and percentage.

- Display the **Pie charter tool** set to 'name and percentage', label the slices and enter the percentages. Ask learners to hold up their sketches and compare them to the chart on the board.

- Explain to the learners that pie charts can be divided into any number of divisions. Introduce them to

Question 3 in the Workbook where the divisions are multiples of 5%.

- Provide each learner with a waffle diagram from Resource sheet 14: Waffle diagrams and ask learners to construct a waffle diagram of the same data. Ask: **What conclusions can you draw?**

- **[T&T] [TWM.07]** Ask questions that compare the two graphs: **What do you think are the strengths of a waffle diagram? The strengths of a pie chart? Which graph do you think displays data more clearly? Why?**

- Discuss the Let's learn section in the Student's Book.

- 🗣 Ask learners to explain to a partner what a pie chart is and the type of data it can be used to represent.

- **[TWM.04]** Introduce the paired activity. Pairs of learners discuss their conclusions from the data.

- Discuss the Guided practice example.

Practice [WB] **[TWM.04]**

- Workbook

Title: Waffle diagrams and pie charts

Pages: 204–207

- Refer to Activity 2 from the Additional practice activities.

Apply 👥 **[TWM.03]**

- Display **Slide 2**. Pairs explain how they worked backwards using the pie chart to deduce the original data.

Review

- Choose learners to present their conclusions from the Workbook to the class.

Assessment for learning

- Give me an example of a statistical question where data could be collected as measurements and represented in a pie chart.

Lesson 4: **Mode, median, mean and range**

Learning objective

Code	Learning objective
6Ss.03	Understand that the mode, median, mean and range are ways to describe and summarise data sets. Find and interpret the mode (including bimodal data), median, mean and range, and consider their appropriateness for the context.

Resources

2 to 9 spinner from Resource sheet 3: Spinner (1) (per pair); pencil and paper clip, for the spinner (per learner)

Revise

Use the activity *Find the median* from Unit 25: *Statistics (A)* in the Revise activities.

Teach 🆂🅱 🖥️ [TWM.04/08]

- Display **Slide 1**. Remind learners that the median is found by arranging data values in order of size and finding the middle value. On the board, write: 12, 16, 10, 18, 13. Explain that the values are the points that Tom scored in different rounds of a video game. Ask: **What is the order, from lowest to highest?** Record: 10, 12, 13, 16, 18. Ask: **What is the median?** (13) Explain that the median is a boundary point where half of the data is below the median and half is above. Ask: **Which of Tom's scores are below the median?** (10, 12) **Which of Tom's scores are above the median?** (16, 18)
- On the board, write: 34, 26, 37, 22, 33, 42. Explain that the values are more of Tom's scores. **[T&T]** Ask: **What do you notice about this set of scores compared to Tom's previous scores?** Take responses. Establish that the number of scores is an even number. Establish that for an even number of events, you find the average of the two middle scores by adding them and then dividing by 2.
- Return to **Slide 1**. Remind learners of the definition of mode. On the board, write: 78 kg, 76 kg, 72 kg, 75 kg, 76 kg, 78 kg, 73 kg, 76 kg, 79 kg, 75 kg. Say: **These are the masses of crates of fruit. What is the mode of the set of measurements?** (76 kg)
- On the board, write: 17, 5, 19, 23, 17, 8, 5, 24, 6. Ask: **What is the mode of this data set?** Give learners time to think then ask them what they notice about the mode. Confirm that there are in fact two modes, 5 and 17 and explain that it is possible for data sets to have two or more modes.
- Read the definition of 'Range' from **Slide 1**. Refer to the list of masses of crates. Ask: **What is the range of this data?** Record the range: 79 kg–72 kg = 7 kg.
- On the board, write: 91, 92, 94, 91, 92, 90, 88, 89, 91, 88, 90, 90, 91, 93, 91, 88. Ask learners, working in pairs, to calculate the mode, median and range of the data. (Mode: 91, Median: 91, Range: 6)
- Read the definition of 'Mean' from **Slide 1** to introduce the term. Write: 2 kg, 8 kg, 11 kg. Ask: **What is the mean?** With help from the learners, record: mean = (2 + 8 + 11) ÷ 3 = 7. Confirm that the mean is 7 kg.
- Write: 9, 5, 6, 11, 3, 14. Ask: **What is the mean of this set of numbers?** (8) **How did you work it out?** (find the sum of the six numbers and divide by 6)

- Display **Slide 2**. Discuss the three types of averages and when they are typically used.
- Display and discuss **Slide 3**. **[TWM.08]**.
- On the board, write: The mean of four numbers is 30 and the mode is 15. **What could the four numbers be?** Elicit that since the mode is 15 at least two of the numbers must be 15. Write: 15   15   ___   ___. Ask: **What does the mean 30 tell us?** Establish that for four numbers to give a mean of 30 their sum must be 120 (4 × 30). Given that two 15s sum to 30, the sum of the remaining two numbers must be 90 (120–30). There are many solutions: (15, 15, x, y) where x + y = 90, e.g. 15, 15, 40, 50.
- Discuss the Let's learn section in the Student's Book.
- **[TWM.04]** Introduce the paired activity and the Guided practice in the Student's Book.

Practice 🆆🅱 [TWM.04]

- Workbook

Title: Mode, median, mean and range

Pages: 208–211

- Refer to Activity 2 (Variation) from the Additional practice activities.

Apply 👥

- Pairs of learners spin the 2 to 9 spinner from Resource sheet 3 20 times, recording the numbers. They then find the mode, mean, median and range.

Review

- Invite 20 learners each to write a random number from 0 to 6 on the board. Ask the class to find the mode, mean, median and range of the set of numbers.

Assessment for learning

- Roll a dice 20 times and record the results. What is the mode, mean, median and range?

Same day intervention
Support

- To understand the differences between median and mode, start by defining the terms. The median is the middle score in a set of given numbers. The mode is the most frequently occurring score in a set of numbers.

Statistics and Probability – Statistics

Additional practice activities

Activity 1 👥 ▲2

Learning objective
• Plan and conduct an investigation and make predictions for a set of related statistical questions, considering what data to collect (categorical, discrete and continuous data).

Resources
paper (per pair); ruler (per pair) (for variation); squared paper (per pair) coloured pencils (per pair)

What to do
• Before starting the activity, write on the board the question: 'If your family has more than one car, are you more likely to be driven to school?'
• Learners work in pairs. Ask them to make a prediction on what they think the results will reveal. Write some of these predictions on the board.
• Ask: **What question will you ask to gather the right information?** Take responses. Establish that a two-part question should be asked: 'Does your family have more than one car? Do you travel to school by car?'
• Have learners construct a tally chart and collect the data from the class.
• They add a third column to the tally chart and find the frequency for each set of data.

• They then use the data to construct a Carroll diagram and answer the question on the board. Confirm that learners create four categories for sorting, for example: 'Have more than 1 car'/'Do not have more than 1 car'; 'Travel to school by car'/'Do not travel to school bay car'.
• Ask learners to draw conclusions from the data represented in the Carroll diagram. Ask: **Was your prediction correct? How do you know?**

Variation
▲2 Change the question to: 'How many vehicles does your family own?' This includes cars, motorbikes, mopeds, bicycles, scooters and skateboards. Ask learners to make predictions about the data. Learners then gather data using a tally chart and record the frequencies. They convert the frequencies to percentages and use these to complete a bar chart. Ask them to draw conclusions from the data represented in the bar chart and decide whether their predictions are correct or not.

Activity 2 👥 ▲2

Learning objective
• Record, organise and represent categorical, discrete and continuous data. Choose and explain which representation to use in a given situation: waffle diagram and pie chart.

Resources
Resource 29: Pie charts (per pair); ruler (per pair); waffle diagram from Resource sheet 14: Waffle diagrams (per pair); coloured pencils (per pair)

What to do
• Working in pairs, learners collect data to investigate the statistical question: 'Which flavour of crisps do most learners in our class prefer?'
• Have them predict which flavour they think will be the most popular and explain their choice.
• They ask 20 learners to make a choice from five flavours and record the data using a tally chart.
• Next, they find the frequency of each flavour chosen and then convert these values to a percentage.

• Lastly, they present the percentage data in a waffle diagram and a pie chart choosing the most appropriate Resource sheet 29: Pie charts 4, 5, 10 or 20 divisions. Choose learners to present their conclusions.

Variation
▲2 Expand the 'favourite flavour of crisps' survey above to include 50 learners and eight flavours: salt and vinegar, prawn cocktail, pickled onion, spicy, cheese and onion, chicken, BBQ, ready salted. Complete the survey and record the results in a frequency table. Find the mean, mode, median and range of the data.

Unit 26: Statistics (B)

Learning objectives

Code	Learning objective
6Ss.01	Plan and conduct an investigation and make predictions for a set of related statistical questions, considering what data to collect (categorical, discrete and continuous data).
6Ss.02	Record, organise and represent categorical, discrete and continuous data. Choose and explain which representation to use in a given situation: [- Venn and Carroll diagrams - tally charts and frequency tables - bar charts - waffle diagrams and pie charts] - frequency diagrams for continuous data - line graphs - scatter graphs - dot plots.
6Ss.04	Interpret data, identifying patterns, within and between data sets, to answer statistical questions. Discuss conclusions, considering the sources of variation and check predictions.

Unit overview

In this unit, the work completed in Unit 25 Statistics (A) is extended to the representation of data in frequency diagrams, line graphs, scatter graphs and dot plots.

Each lesson in Units 25 and 26 focuses on a different way of representing data. Within each lesson, the statistical question investigated and the type of data collected match this representation. Learners make predictions before conducting the investigation, and then analyse and interpret the data, identifying patterns, within and between data sets. They draw conclusions and discuss them, considering sources of variation.

Prerequisites for learning

Learners need to:

- be able to suggest appropriate statistical questions to investigate
- be able to decide on a suitable way to collect data and how to represent it
- know how to interpret data represented in various charts and graphs and draw conclusions related to the statistical question that founded the investigation
- know how to interpret and present continuous data in frequency diagrams
- be able to solve a problem by interpreting and presenting continuous data in simple line graphs and dot plots.

Vocabulary

statistical question, frequency diagram, line graph, scatter graph, line of best fit, dot plot, frequency, peak, cluster, gap

Common difficulties and remediation

Teachers should encourage construction of tables and graphs on squared paper. This encourages learners to construct accurate columns, lines, crosses and dots, and allows them to focus on the plotted data and makes interpreting data and identifying patterns easier.

Supporting language awareness

If appropriate, when a new key word is introduced, ask learners to write a definition in their books, drawing a box around it for emphasis. Encourage learners to write the definition in their own words, for example: 'A scatter graph can be used to try to find a relationship between two measures.'

Promoting Thinking and Working Mathematically

Opportunities to develop all four pairs of TWM characteristics are provided throughout the unit.

In Lesson 1, learners conjecture (TWM.03) when they suggest how they would change an investigation to test a prediction based on the comparison of two data sets: 'I believe that boys have bigger hand spans than girls.'

In Lesson 2, learners improve (TWM.08) when they decide on the most suitable format for a data collection sheet.

Learners characterise (TWM.05) in Lesson 4 when they comment on the features of a dot plot, including peaks, clusters and gaps.

Success criteria

Learners can:

- suggest appropriate statistical questions to investigate and make predictions about what they think the data will show
- construct and interpret frequency diagrams, line graphs and dot plots
- say what data is best represented in a scatter graph
- construct a line of best fit on a scatter diagram
- interpret a scatter diagram using understanding of correlation.

Unit 26 Statistics (B)

Statistics and Probability – Statistics

Lesson 1: Frequency diagrams

Learning objectives

Code	Learning objective
6Ss.01	Plan and conduct an investigation and make predictions for a set of related statistical questions, considering what data to collect (categorical, discrete and continuous data).
6Ss.02	Record, organise and represent categorical, discrete and continuous data. Choose and explain which representation to use in a given situation: frequency diagrams for continuous data.
6Ss.04	Interpret data, identifying patterns, within and between data sets, to answer statistical questions. Discuss conclusions, considering the sources of variation and check predictions.

Resources

paper (per learner); ruler (per learner); paper or mini whiteboard and pen (per learner); several sheets of squared paper or Resource sheet 30: 2 cm squared paper (per learner); coloured pencil (per learner)

Revise

Use the activity *Time before bed* from Unit 26: *Statistics (B)* in the Revise activities.

Teach 📘 🖥 [TWM.03/06]

- Display and discuss **Slide 1**.
- Say: **We are going to gather data to answer a statistical question: 'Which hand span size is most common in our class?'** Explain that 'hand span' is a measure of distance from the tip of the thumb to the tip of the little finger when the hand is fully extended.
- Say: **We will collect grouped data arranged in intervals. [T&T] [TWM.06]** Ask: **Why do you think it is better to collect grouped data such as 26 to 27 cm rather than individual measurements such as 27·1, 27·2 cm, and so on?** Take responses. Elicit that because hand spans can take on a wide range of values, it is easier to organise and record the values as groups arranged in intervals. This makes the data more readable and easier to interpret.
- Provide paper and rulers and ask learners to draw around their right hand with their little finger and thumb fully outstretched. **[TWM.03]** Ask: **Why do you think you are all measuring your right hands, not a mix of right and left hands?** Elicit that if you vary right and left hands, then you are not comparing like with like.
- Ask learners to measure their hand span from the tip of the thumb to the tip of the little finger to the nearest millimetre, for example, 28·3 cm. Ask about a third of the learners in the class to call out their measurements and list these on the board. **[T&T]** Ask: **These are a sample of hand span widths. How should you group them into intervals?** Take responses. Decide on suitable intervals, such as 1 cm intervals: 25 to 26, 26 to 27 cm, and so on. Draw a data collection table on the board with two columns headed 'hand span width (cm)' and 'frequency'. List the width intervals down the first column.
- Ask learners to copy the table on the board. Ask them to make predictions about which intervals will be the most frequent and explain why.
- Begin the survey. Learners move around the class and collect the hand span data, making tally marks next to the appropriate intervals. Remind them that the end

numbers in each range should be included in the lower value group. Count these to complete the frequencies.

- Learners then use their data to construct a frequency diagram. Ask questions: **What is the most common width interval? Was your prediction correct? If not, can you explain why? Tell me one conclusion that you can draw from the results.**
- 🗣 **[TWM.03]** Ask: **How would you alter the investigation to test the prediction: 'I believe that boys have bigger hand spans than girls'?** Discuss.
- Learners should now be ready to suggest their own statistical question to investigate.
- **[TWM.04]** Discuss Let's learn, the paired activity and Guided practice in the Student's Book.

Practise 📒 [TWM.03/04]

- Workbook

Title: Frequency diagrams

Pages: 212–214

- Refer to Activity 1 from the Additional practice activities.

Apply 👥 🖥 [TWM.04]

- Display **Slide 2**.

Review

- Invite pairs of learners to share their statistical question, method of data collection, and results including their completed frequency diagram. What conclusions can they draw? Does it answer the statistical question?

Assessment for learning

- Give me an example of a statistical question that you could test at home or in class.

Same day intervention

Support

- Explain that grouping data enables graphs and frequency diagrams to be drawn showing class intervals rather than individual values. Demonstrate that graphing data with multiple values on the x-axis is time-consuming and challenging.

Lesson 2: **Line graphs**

Learning objectives

Code	Learning objective
6Ss.01	Plan and conduct an investigation and make predictions for a set of related statistical questions, considering what data to collect (categorical, discrete and continuous data).
6Ss.02	Record, organise and represent categorical, discrete and continuous data. Choose and explain which representation to use in a given situation: line graphs
6Ss.04	Interpret data, identifying patterns, within and between data sets, to answer statistical questions. Discuss conclusions, considering the sources of variation and check predictions.

Resources

several sheets of squared paper or Resource sheet 30: 2 cm squared paper (per learner/pair); ruler (per learner)

Revise

Use the activity *Greenhouse temperature* from Unit 26: *Statistics (B)* in the Revise activities.

Teach [SB] 🖵 [TWM.03/08]

- Say: **Phoebe records the time it takes for a beaker of warm water to cool down. The data values range from 50°C to 23°C. Which type of graph should she use to represent this data? Why?** Take responses. Establish that a line graph is a good choice for showing changes over time.
- [T&T] Ask: **What do you think the graph of water cooling over time would look like?** Take responses. Record some of these predictions on the board. Display **Slide 1**. Explain that the table is data collected from a water-cooling experiment. Ask: **If you were going to plot the data as a line graph, how would you label each axis?** Establish that the horizontal axis would be time in minutes marked in intervals of five minutes. Ask: **For the vertical axis, would intervals of one degree be the best choice?** Take responses. Elicit that intervals of two or five degrees would be best. The intervals should make best use of the vertical space, whilst ensuring that the range of tick marks on the vertical scale includes all of the data that has been plotted.
- Organise learners into pairs and provide them with squared paper and a ruler. Ask them to draw a graph using the data on the slide. Choose learners to share their graphs and explain what information the graph reveals. Does it support their predictions? If not, can they explain why?
- Invite learners to suggest their own statistical questions. These should be questions that can be answered by collecting continuous data that will be represented in a line graph. Remind them that a useful way of framing a question is to begin with 'I wonder…'. Take responses. Confirm with the class that each suggestion follows the definition of a statistical question, particularly that the data collected will be varied.
- Ask learners to predict what they think the data will show. Have them work in pairs to devise a recording

method. [TWM.08] Choose pairs to explain their method and invite the class to comment on its strengths and weaknesses, suggesting possible improvements.
- Ask learners to conduct an investigation to collect data to answer the question they have chosen. Have them represent the data in a graph. Once complete, ask learners to state one conclusion they can draw from the results and to comment on how well they did with their predictions.
- 🗣 Say: **Finn plans to investigate his prediction that: 'Learners drink more glasses of water at home than they do at school.' What question should he ask the learners in his class? How should Finn collect and represent his data? What should he do to find out if his prediction is correct or not?**
- Discuss the Let's learn section in the Student's Book.
- [TWM.03] Introduce the paired activity in the Student's Book.
- Discuss Guided practice in the Student's Book.

Practise [WB] [TWM.03/04]

- Workbook

Title: Line graphs

Pages: 215–217

- Refer to Activity 1 (Variation and accompanying NOTE) from the Additional practice activities.

Apply 👥 🖵

- Distribute squared paper and ask pairs of learners to use the information on **Slide 2** to make a line graph. They should then share their conclusions.

Review 🖵

- Display **Slide 3**. Pose questions that compare and contrast the data, for example: **How do the company profits in 2013 compare to those in 2010?**

Assessment for learning

- For a graph that plots profits against time, what does a downward sloping line indicate?

Unit 26 Statistics (B)

Statistics and Probability – Statistics

Lesson 3: **Scatter graphs**

Learning objectives

Code	Learning objective
6Ss.01	Plan and conduct an investigation and make predictions for a set of related statistical questions, considering what data to collect (categorical, discrete and continuous data).
6Ss.02	Record, organise and represent categorical, discrete and continuous data. Choose and explain which representation to use in a given situation: scatter graphs.
6Ss.04	Interpret data, identifying patterns, within and between data sets, to answer statistical questions. Discuss conclusions, considering the sources of variation and check predictions.

Resources

paper (per pair); measuring equipment such as measuring tape (per pair); several sheets of squared paper or Resource sheet 30: 2 cm squared paper (per learner/pair); ruler (per learner)

Revise

Use the activity *Is there a relationship?* from Unit 26: *Statistics (B)* in the Revise activities.

Teach 🔲 💻 [TWM.07]

- Display **Slides 1** and **2**. Discuss how to construct a scatter graph and draw a line of best fit. Say: **We are now going to draw a scatter graph to see if there is a relationship between two sets of data. [T&T]** Ask: **Is there a connection between a person's height and another body measurement?** Take responses. Prompt learners to think about the body measurements they have already investigated, such as hand span or arm span. Some learners might suggest shoulder width or shoe size. Decide on the two measurements that will be compared.

- Provide paper and suitable measuring equipment to pairs of learners and ask them to collect two measurements from each learner in the class. They need to decide on a recording format that will ensure the two sets of data are recorded alongside each other but remain separate. **[TWM.07]** Ask learners to share their data collection sheets and discuss them as a class, commenting on any strengths and weaknesses. Refine the format until everyone is agreed on the best method.

- Learners collect the data and then represent it in a scatter graph. Provide squared paper and a ruler and remind learners that they need to draw an axis for each measurement, deciding on suitable scales for both. They plot the data and draw a line of best fit. Choose learners to share their graphs and confirm that they are constructed correctly and the line passes through the middle of the plotted points.

- Ask questions such as: **What does the line tell you about the data? If you found it difficult to draw a line of best fit, what does that tell you about the two sets of data? If you have a learner with a height of 140 cm, what arm span would you expect them to have?**

- Repeat the comparison process for a different pair of data sets. What do the learners discover?

- 🗣 Say: **Sara plotted the height of a plant against the number of leaves. What do you predict the scatter plot to show? Why?**

- Discuss the Let's learn section in the Student's Book.

- Introduce the paired activity in the Student's Book. Choose pairs of learners to read the questions they wrote and explain how they used the information in the graph to answer them.

- Discuss Guided practice in the Student's Book.

Practise 📖 [TWM.03/04]

- Workbook

Title: Scatter graphs

Pages: 218–220

- Refer to Activity 2 from the Additional practice activities.

Apply 👥 💻 [TWM.04]

- Display **Slide 3**. Choose pairs to explain whether Toby's prediction is correct or not.

Review 💻

- Referring to the appropriate questions in the Workbook, choose individual learners to present their statistical questions, predictions and conclusions to the class.

Assessment for learning

- Give me an example of a statistical question where data could be collected as measurements and represented in a scatter graph.

Same day intervention
Support

- When drawing a line of best fit, a learner might suggest connecting all the plotted points on the graph. Elicit that doing so would result in a sequence of line segments rather than a single line of best fit.

Lesson 4: **Dot plots**

Learning objectives

Code	Learning objective
6Ss.01	Plan and conduct an investigation and make predictions for a set of related statistical questions, considering what data to collect (categorical, discrete and continuous data).
6Ss.02	Record, organise and represent categorical, discrete and continuous data. Choose and explain which representation to use in a given situation: Dot plots.
6Ss.04	Interpret data, identifying patterns, within and between data sets, to answer statistical questions. Discuss conclusions, considering the sources of variation and check predictions.

Resources

several sheets of Resource sheet 30: 2 cm squared paper (per pair); ruler (per learner)

Revise

Use the activity *Siblings* from Unit 26: *Statistics (B)* in the Revise activities.

Teach 🆂🅱 💻 [TWM.03/05]

- Ask: **Who can remember what a dot plot is?** Take responses. Display **Slide 1**. Discuss the data given in the dot plot. Remind learners that a dot plot is used to illustrate frequency and is best used for small data sets. Explain that the vertical axis does not need to be labelled in this case because every dot represents just one data value and they can easily be counted.
- Pose questions that require analysis of the scatter graph on the slide, for example: **How many books did most learners read this month? (2) What was the lowest number of books read? (0) How many learners read more than six books this month?** Take responses. Establish that the answer is found by totalling all the frequencies from seven books to 12 books. (23) Discuss where the data peaks (two books and 12 books) and clusters (two ranges: 0 to 4; 7 to 12 books), and where there is a gap (five and six books). **[TWM.03/05]** Ask learners to comment on these features of the graph and prompt them to offer possible explanations.
- Display **Slide 2**. Provide squared paper and a ruler. Ask learners to work in pairs to construct a dot plot from the data. Choose pairs to discuss their conclusions.
- **[T&T] [TWM.03]** Invite pairs of learners to suggest their own statistical question. Explain that the question must be answered by gathering data that is suitable for representing in a dot plot. Invite learners to share their ideas and confirm as a class that the question is valid and the corresponding data gathered can be effectively represented in a dot plot. Once confirmed, give learners time to make predictions about what they think the data will show, and time to gather data and use it to construct a dot plot. Once completed, choose pairs of learners to explain the conclusions they have drawn from the data and whether it supports their predictions.

- Discuss the Let's learn section in the Student's Book.
- **[TWM.03]** Introduce the paired activity in the Student's Book. Remind learners of the meaning of the terms 'peak', 'cluster' and 'gap'. Choose learners to share their conclusions.
- Discuss the Guided practice example.

Practise 🆆🅱 [TWM.04]

- Workbook

Title: Dot plots

Pages: 221–223

- Refer to Activity 2 (Variation) from the Additional practice activities.

Apply 👥 💻

- Display **Slide 3**. Choose pairs to explain whether the teacher's prediction is correct or not.

Review 💻

- As a class, decide upon a statistical question to investigate. The question should generate a discrete data set where the range of values is small, for example: 'How many people live in your home?' Make predictions, collect the data and then use the **Dot plot tool** to represent the data. What conclusions can be drawn from the results? Where are the peaks, clusters and gaps, if any? Does it answer the statistical question? How?

Assessment for learning

- Describe to your partner what is meant by the terms 'peak', 'cluster' and 'gap' when used to describe a dot plot.

Same day intervention
Support

- Learners may find plotting dots at equal intervals above the scale difficult. Provide grid or graph paper to help learners place one dot in each block or square.

Statistics and Probability – Statistics

Additional practice activities

Activity 1

Learning objective
- Plan and conduct an investigation and make predictions for a set of related statistical questions, considering what data to collect (categorical, discrete and continuous data).

Resources
paper (per pair); squared paper or Resource sheet 30: 2 cm squared paper (per pair); ruler (per pair); coloured pencil (per pair).

For the variation, as well as the above: large sheet of poster paper (optional) (per class); thermometer (per class); cup of warm water (optional) (per class)

What to do
- Prior to the activity, write on the board the question: 'What is the most common foot length in our class?' Have learners make predictions and record some of them on the board. Provide rulers and ask each learner to measure the length of his/her foot.
- Learners work in pairs. They construct a tally chart and collect the data from the class responding to the question: 'What is your foot length?' The results should be given as a whole number or decimal.
- Learners construct a frequency table for the data deciding on suitable intervals for the horizontal (foot length) axis, such as 16–18 cm, 18–20 cm, 20–22 cm, 22–24 cm, and so on. Remind learners that numbers at the end of each interval go in the lower value group.
- They then use the data to construct a frequency diagram and answer the question on the board.

Discuss the predictions and whether they were correct.

Variation
NOTE: This Variation is designed to be undertaken as part of Lesson 2: Practise. The day **prior** to this lesson, undertake the first part of the activity described below including the data collection.

Write on the board the question: 'How does the temperature outside our classroom change over time?' Have learners make predictions and record some of them on the board (or a large sheet of poster paper to be displayed and used later). If you live in a hot country, then take the temperature of a cooling cup of warm water over time.

Work as a class to measure and record (on the board or large sheet of poster paper) the temperature outside the classroom every hour throughout the school day.

On the day that Lesson 2 is taught, learners construct a line graph for the data deciding on suitable intervals for the vertical (temperature) axis.

They then use the data to answer the question on the board. Discuss the predictions and whether they were correct.

Activity 2

Learning objective
- Interpret data, identifying patterns, within and between data sets, to answer statistical questions. Discuss conclusions, considering the sources of variation and check predictions.

Resources
paper (per pair); squared paper or Resource sheet 30: 2 cm squared paper (per pair); ruler (per pair)

What to do
- Working in pairs, learners collect data to investigate the statistical question: 'Is there a relationship between the number of hours learners spend online and the number of hours of sleep learners get?'
- Have them predict what the data will show.
- Collect data as a class and record the results on the board.

- Working in pairs, learners construct a scatter graph to represent the data and draw a line of best fit.
- Lastly, they draw conclusions from the graph and decide whether their predictions were correct.

Variation
Ask learners to focus on the data collected for the number of hours of sleep. They use it to construct a dot plot. Ask them questions about the dot plot, including where any peaks, clusters or gaps are. Can learners offer an explanation for these features?

Unit 27: Probability

Collins International Primary Maths
Recommended Teaching and
Learning Sequence Term 3, Week 10

Learning objectives

Code	Learning objective
6Sp.01	Use the language associated with probability and proportion to describe and compare possible outcomes.
6Sp.02	Identify when two events can happen at the same time and when they cannot, and know that the latter are called 'mutually exclusive'.
6Sp.03	Recognise that some probabilities can only be modelled through experiments using a large number of trials.
6Sp.04	Conduct chance experiments or simulations, using small and large numbers of trials. Predict, analyse and describe the frequency of outcomes using the language of probability.

Unit overview

In this unit, learners describe and predict outcomes from data, using the language of probability. They discuss the relative chances of an event taking place and describe and compare possible outcomes using language associated with probability and proportion. For example, using a spinner marked 1 to 4, they give the likelihood of spinning a '3' as 1 in 4 chance or 25%, spinning a number less than 3 as a 1 in 2 chance or 50%, and a number less than 5 as 100% ('certain').

Learners are introduced to the term 'mutually exclusive' through everyday situations, for example, turning left and turning right are mutually exclusive (you can't do both at the same time); whereas turning left and scratching your head can happen at the same time. They understand that independent events are ones that have no effect on the probability of each other happening. Independent events therefore can happen at the same time.

In Lessons 3 and 4, learners conduct probability experiments and predict, analyse and describe the frequency of outcomes using the language of probability. They understand that some probabilities can only be modelled through large numbers of trials.

Prerequisites for learning

Learners need to:
• be able to express a proportion as a percentage
• be able to use vocabulary associated with chance
• know how to order the vocabulary that describes the probability of events on the correct position on a scale.

Vocabulary

probability, proportion, percentage, possible outcomes, event, mutually exclusive, independent, trial, outcome, probability experiment, predicted probability, experimental probability

Common difficulties and remediation

When teaching learners to expect empirical (experimental) probability to follow theoretical (predicted) probability with a large number of experimental

trials, be careful that learners do not assume that two probabilities will converge without question. It is possible to have an empirical probability that moves closer and then moves away from the theoretical probability, even after a large number of trials. This misconception is strengthened if there is an expectation that the 'convergence' should be rapid. Teachers should avoid using 'definites' in their discussion of probability: that the experimental probability *will* be close to the theoretical probability with a larger number of trials (this could be just after 50 or 100 trials when a 1000 might be more effective) or using phrases such as 'gets closer and closer to the probability we calculated'.

Supporting language awareness

Encourage learners to write the definition in their own words of any new key words, for example: 'Two events are mutually exclusive if you can't do both at the same time.'

Promoting Thinking and Working Mathematically

In Lesson 3, learners specialise and convince (TWM.01/04) when they compare the predicted probability of an outcome to the experimental probability, and suggest why there might be a disparity.

Learners critique (TWM.07) in Lesson 4 when they devise and conduct a probability experiment to test a prediction.

Success criteria

Learners can:
• use the language of probability and proportion
• identify two events that are mutually exclusive and explain why they are mutually exclusive
• identify two independent events and explain why they are independent
• make predictions about the frequency of outcomes using the language of probability and proportion
• explain whether the experimental probability of an outcome matches that of a theoretical probability and give reasons for why the two may be different.

Statistics and Probability – Probability

Lesson 1: **Describing and comparing outcomes**

Statistics and Probability – Probability

Learning objective

Code	Learning objective
6Sp.01	Use the language associated with probability and proportion to describe and compare possible outcomes.

Resources

one 8-sector blank spinner from Resource sheet 31: 8-sector blank spinner (per pair)

Revise

Use the activity *Probability scale* from Unit 27: *Probability* in the Revise activities.

Teach [SB] [📊] [TWM.02/04]

- Display the **Spinner tool** set to four segments (blue, orange, red, yellow) with labels removed. Say: **I am going to randomly spin a colour. Give me an outcome that would be impossible.** Take responses. Expect: a colour that is not on the spinner. Ask: **If you were to give the probability of spinning 'black' what percentage would this be?** (0%) Ask: **What probability word would you use to describe the likelihood of spinning a colour?** (certain) **Give me the probability as a percentage.** (100%)

- Explain that you can describe the probability of an event taking place as a proportion or a percentage, where an event that is certain has a probability of '1 in every 1' or 100%, and an event that is impossible has '0 in every 1' or 0% of occurring.

- **[T&T]** Ask: **What is the probability of spinning 'blue'?** Take responses. Ask questions to develop thinking: **How many possible different outcomes are there?** (4 colours: 4 possible outcomes) **How many outcomes result in 'blue' being spun?** (1) Explain that the probability of spinning blue is therefore '1 in 4'. Ask: **How do you express the proportion '1 in 4' as a percentage?** (25%) **[TWM.02]** Ask: **What other events have an equal probability to spinning 'blue'?** (spinning orange, red or yellow) **How do you know?** Establish that each event is one outcome out of four possible outcomes.

- Ask: **What is the probability of spinning red or yellow?** Take responses. Ask questions to develop thinking: **How many possible different outcomes are there?** (4 possible outcomes) **How many outcomes result in red or yellow being spun?** (2) Ask: **What is the probability as a proportion?** (2 in 4) **How can this proportion be simplified?** (1 in 2) **What is the probability as a percentage?** (50%) **What probability word do you use to describe an event that has a 50% chance of taking place?** (even chance)

- Set the spinner to 10 segments labelled 1 to 10. Ask: **What is the probability of spinning a '3'? How do you know?** Ask learners to express their answers as both a proportion and as a percentage. Establish that spinning a '3' is one outcome out of a possible 10 outcomes and therefore the probability is 1 in 10 or 10%. Ask: **Would the probability of spinning a**

'7' be a greater percentage? How do you know? (no, they are both 10%) Ask: **Would the probability of spinning a number less than 6 be higher or lower? Why?**

- 📱 Say: **Ali rolls an 8-sided dice numbered 1 to 8. What is the probability that he will spin a number less than 7?** (6 out of 8 or 3 out of 4; 75%)

- Discuss the Let's learn section in the Student's Book.

- Introduce the paired activity in the Student's Book.

- Discuss the Guided practice example.

Practise [WB] [TWM.01]

- Workbook

Title: Describing and comparing outcomes

Pages: 224–225

- Refer to Activity 1 from the Additional practice activities.

Apply [👥] [🖥️] [TWM.01]

- Display **Slide 1**. Choose learners to demonstrate their spinners and confirm they are correctly labelled.

Review

- Describe a probability experiment: ten jellybeans of different colours are placed in a jar and one is chosen at random. Ask learners to describe the colours of jelly beans in the jar for the following probability scenarios: a) There is a 50% chance of picking an orange bean; b) The probability of picking a red bean is 20% greater than the probability of picking a yellow bean; c) There is a 4 in 5 chance of picking a blue jellybean.

Assessment for learning

- A spinner is divided into eight sectors, five are marked 'A', two 'B' and one 'C'. Describe the probability of spinning each letter.

Same day intervention
Support

- Some learners may have difficulty establishing the meaning of probability words, such as 'possible', 'unlikely' and 'certain', for example, equating 'rare' with 'impossible'. Provide plenty of opportunities for learners to discuss the likelihood of everyday events happening and model the correct use of the language of probability.

Lesson 2: **Independent and mutually exclusive events**

Learning objective

Code	Learning objective
6Sp.02	Identify when two events can happen at the same time and when they cannot, and know that the latter are called 'mutually exclusive'.

Resources

paper or mini whiteboard and pen (per pair); one 6-sector blank spinner from Resource sheet 32: 6-sector blank spinner (per pair)

Revise

Use the activity *Likely thumbs* from Unit 27: *Probability* in the Revise activities.

Teach 🔲 🖥 [TWM.01/05/06]

- Display **Slide 1**. Read the first pair of events based on the outcome of a football match. **[T&T]** Ask: **Can these events take place at the same time?** Take responses. Establish that the football team can't both win and lose a match. Explain that you call events that cannot happen at the same time 'mutually exclusive'.

- Read the second pair of events based on the outcome of flipping a coin. Ask: **Can these events take place at the same time?** Take responses. Establish that a flipped coin can't land on both 'heads' and 'tails'. Explain that when two events (call them 'A' and 'B') are 'mutually exclusive', it is impossible for them to happen together.

- Discuss the independent events on **Slide 2**. Explain that events are independent if the outcome of one event does not affect the outcome of a second event. For example, if you choose a number card from a pack of cards and then flip a coin, the number chosen from the pack does not affect the result you get on the coin.

- **[TWM.01/05]** Have learners work in pairs and ask them to record two pairs of events. Events A and B should be mutually exclusive and events C and D should be independent. Choose learners to share their events and, as a class, confirm that they are mutually exclusive or independent. Record some of the suggestions on the board.

- 🔲 Present learners with the following problem: **Which pair of events is independent and which is mutually exclusive – throwing a number less than 4 with a 1 to 6 dice/throwing a number greater than 4 with a 1 to 6 dice; eating cereal for breakfast/eating sweetcorn for dinner? Explain your answer.**

- Discuss the Let's learn section in the Student's Book and address any misunderstandings.

- **[TWM.06]** Introduce the paired activity in the Student's Book. Choose pairs of learners to explain how they decided which pairs of events are mutually exclusive.

- Discuss the Guided practice example in the Student's Book.

Practise 📖 [TWM.02/06]

- Workbook

Title: Independent and mutually exclusive events

Pages: 226–227

- Refer to Activity 1 (Variation) from the Additional practice activities.

Apply 👥 🖥 [TWM.01]

- Display **Slide 3**. Learners label each sector of a 6-sector spinner with letters so that the probability of spinning a letter from one three-letter word is mutually exclusive from spinning a letter from a second three-letter word. If learners need help, then suggest two three-letter animals (D-O-G, C-A-T). Choose learners to demonstrate their spinners and confirm they are correctly labelled.

Review

- Ask learners to say whether each of the following pairs of events are mutually exclusive or not: 'Rolling a 1 to 6 dice and getting a prime number and an even number'; 'Getting an odd number and a number greater than 7 when rolling 1 to 8 dice'; 'With two rolls of a 1 to 6 dice, rolling a 5 twice.'

Assessment for learning

- Describe to your partner, using an example, what is meant by a pair of mutually exclusive events.
- Describe to your partner, using an example, what is meant by a pair of independent events.

Same day intervention

Support

- Look out for learners who have difficulty with the notions of independent and mutually exclusive events; they very often confuse the two. Provide other examples of each type of event. Ensure learners understand that if two events are mutually exclusive, it does not necessarily mean they are independent.

Statistics and Probability – Probability

269

Unit **27** Probability

Statistics and Probability – Probability

Lesson 3: **Event probability and the number of trials (1)**

Learning objectives

Code	Learning objective
6Sp.03	Recognise that some probabilities can only be modelled through experiments using a large number of trials.
6Sp.04	Conduct chance experiments or simulations, using small and large numbers of trials. Predict, analyse and describe the frequency of outcomes using the language of probability.

Resources

coin clearly showing 'heads' or 'tails' (per pair); one 10-sector blank spinner from Resource sheet 33: 10-sector blank spinner (per pair); pencil and paper clip, for the spinner (per pair)

Revise

Use the activity *Independent or mutually exclusive?* from Unit 27: *Probability* in the Revise activities.

Teach [SB] [📊] [TWM.01/04]

- Display the **Spinner tool** set to two quadrants coloured red and blue. Ask: **Which colour do you think the spinner will land on?** Spin the spinner and discuss the result. Repeat several times, giving learners several opportunities to predict the colour spun. [TWM.04] [T&T] Ask: **If the colour spinner was a device used in a two-player game, would it be a fair game? Why?** Discuss what makes a fair spinner. Ask: **Do both players have an equal chance of winning?** (yes) Ask: **What fraction of the spinner is blue?** ($\frac{1}{2}$) **What fraction is red?** ($\frac{1}{2}$) **What is the probability of spinning blue?** (1 in 2 or 50%) **If I spin the spinner 100 times, how many times would you expect to spin blue?** (50)
- Perform the probability experiment for ten spins. Explain that a spin is an example of a trial, one particular performance of a probability experiment. In this case, you perform ten trials (ten spins). Discuss the outcomes of the spins and explain that an outcome is the result of a trial.
- Populate the table in the **Spinner tool** with sample data by selecting '100 spins' in the drop-down menu and clicking 'spin'. Invite learners to compare the experiment results with the predicted results. [TWM.01/04] Ask: **How close was the predicted probability to the actual results? Why do you think there is a difference between the predicted probability and the experimental probability?** Explain that probability gives you an indication of the likelihood of an event happening. You expect to spin 'red' one in every two spins or 50% of the time. [TWM.04] Ask: **If I spin 'red' seven times in a row, is it more likely that I will spin 'red' on the eighth spin?** Emphasise that this is a prediction and the likelihood of spinning red remains 50% for every spin. The probability does not change as it is independent of the results of previous spins.
- Repeat the previous probability experiment with a spinner set to four sectors: three sectors yellow and one sector pink. Ask: **What is the probability of spinning pink?** (1 in 4 or 25%) **What is the probability of spinning yellow?** (3 in 4 or 75%)

Perform the probability experiment for ten spins. Discuss what the data reveals. Ask: **If you spin 100 times, how many times should you expect to spin yellow?** (75) **Pink?** (25) Populate the table in the **Spinner tool** with sample data by selecting '100 spins' in the drop-down menu and clicking 'Spin'. Invite learners to compare the experiment results with the predicted results.

- 🗣 Say: **A five-sector spinner has four sectors orange and one sector red. Over 100 trials, how many times would you expect to spin orange? Red? Why?** (80, 20)
- Discuss the Let's learn section in the Student's Book.
- Introduce the paired activity in the Student's Book.
- Discuss the Guided practice example.

Practise [WB] [TWM.03]

- Workbook

Title: Event probability and the number of trials (1)

Pages: 228–229

- Refer to Activity 2 from the Additional practice activities.

Apply [👥] [🖥] [TWM.04]

- Display **Slide 1**. Ask learners to explain their expected frequencies and their results and why they may be different.

Review

- Say: **Tammy rolls a 1 to 20 dice. What is the percentage probability that she will spin an odd number?** (50%) **A number greater than 15?** (25%) **A prime number?** (40%) **What frequency of each outcome should Tammy expect to record after 80 rolls of the dice?** (40, 20, 32)

Assessment for learning

- How many spins should Liam expect to record 'green' if he spins 50 times? (30)

Same day intervention
Support

- Ensure learners understand that rolling a single dice or flipping a coin several times results in independent events. One flip of a coin does not affect the outcome of subsequent flips.

Lesson 4: **Event probability and the number of trials (2)**

Learning objectives

Code	Learning objective
6Sp.03	Recognise that some probabilities can only be modelled through experiments using a large number of trials.
6Sp.04	Conduct chance experiments or simulations, using small and large numbers of trials. Predict, analyse and describe the frequency of outcomes using the language of probability.

Resources

letters A to E from Resource sheet 34: Letter cards (per class)

Revise

Use the activity *Higher or lower?* (2) from Unit 27: *Probability* in the Revise activities.

Teach [SB] [📊] [TWM.02/03/04/08]

- Shuffle the five cards A to E from Resource sheet 34. Fan them out and explain to the class that each card displays a different letter from A to E. Ask: **What is the probability of picking a 'D'?** (20%) Draw a frequency table on the board with rows for each letter. Invite a learner to pick one card and record a tally mark next to the outcome on the board. Repeat this 49 times more. Remember to replace the card each time and reshuffle before fanning the cards out once again.
- [T&T] [TWM.02] Ask: **How do the experimental frequencies compare to the predicted frequencies?** Discuss how close the two probabilities are. Ask learners to make suggestions as to why the two frequencies might be different. Ensure learners understand that knowing the probability of an outcome does not indicate the outcome is definite. Say: **Probabilities do not tell you what is going to happen, they simply tell you what is *likely* to happen.** Explain that it is unlikely that you will pick a letter 'B' exactly ten times out of 50 every time you conduct the experiment; however, if enough people pick cards for long enough, then this may well happen. Even then, it is still not definite. Say: **When we conduct probability experiments, the greater the number of trials completed, in this case the greater the number of cards picked, the more likely the experimental probability will agree with the predicted probability.**
- Display the **Spinner tool** set to four sectors: three sectors yellow and one sector pink. Remind learners that this is the spinner combination used in Lesson 3 when 100 trials were conducted. Remind them of the probabilities they calculated: 25% chance of spinning pink and 75% chance of spinning yellow. Say: **We performed 100 spins last time, now let's perform 1000 spins. [TWM.03]** Ask: **How might a tenfold increase in the number of trials affect the results?** Take responses. Conduct the experiment by selecting '1000 spins' and ask learners to compare the experiment results with the

predicted results. Ask: **Has the tenfold increase in the number of trials brought the predicted and experimental frequencies closer?**
- Discuss the Let's learn section in the Student's Book.
- **[TWM.04/08]** Introduce the paired activity in the Student's Book. Choose pairs of learners to discuss what the experiment revealed.
- Discuss Guided practice in the Student's Book.

Practise [WB] [TWM.04]

- Workbook

Title: Event probability and the number of trials (2)

Pages: 230–232

- Refer to Activity 2 (Variation: **Slide 1**). from the Additional practice activities.

Apply 👥 🖥️

- Referring to **Slide 1**, ask learners to work out the predicted probabilities for each of the following outcomes:
a) picking a '3' b) picking a '7' c) picking a '9'.

Review

- Working in pairs, learners conduct their own probability experiment. This might involve a dice, a spinner or a coin. Have learners make a prediction about the probability of one outcome and the frequency of this outcome over a certain number of trials. Give them time to conduct the experiment and record the results. How do the predicted and experimental probabilities compare? Have them increase the number of trials and comment on the same question.

Assessment for learning

- How might the number of trials conducted in a probability experiment affect the outcome? Why?

Same day intervention
Support

- Discuss the likelihood of each outcome in a probability event and encourage learners to work out and design predicted probabilities.

Additional practice activities

Activity 1 👥 ⚠2

Learning objective
• Use the language associated with probability and proportion to describe and compare possible outcomes.

Resources
• one 10-sector blank spinner from Resource sheet 33: 10-sector blank spinner (per learner); coloured pencils in blue, green, red and yellow (per learner)

What to do
• Without their partner seeing, each learner lightly marks the sectors of their spinner with the letters 'B', 'G', 'R' and 'Y' to represent the colours blue, green, red and yellow. They may choose any combination of letters and sectors.
• Learners then write down the probability rules for designing their spinner, for example, 'The probability of spinning red is 3 in 5' or 'The probability of spinning green is 30%'.
• They erase the letters on their spinners and then swap probability rules.

• Learners then colour their spinners to satisfy the probability rules.
• They return their spinners to confirm they have been coloured correctly.

Variation
⚠2 Learners take turns to describe a probability event to each other, for example, spinning a red-green-blue spinner, rolling a 1 to 8 dice or throwing a dart at a dartboard numbered 1 to 25.

Learner A describes one possible outcome and then says 'Mutually exclusive' or 'Independent'. Learner B then must describe a related outcome. For example, if learner A says, 'A red-green-blue spinner is spun and the result is blue' and 'Independent' then Learner B might say, 'A second spinner is spun and the result is blue' or 'A 1 to 8 dice is rolled and the result is 3'.

Activity 2 👥👥 ⚠2

Learning objectives
• Recognise that some probabilities can only be modelled through experiments using a large number of trials.
• Conduct chance experiments or simulations, using small and large numbers of trials. Predict, analyse and describe the frequency of outcomes using the language of probability.

What to do
• On the board, write the numbers: 8, 2, 8, 1, 4. Say: **These are the numbers on the faces of a five-sided dice.**
• Ask: **If you were to roll the dice, what is the probability of each of the following events:**
  – **Rolling an '8'?** (2 in 5 or 40%)
  – **Rolling an odd number?** (1 in 5 or 20%)
  – **Rolling an even number?** (4 in 5 or 80%)
• Ask: **If you were to roll the dice 40 times, how many times would you expect to roll these outcomes?** (16, 8, 32)

Variation [TWM.04]
⚠2 Display Stage 6, Unit 27, Lesson 4, **Slide 1** (Apply). Read and discuss the experiment.

Ask learners to comment on how the predicted and experimental probabilities compare for each set of trials, 100 and 1000.

[TWM.04] Ask: What does this tell you about how the number of trials might influence the results of a probability experiment?

1–20 number cards

1	2	3	4
5	6	7	8
9	10	11	12
13	14	15	16
17	18	19	20

0–100 number cards

0	1	2	3	4	5	6
7	8	9	10	11	12	13
14	15	16	17	18	19	20
21	22	23	24	25	26	27

0–100 number cards

34	41	48	55
33	40	47	54
32	39	46	53
31	38	45	52
30	37	44	51
29	36	43	50
28	35	42	49

0–100 number cards

62	69	76	83
61	68	75	82
60	67	74	81
59	66	73	80
58	65	72	79
57	64	71	78
56	63	70	77

0–100 number cards

88	93	98	100
87	92	97	99
86	91	96	
85	90	95	
84	89	94	

Spinner (1)

How to use the spinner

Hold the paper clip in the centre of the spinner using the pencil and gently flick the paper clip with your finger to make it spin.

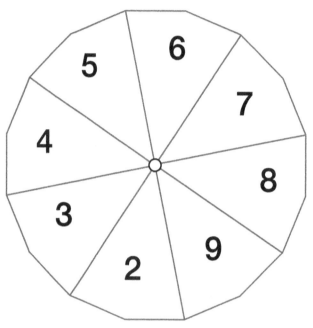

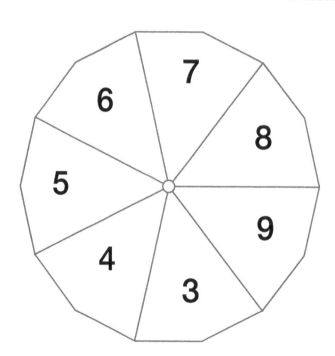

Gameboard

1	2	3	4	5	6
8	9	10	15	16	18
20	24	25	30	32	36

1	2	3	4	5	6
8	9	10	15	16	18
20	24	25	30	32	36

1–6 and 1–9 spinners

How to use the spinner

Hold the paper clip in the centre of the spinner using the pencil and gently flick the paper clip with your finger to make it spin.

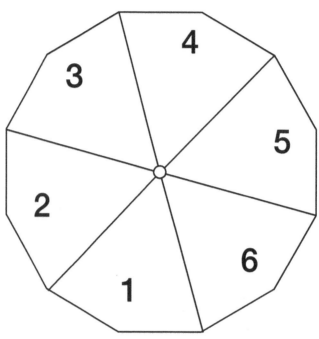

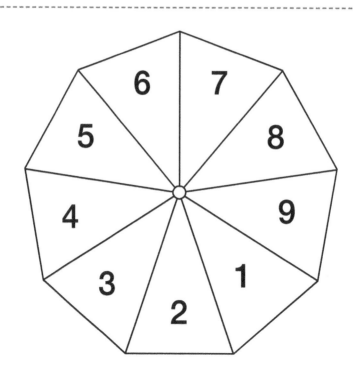

10–100 number cards

16	23	30	37
15	22	29	36
14	21	28	35
13	20	27	34
12	19	26	33
11	18	25	32
10	17	24	31

10–100 number cards

44	51	58	65
43	50	57	64
42	49	56	63
41	48	55	62
40	47	54	61
39	46	53	60
38	45	52	59

10–100 number cards

72	79	86	93
71	78	85	92
70	77	84	91
69	76	83	90
68	75	82	89
67	74	81	88
66	73	80	87

10–100 number cards

98

97

96

95

100

94

99

3–99 number cards

9	16	23	30
8	15	22	29
7	14	21	28
6	13	20	27
5	12	19	26
4	11	18	25
3	10	17	24

3–99 number cards

37	43	50	57
36	42	49	56
35	41	48	55
34	40	47	54
33	40	46	53
32	39	45	52
31	38	44	51

3–99 number cards

64	71	78	85
63	70	77	84
62	69	76	83
61	68	75	82
60	67	74	81
59	66	73	80
58	65	72	79

3–99 number cards

92	99
91	98
90	97
89	96
88	95
87	94
86	93

Place value counters

100	100	100	100	100	100	100	100	100
100	100	100	100	100	100	100	100	100
100	100	100	100	100	100	100	100	100
10	10	10	10	10	10	10	10	10
10	10	10	10	10	10	10	10	10
10	10	10	10	10	10	10	10	10
1	1	1	1	1	1	1	1	1
1	1	1	1	1	1	1	1	1
1	1	1	1	1	1	1	1	1

Spinner (2)

How to use the spinner

Hold the paper clip in the centre of the spinner using the pencil and gently flick the paper clip with your finger to make it spin.

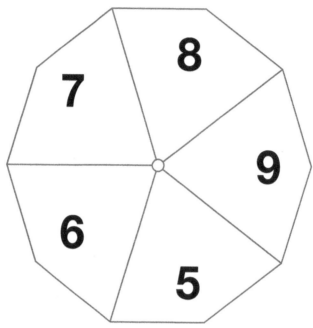

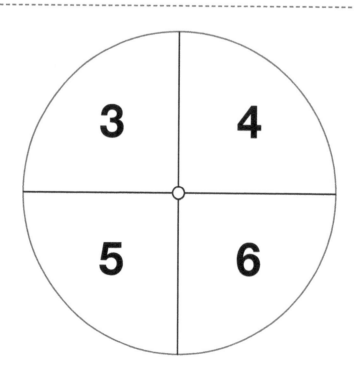

Division number cards (1)

12	**13**	**19**
23	**7**	**6**
5	**4**	**84**
78	**95**	**92**

Operator cards

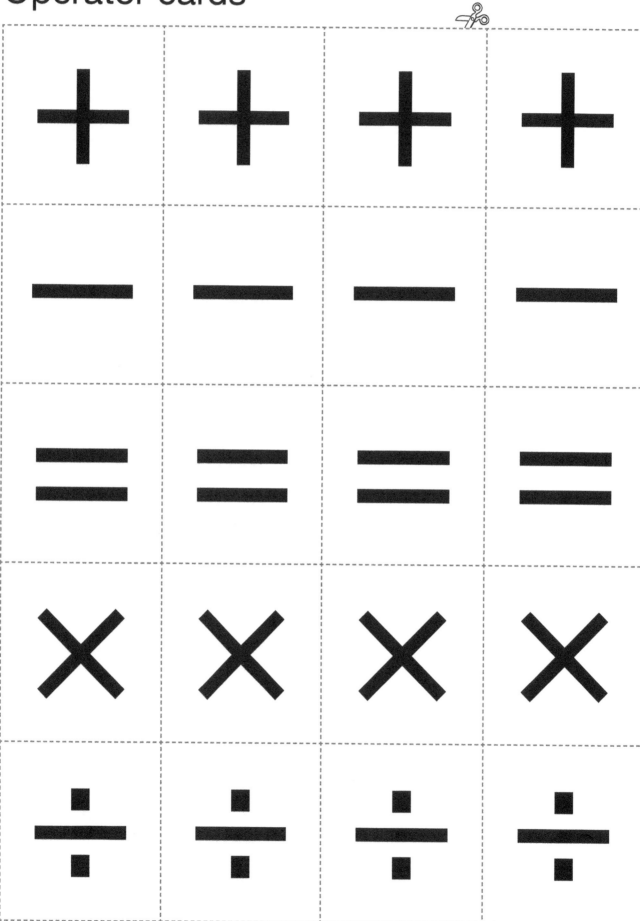

Division number cards (2)

14	16	18	22
23	25	28	31
434	448	450	506

0–9 digit cards

4

9

3

8

2

7

1

6

0

5

Waffle diagrams

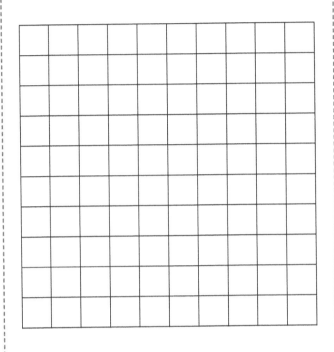

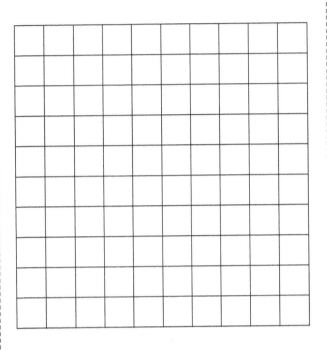

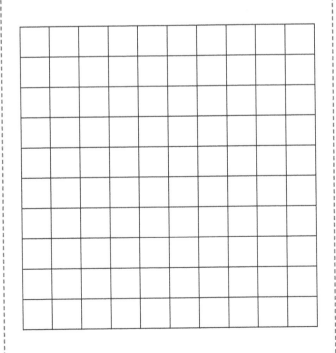

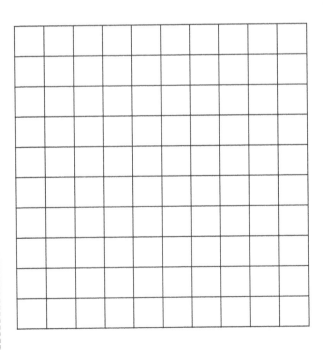

Multiples of 5% cards

5%	10%	15%	20%
25%	30%	35%	40%
45%	50%	55%	60%
65%	70%	75%	80%
85%	90%	95%	

Multiples of 20 cards

120	140	160	180
200	220	240	260
280	300	320	340
360	380	400	

Square dot paper

Circle

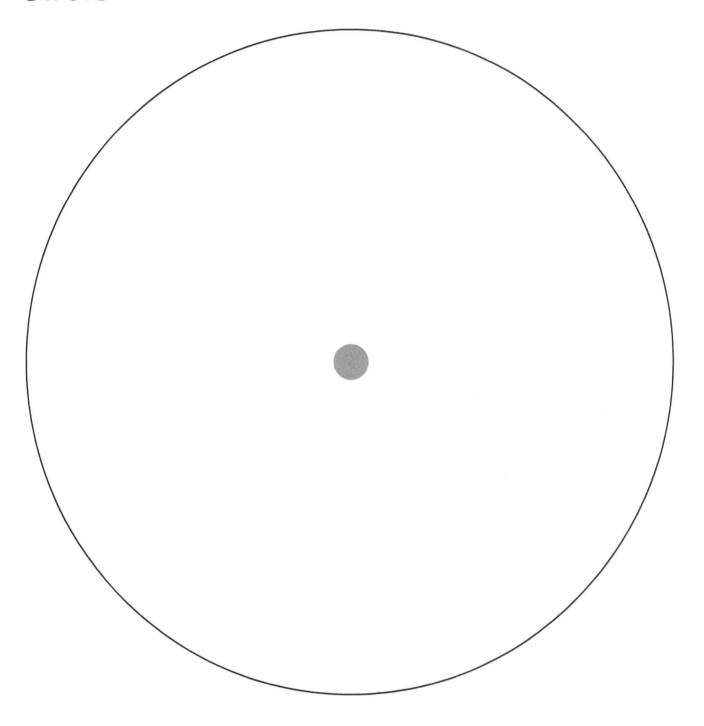

Triangular dot paper

Nets of pyramids

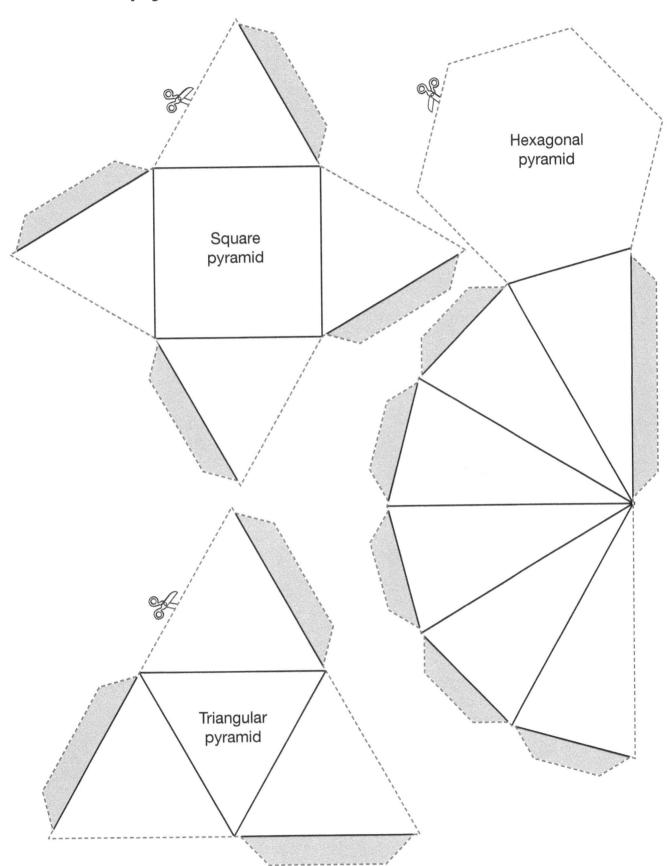

Square pyramid

Hexagonal pyramid

Triangular pyramid

Nets of prisms

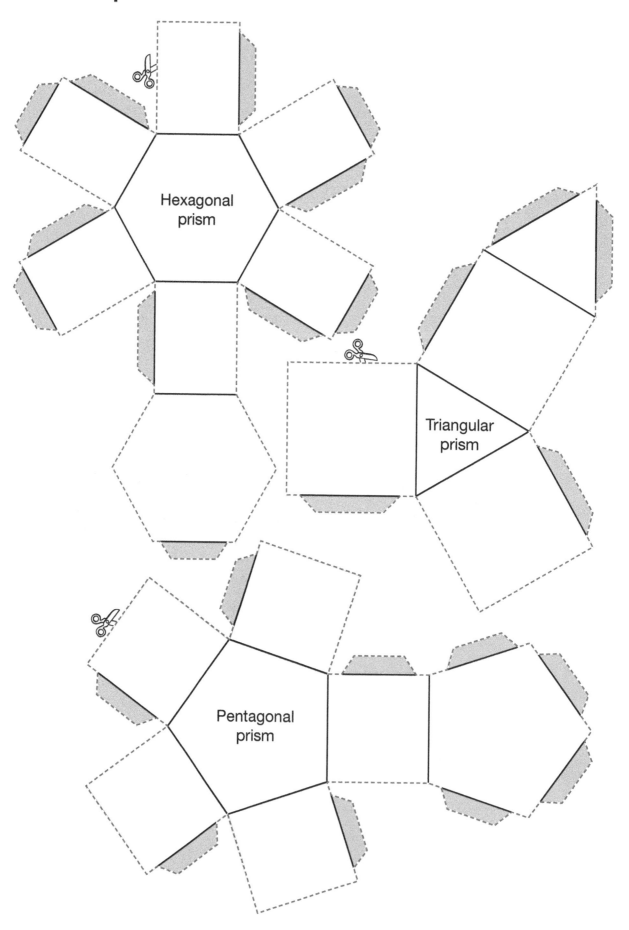

Pyramid properties

	Triangular pyramid	Square pyramid	Pentagonal pyramid
Faces			
Edges			
Vertices			

Angles

a)

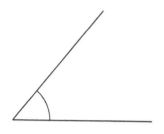

This angle is _____ degrees.

b)

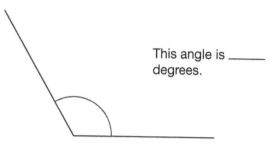

This angle is _____ degrees.

c)

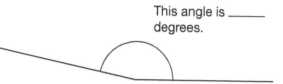

This angle is _____ degrees.

d)

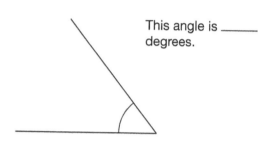

This angle is _____ degrees.

e)

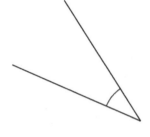

This angle is _____ degrees.

f)

This angle is _____ degrees.

g)

This angle is _____ degrees.

h)

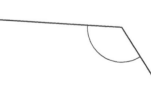

This angle is _____ degrees.

Missing angle problems

- Turn the drawing into a missing angle problem to give to another pair of learners to solve. Work out the minimum number of angles that must be labelled to make it possible to find all the remaining angles.

- Label the angles and swap with another pair to solve.

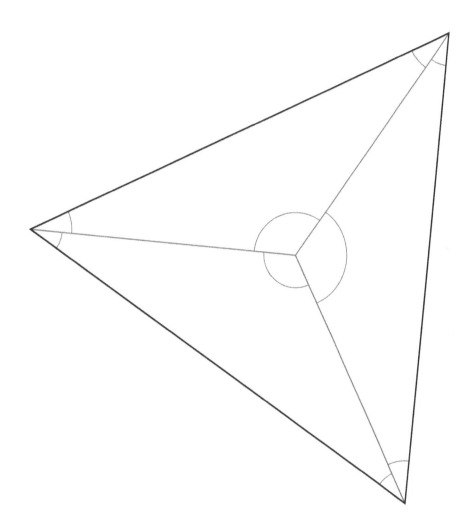

Fraction cards

$\frac{1}{2}$	$\frac{1}{4}$	$\frac{3}{4}$	$\frac{1}{10}$
$\frac{2}{10}$	$\frac{3}{10}$	$\frac{4}{10}$	$\frac{5}{10}$
$\frac{6}{10}$	$\frac{7}{10}$	$\frac{8}{10}$	$\frac{9}{10}$

$\frac{1}{2}$	$\frac{1}{4}$	$\frac{3}{4}$	$\frac{1}{10}$
$\frac{2}{10}$	$\frac{3}{10}$	$\frac{4}{10}$	$\frac{5}{10}$
$\frac{6}{10}$	$\frac{7}{10}$	$\frac{8}{10}$	$\frac{9}{10}$

Four quadrant grid

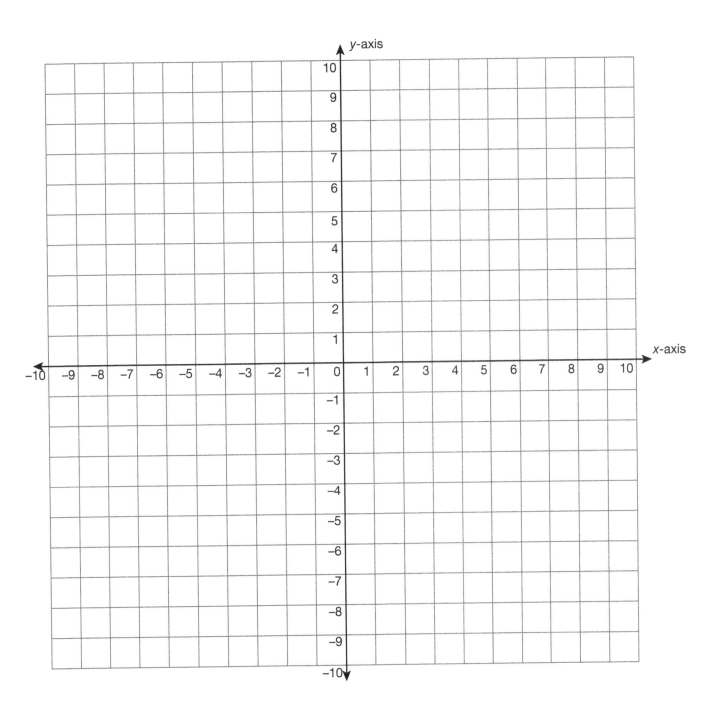

Spinner (3)

How to use the spinner

Hold the paper clip in the centre of the spinner using the pencil and gently flick the paper clip with your finger to make it spin.

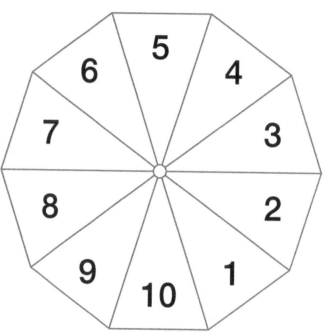

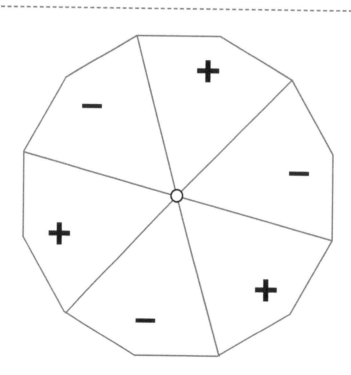

Spinners (4)

How to use the spinner

Hold the paper clip in the centre of the spinner using the pencil and gently flick the paper clip with your finger to make it spin.

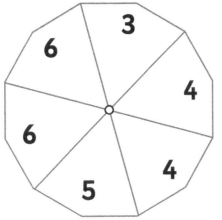

How to use the spinner

Hold the paper clip in the centre of the spinner using the pencil and gently flick the paper clip with your finger to make it spin.

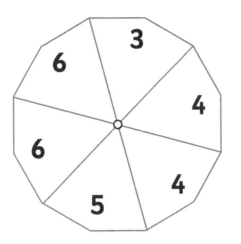

Pie chart (4 divisions)

Pie chart (5 divisions)

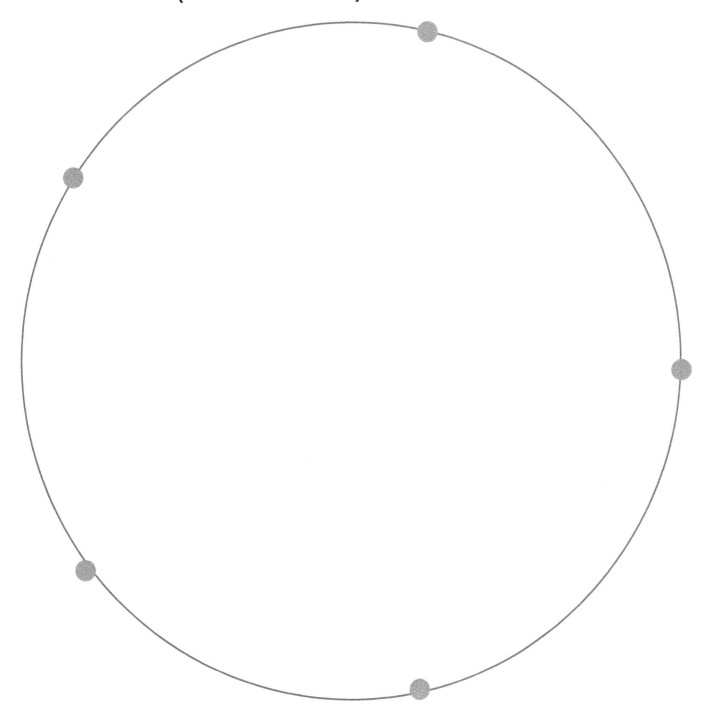

Pie chart (10 divisions)

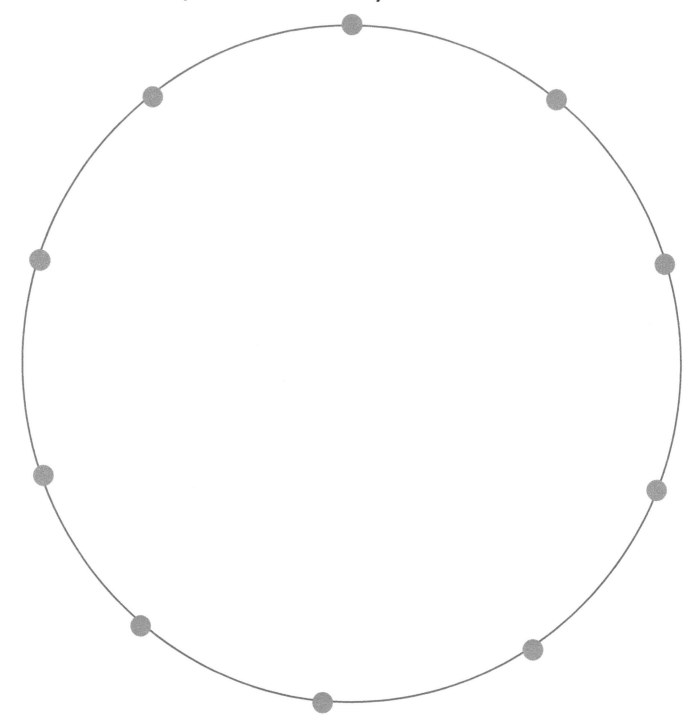

Pie chart (20 divisions)

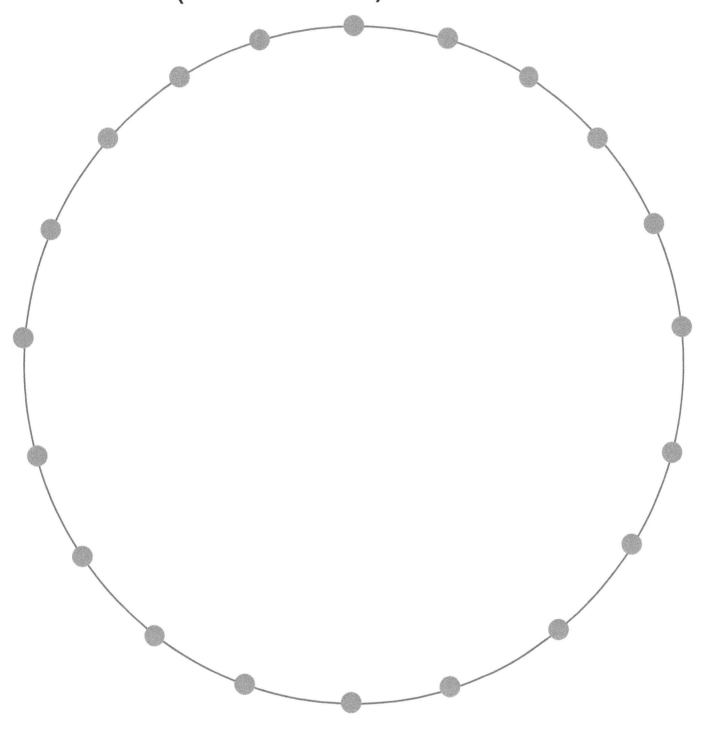

2 cm squared paper

8-sector blank spinner

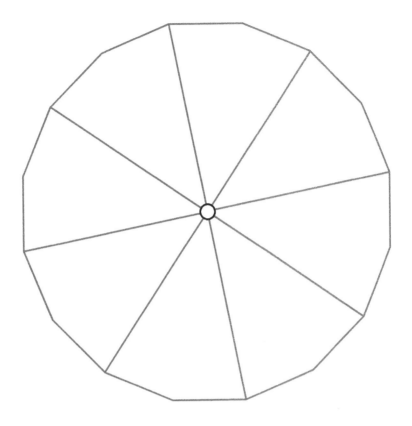

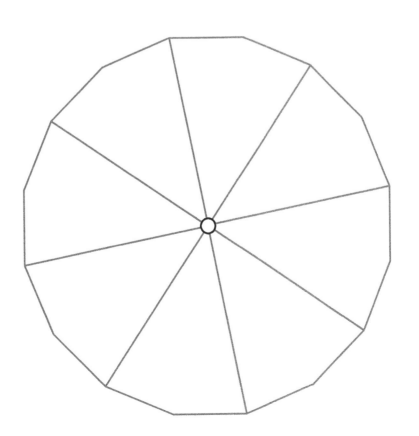

6-sector blank spinner

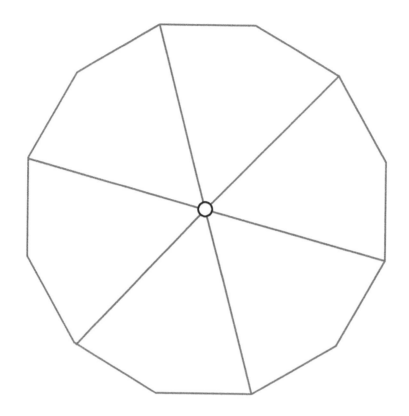

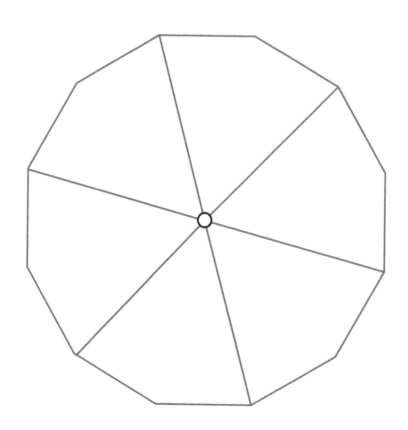

10-sector blank spinner

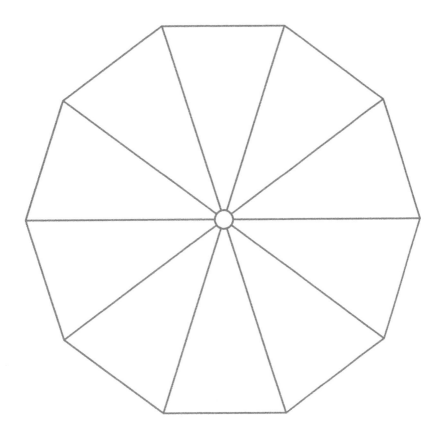

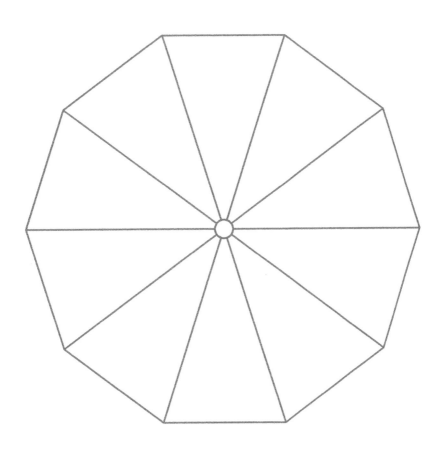

Letter cards

A

B

C

D

E

Operations spinner

How to use the spinner

Hold the paper clip in the centre of the spinner using the pencil and gently flick the paper clip with your finger to make it spin.

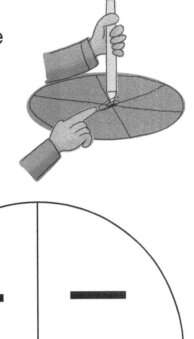

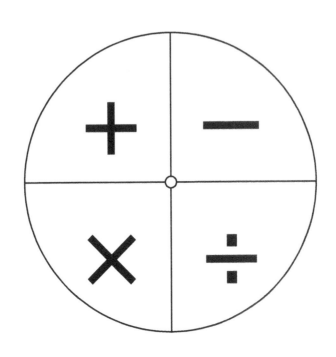

Symmetrical letters

Which of these letters have rotational symmetry?

If you answer yes, write the order of rotational symmetry for the letter.

(i)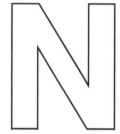

(ii)

(iii)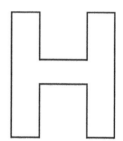

(iv)

(v)

(vi)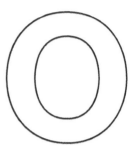

(vii)

Unit 1

Lesson 1: Counting on and back in fractions and decimals

Challenge ❶

1 a 2·7, 2·9, 3·1, 3·5, 3·7, 4·1, 4·3

b 5·21, 5·2, 5·19, 5·17, 5·16, 5·14, 5·13

c $6\frac{2}{3}$, 7, $7\frac{1}{3}$, 8, $8\frac{1}{3}$, 9, $9\frac{1}{3}$

d 8, $7\frac{1}{2}$, 7, 6, $5\frac{1}{2}$, $4\frac{1}{2}$, 4

2 a 7·1, 7·5, 7·9, 8·7, 9·1, 9·9, 10·3

b 4·52, 4·49, 4·46, 4·4, 4·37, 4·31, 4·28

c 2·164, 2·214, 2·264, 2·364, 2·414, 2·514, 2·564

d $6\frac{2}{3}$, 6, $5\frac{1}{3}$, 4, $3\frac{1}{3}$, 2, $1\frac{1}{3}$

e $1\frac{4}{5}$, $2\frac{2}{5}$, 3, $4\frac{1}{5}$, $4\frac{4}{5}$, 6, $6\frac{3}{5}$

Challenge ⚠

2

Start balance	$20	–$15	$1.40	–$2.30	–$3.10
Daily increase or decrease	Decrease $7 per day	Increase $4 per day	Decrease $0.30 per day	Increase $0.40 per day	Increase $0.80 per day
Number of days	4	9	6	7	8
Final bank balance	–$8	$21	–$0.40	$0.50	$3.30

3

Start temperature (°C)	1·5	–2·3	0·8	–3·1	–2·7
Temperature rise/fall	Fall 0·4 degrees per hour	Rise 0·5 degrees per hour	Fall 0·7 degrees per hour	Rise 0·6 degrees per hour	Rise 0·4 degrees per hour
Number of hours	6	8	6	7	7
End temperature (°C)	–0·9	1·7	–3·4	1·1	0·1

4 $1\frac{1}{10}$, $\frac{8}{10}$, $\frac{5}{10}$, $\frac{2}{10}$, $\frac{-1}{10}$, $\frac{-4}{10}$, $\frac{-7}{10}$

Challenge ❸

5 a –3·2 m **b** –5·4 m **c** –6·5 m

Lesson 3: Finding the position-to-term rule

Challenge ❶

1 a

Position	1	2	3	4	5	6	7	8
Term	2	4	6	8	10	12	14	16

   i 18 and 20 **ii** 12 **iii** Explanations will vary

b

Position	1	2	3	4	5	6	7	8
Term	4	8	12	16	20	24	28	32

   i 36 and 40 **ii** 11

Challenge ⚠

3 a 2·15, 2·19, 2·27, 2·35, 2·39, 2·43, 2·51, 2·59, 2·63

b 6·78, 6·63, 6·6, 6·39, 6·33

c $3\frac{1}{8}$, $3\frac{4}{8}$, $4\frac{2}{8}$, $4\frac{5}{8}$, $5\frac{3}{8}$, $5\frac{6}{8}$, $6\frac{1}{8}$, $7\frac{5}{8}$, 8

4 16·17 m

Challenge ❸

5 a 4·1, 4·7, 5·9

b 16·53, 16·45, 16·41

c $16\frac{7}{8}$, 16, $14\frac{2}{8}$

Lesson 2: Counting on and back beyond zero

Challenge ❶

1 a Sequence: 10, 8, 6, 4, 2, 0, –2, –4, –6, –8, –10

b Sequence: 1, 0·8, 0·6, 0·4, 0·2, 0, –0·2, –0·4, –0·6, –0·8, –1

c Sequence: –3, $-2\frac{1}{4}$, $-1\frac{2}{4}$, $-\frac{3}{4}$, 0, $\frac{3}{4}$, $1\frac{2}{4}$, $2\frac{1}{4}$, 3

Challenge ⚠

2 a

Position	1	2	3	4	5	6
Term	5	10	15	20	25	30

   i Multiply by 5 **ii** 45 **iii** 20

b

Position	1	2	3	4	5	6
Term	12	24	36	48	60	72

   i Multiply by 12 **ii** 84 **iii** 10

3 a

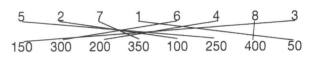

5   2   7   1   6   4   8   3

150 300 200 350 100 250 400 50

b Multiply by 50 **c** 500

Challenge 3

4 a i 50 **ii** 225 **iii** 350 **iv** 475

 b i 48 **ii** 144 **iii** 192 **iv** 276

Lesson 4: Finding terms of a square number sequence

Challenge 1

1

Position	1	2	3	4	5
Pattern	•	• •	• • • • • • • • •	• • • • • • • • • • • • • • • •	• • • • • • • • • • • • • • • • • • • • • • • • •
Term	1×1 $= 1$	2×2 $= 4$	3×3 $= 9$	$4 \times 4 =$ 16	$5 \times 5 =$ 25

Challenge 2

2

Position	Calculation	Value
1	1^2	1
2	2^2	4
3	3^2	9
4	4^2	16
5	5^2	25
6	6^2	36
7	7^2	49
8	8^2	64
9	9^2	81
10	10^2	100

3 a 36 **b** 81 **c** 169

Challenge 3

4 a Example answers: $9 + 16 = 25$ or $36 + 64 = 100$ or $81 + 144 = 225$

 b Example answers: $4 + 9 + 36 = 49$ or $1 + 16 + 64 = 81$

 c Explanations will vary, e.g. it's more methodical and less random

Unit 2

Lesson 1: Adding positive and negative numbers (1)

Challenge 1

1 a $-4 + 3 = -1$ **b** $-6 + 2 = -4$

 c $-8 + 6 = -2$ **d** $-7 + 5 = -2$

 e $-9 + 8 = -1$ **f** $-8 + 7 = -1$

 g $-8 + 14 = 6$ **h** $-5 + 15 + 10$

 i $-9 + 18 = 9$ **j** $-10 + 18 = 8$

 k $-9 + 19 = 10$ **l** $-2 + 9 = 7$

Challenge 2

2 a $-11 + 8 = -3$ **b** $-14 + 12 = -2$

 c $-13 + 5 = -8$ **d** $-15 + 14 = -1$

 e $-16 + 15 = -1$ **f** $-14 + 13 = -1$

 g $-17 + 28 = 11$ **h** $-18 + 31 = 13$

 i $-19 + 29 = 10$ **j** $-20 + 39 = 19$

 k $-6 + 14 = 8$ **l** $-9 + 19 = 10$

3

Starting balance	Money in	New balance	Calculation
−$16	$23	$7	$-16 + 23 =$
−$18	$26	$8	$-18 + 26 = 8$
−$11	$24	$13	$-11 + 24 = 13$
−$14	$32	$18	$-14 + 32 = 18$
−$4	$35	$31	$-4 + 35 = 31$
−$2	$37	$35	$-2 + 37 = 35$

Challenge 3

4 Learner's calculations, following the example given

Lesson 2: Adding positive and negative numbers (2)

Challenge 1

1 a −8°C **b** −11°C **c** −4°C

 d 2°C **e** 10°C **f** 10°C

Challenge 2

2 a −7 m **b** −1 m **c** −1 m

 d 13 m **e** 18 m

3 a $-17 + 24 = 7$ **b** $-19 + 26 = 7$

 c $-22 + 19 = -3$ **d** $-23 + 26 = 3$

 e $-18 + 34 = 16$ **f** $-24 + 42 = 18$

 g $-44 + 52 = 8$ **h** $-52 + 67 = 15$

 i $-55 + 73 = 18$ **j** $-58 + 84 = 26$

 k $-46 + 75 = 29$ **l** $-37 + 83 = 46$

Challenge 3

4

Temperature (°C)	Temperature rise	New temperature (°C)
−36	77	41
−27	64	37
−32	80	48
−24	86	62
−27	64	37
−42	75	33
−31	58	27

Lesson 3: Identifying values for symbols in addition calculations

Challenge ❶

1

= $7   = $2   = $3

= $6   = $4

Challenge ⚠

2

a	36 + a = 76	a = 40
b	b + 40 = 83	b = 43
c	c + 17 = 55	c = 38
d	43 + d = 69	d = 26
e	12 + e = 52	e = 40
f	f + 30 = 53	f = 23
g	73 = 29 + g	g = 44
h	h + 33 = 51	h = 18
i	41 + i = 89	i = 48
j	71 = j + 36	j = 35
k	k + 53 = 72	k = 19
l	57 + l = 95	l = 38

3 a 16 °C **b** 47 kg   **c** 29 litres

Challenge ❸

4

a	133 + a = 224	a = 91
b	b + 156 = 235	b = 79
c	c + 143 = 351	c = 208
d	234 + d = 463	d = 229
e	467 + e = 524	e = 57
f	f + 376 = 611	f = 235
g	335 = 177 + g	g = 158
h	h + 566 = 723	h = 157
i	654 + i = 812	i = 158
j	813 = j + 687	j = 126
k	k + 765 = 921	k = 156
l	856 + l = 923	l = 67

Lesson 4: Identifying values for symbols in subtraction calculations

Challenge ❶

1 a 600 ml **b** 250 ml **c** 345 ml **d** 135 ml

Challenge ⚠

2

a	65 – a = 32	a = 33
b	b – 27 = 24	b = 51
c	c – 18 = 43	c = 61
d	57 – d = 18	d = 39
e	71 – e = 36	e = 35
f	f – 59 = 28	f = 87
g	48 = 72 – g	g = 24
h	h – 44 = 32	h = 76
i	94 – i = 57	i = 37
j	27 = j – 66	j = 93
k	k – 65 = 76	k = 141
l	121 – l = 69	l = 52

3 a 91 **b** 78

4 a 23 °C **b** $55 **c** 56

Challenge ❸

5

a	246 – a = 178	a = 68
b	b – 197 = 76	b = 273
c	c – 188 = 165	c = 353
d	312 – d = 186	d = 126
e	421 – e = 255	e = 166
f	f – 243 = 229	f = 472
g	473 = 625 – g	g = 152
h	h – 448 = 377	h = 825
i	867 – i = 594	i = 273
j	294 = j – 618	j = 912
k	k – 264 = 677	k = 941
l	915 – l = 693	l = 222

Unit 3

Lesson 1: Subtracting positive and negative integers

Challenge ❶

1 a	4	**b**	5	**c**	5	**d**	12
e	7	**f**	8	**g**	16	**h**	16
i	16	**j**	17	**k**	20	**l**	8

Challenge ⚠

2 18 degrees

3 a	13	**b**	15	**c**	17	**d**	16
e	19	**f**	21	**g**	30	**h**	30
i	33	**j**	37	**k**	24	**l**	31

4

Starting temperature	Final temperature	Temperature change
−13 °C	9 °C	22 degrees
14 °C	−12 °C	26 degrees
−15 °C	11 °C	26 degrees
17 °C	−8 °C	25 degrees
−16 °C	13 °C	29 degrees
19 °C	−17 °C	36 degrees

Challenge 3

5 a 11 **b** −2 **c** 8
d −6 **e** 4 **f** −3
g 11 **h** −14 **i** 11
j −12 **k** 21 **l** −22

6 Explanations will vary

Lesson 2: Subtracting two negative integers

Challenge 1

1 a −7 m **b** Explanations will vary **c** −5 m
d i 2m **ii** 3 m **iii** 5 m **iv** 2 m

Challenge 2

2 a 3 **b** 4 **c** 4
d 4 **e** 7 **f** 4
g 5 **h** 5 **i** 9
j 9 **k** 3 **l** 13
3 a 5 **b** $12 **c** 16 **d** 16
4 a 7 **b** 7 **c** 8
d 12 **e** 15 **f** 13
g 27 **h** 24 **i** 39

Challenge 3

5

	Lower depth (m)	Upper depth (m)	Height moved (m)
Monday	−52	−16	36
Tuesday	−63	−18	45
Wednesday	−55	−7	48
Thursday	−66	−31	35
Friday	−71	−13	58
Saturday	−91	−14	77

Lesson 3: Identifying values of variables in calculations (1)

Challenge 1

1 (1, 5), (2, 4), (3, 3), (4, 2), (5, 1)

Challenge 2

2

p	0	1	2	3	4	5	6	7
q	7	6	5	4	3	2	1	0

3

m	7	8	9	10	11	12	13
n	13	12	11	10	9	8	7

4 a $x + y = 14$ (letters will vary)
(3, 11), (4, 10), (5, 9), (6, 8), (7, 7), (8, 6), (9, 5), (10, 4), (11, 3)
b $a + b = 22$ (letters will vary)
(6, 16), (7, 15), (8, 14), (9, 13), (10, 12), (11, 11), (12, 10), (13, 9), (14, 8), (15, 7), (16, 6)
c $c + d = 18$ (letters will vary)
(4, 14), (5, 13), (6, 12), (7, 11), (8, 10), (9, 9), (10, 8), (11, 7), (12, 6), (13, 5), (14, 4)

Challenge 3

5 a $x − y = 7$ (letters will vary)
(37, 30), (36, 29), (35, 28), (34, 27), (33, 26), (32, 25)
b $2x + 5 = y$
(3, 11), (4, 13), (5, 15), (6, 17), (7, 19)

Lesson 4: Identifying values of variables in calculations (2)

Challenge 1

1 a $14 + 3 = 17$
b $14 + 4 = 18$
c $14 + 6 = 20$

Challenge 2

2

T ($)	9	15	23	38	46	57
E ($)	104	110	118	133	141	152

3

Side length (cm)	7	14	23	37	58	77
Perimeter (cm)	28	56	94	148	232	308

4

Side length (cm)	4	6	7	9	12	20
Area (cm²)	16	36	49	81	144	400

5

a (cm)	6	8	18	16	23	39
b (cm)	23	25	35	33	40	56

Challenge 3

6

W ($)	42	74	63	101	76	112
H ($)	25	57	46	84	59	95

7 Formula: $P = 2l + 2w$
a $l = 1$, $w = 3$; $l = 2$, $w = 2$; $l = 3$, $w = 1$
b $l = 1$, $w = 5$; $l = 2$, $w = 4$; $l = 3$, $w = 3$; $l = 4$, $w = 2$; $l = 5$, $w = 1$
c $l = 1$, $w = 9$; $l = 2$, $w = 8$; $l = 3$, $w = 7$; $l = 4$, $w = 6$; $l = 5$, $w = 5$; $l = 6$, $w = 4$; $l = 7$, $w = 3$; $l = 8$, $w = 2$; $l = 9$, $w = 1$

Unit 4

Lesson 1: Common multiples

Challenge ❶

1 **a** Multiples of 4

4	8	12	16	20	24	28	32	36	40
× 1	× 2	× 3	× 4	× 5	× 6	× 7	× 8	× 9	× 10

b Multiples of 6

6	12	18	24	30	36	42	48	54	60
× 1	× 2	× 3	× 4	× 5	× 6	× 7	× 8	× 9	× 10

c Multiples of 9

9	18	27	36	45	54	63	72	81	90
× 1	× 2	× 3	× 4	× 5	× 6	× 7	× 8	× 9	× 10

Challenge ⚠

2 **a** 4: 4, 8, 12, 16, 20, 24, 28, 32, 36, 40, 44, 48, 52, 56, 60
5: 5, 10, 15, 20, 25, 30, 35, 40, 45, 50, 55, 60, 65, 70, 75
The common multiples of 4 and 5 include: 20, 40, 60

b 2: 2, 4, 6, 8, 10, 12, 14, 16, 18, 20, 22, 24, 26, 28, 30
5: 5, 10, 15, 20, 25, 30, 35, 40, 45, 50, 55, 60, 65, 70, 75
The common multiples of 2 and 5 include: 10, 20, 30

3 **a** 30   **b** Explanations will vary

Challenge ❸

4 **a** 84, 168
b 99, 198
c 104, 208
d 156, 312

Lesson 2: Common factors

Challenge ❶

1 **a** 16: 1, 2, 4, 8, 16
b 24: 1, 2, 3, 4, 6, 8, 12, 24
c 30: 1, 2, 3, 5, 6, 10, 15, 30
d 36: 1, 2, 3, 4, 6, 9, 12, 18, 36
e 50: 1, 2, 5, 10, 25, 50
f 34: 1, 2, 17, 34

Challenge ⚠

2

18, 28	18: 3, 6, 9, 18
28: 4, 7, 14, 28	Intersection: 1, 2,
28, 42	28: 4, 28
42: 3, 6, 21, 42	Intersection: 1, 2, 7, 14
44, 56	44: 11, 22, 44
56: 7, 8, 14, 28, 56	Intersection: 1, 2, 4
60, 72	60: 5, 10, 15, 20, 30, 60
72: 8, 9, 18, 24, 36, 72	Intersection: 1, 2, 3, 4, 6, 12

3 4 trays

Challenge ❸

4 **a** 66 and 84: 1, 2, 3, 6
b 76 and 92: 1, 2, 4
c 66 and 99: 1, 3, 11, 33
d 84 and 108: 1, 2, 3, 4, 6, 12

Lesson 3: Tests of divisibility by 3, 6 and 9

Challenge ❶

1 **a** 12   **b** 27   **c** 24   **d** 45
e 18   **f** 21   **g** 6   **h** 9
i 9   **j** 6   **k** 7   **l** 9

2 144, 454, 986, 346, 888, 432

Challenge ⚠

3 **a** 981, 228, 423, 633, 831, 483, 834, 201
b 636
c 459

4 Learner's choices. Check their numbers meet the criteria.

Challenge ❸

5 Learner's choices. Check their numbers meet the criteria.

6 **a** Answers may vary, for example: 1764
b Answers may vary, for example: 7110
c Answers may vary, for example: 6516

Lesson 4: Cube numbers

Challenge ❶

1 **a** 8   **b** 64   **c** 1
d 27   **e** 125

Challenge ❷

2 **a** 27   **b** 1   **c** 125
d 8   **e** 64

3 **b** The cube of 4 or **4³** is **64**. **64** is a cube number.
c The cube of 2 is **2³** is **8**. **8** is a cube number.
d The cube of 5 is **5³** is **125**. **125** is a cube number.
e The cube of 3 is **3³** is **27**. **27** is a cube number.

4

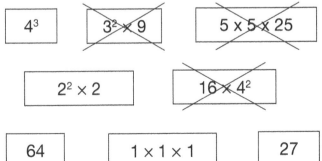

5 a 3   b 5   c 64   d 2
   e 216   f 512   g 10   h 729
   i 343

Challenge 3
6 a 25   b 4   c 27   d 4
   e 1000   f 81   g 343   h 512

Unit 5

Lesson 1: Simplifying calculations (1)
Challenge 1
1 a 5   b 32   c 60   d 2

Challenge 2
2 a 100 (commutative and associative)
   b 720 (commutative and associative)
   c 311 (distributive)
   d 635 (commutative and associative)
   e 360 (commutative and associative)
   f 475 (distributive)
3 Learner's problems and solutions

Challenge 3
4 a 18 248 b 953   c 2030   d 4000

Lesson 2: Simplifying calculations (2)
Challenge 1
1 a 52   b 40

Challenge 2
2 a 680   b 240   c 1340   d 4800
3 a 2348   b 1180

Challenge 3
4 a 4527   b 5580   c 5510 kg

Lesson 3: Using brackets (1)
Challenge 1
1 a 6, 12   b 5, 2   c 9, 45
   d 22, 110 e 70, 10 f 66, 198

Challenge 2
2 a 35   b 3   c 27   d 77
   e 10   f 60
3 a 126   b 32   c 98   d 312
   e 16   f 1665
4 Learner's problems and solutions, e.g. (3 + 4) × 18 = 126

Challenge 3
5 a 3630   b 6876   c 82   d 97
6 a 12   b 22   c 5   d –2
   e 8   f 85   g 20   h 15
   i 65   j 120

Lesson 4: Using brackets (2)
Challenge 1
1 a 16 x (12 – 5)
   b Explanations will vary, for example: She multiplied first, then subtracted 5.

Challenge 2
2 a 32   b 29   c 315 km d 777

Challenge 3
3 81
4 37
5 6375 kg

Unit 6

Lesson 1: Multiplying by 1-digit numbers (1)
Challenge 1
1 a 304   b 464   c 702   d 2280

Challenge 2
2 a 10 875   b 25 732
   c 47 076   d 36 752
3 a 9868   b 44 082

Challenge 3
4 A: 17 265 km, B: 17 511 km, C: 20 846 km, D: 17 076 km
   Order: D, A, B, C

Lesson 2: Multiplying by 1-digit numbers (2)
Challenge 1
1 a 3729   b 21 696
   c 27 936   d 62 936

Challenge 2
2 a 144   b 584   c 3402
3 a 10 392 b 29 900 c 44 506

Challenge 3
4 Week 1: 13 956 m, Week 2: 13 936 m
   Furthest distance is Week 1.

Lesson 3: Multiplying by 2-digit numbers (1)
Challenge 1
1 a 1296   b 4644

Challenge 2
2 a 149 360   b 351 540
3 a $250 260   b $558 914

Challenge 3
4 a 30 284   b 396 082

Lesson 4: Multiplying by 2-digit numbers (2)

Challenge ❶

1 a 2668 **b** 2376 **c** 5112 **d** 11328

Challenge ⚠

2 a 792 **b** 1855 **c** 2898 **d** 5216
e 9604 **f** 16344

3 a 39882 **b** 39844 **c** 82145 **d** 264768

Challenge ❸

4 a 35076 **b** 69168 **c** 162949 **d** 188117

Unit 7

Lesson 1: Dividing 2-digit numbers by 1-digit numbers (1)

Challenge ❶

1 a 27 **b** 16 **c** 13 **d** 14

Challenge ⚠

2 a 24 r 3 **b** 14 r 5
c 13 r 2 **d** 12 r 1

Challenge ❸

3 9 in each pile, remainder: 8

Lesson 2: Dividing 2-digit numbers by 1-digit numbers (2)

Challenge ❶

1 a 28 **b** 24 **c** 17 r 2 **d** 14 r 5

Challenge ⚠

2 a 22 r 1 **b** 19 r 3 **c** 12 r 3 **d** 23 r 3

Challenge ❸

3 a $19\frac{2}{5}$ **b** $11\frac{3}{4}$

Lesson 3: Dividing 3-digit numbers by 1-digit numbers (1)

Challenge ❶

1 a 59 **b** 64 **c** 44 **d** 64

Challenge ⚠

2 a 77 r 3 **b** 44 r 1 **c** 84 r 2 **d** 89 r 4
3 a 74 **b** 97 **c** 52 **d** 26
e 99 **f** 46 **g** 91 r 1 **h** 98 r 1
i 88 r 2 **j** 63 r 1 **k** 48 r 4 **l** 65 r 3

Challenge ❸

4 Examples: 148, 208, 274

Lesson 4: Dividing 3-digit numbers by 1-digit numbers (2)

Challenge ❶

1 a 46 **b** 93 **c** 87 r 2 **d** 82 r 2

Challenge ⚠

2 a 75 r 3 **b** 58 r 1 **c** 85 r 7 **d** 77 r 2

Challenge ❸

3 a $72 **b** $52\frac{1}{9}$ litres

Unit 8

Lesson 1: Dividing 2-digit numbers by 2-digit numbers (1)

Challenge ❶

1 a 51 **b** 92 **c** 90 **d** 78
e 76 **f** 78 **g** 96 **h** 96
i 84 **j** 69 **k** 87 **l** 85

Challenge ⚠

2 a 4 **b** 3
3 a 3 **b** 3

Challenge ❸

4 a 3 pages **b** 3 strips

Lesson 2: Dividing 2-digit numbers by 2-digit numbers (2)

Challenge ❶

1 a 4 **b** 3

Challenge ⚠

2 a 8 **b** 3
3 a 2 **b** 4

Challenge ❸

4 a 6 **b** 52

Lesson 3: Dividing 3-digit numbers by 2-digit numbers (1)

Challenge ❶

1 a 414 ÷ 18 18 × 10 ⟨18 × 20⟩ 18 × 30 18 × 40
b 782 ÷ 23 23 × 10 23 × 20 ⟨23 × 30⟩ 23 × 40
c 704 ÷ 16 16 × 10 16 × 20 16 × 30 ⟨16 × 40⟩
d 918 ÷ 27 27 × 10 27 × 20 ⟨27 × 30⟩ 27 × 40
e 576 ÷ 32 32 × 10 ⟨32 × 20⟩ 32 × 30 32 × 40
f 598 ÷ 46 ⟨46 × 10⟩ 46 × 20 46 × 30 46 × 40

Challenge ⚠

2 a 18 **b** 26
3 a 26 **b** 18

Challenge ❸

4 a 24 games **b** 19 boxes

Lesson 4: Dividing 3-digit numbers by 2-digit numbers (2)

Challenge ❶

1 a 34 **b** 53

Challenge ⚠

2 a 46 **b** 26
3 a 25 **b** 56

Challenge ❸

4 a 162 **b** 513

Unit 9

Lesson 1: Decimal place value

Challenge ❶

1 a Top row: 0·013, 0·034, 0·069, 0·096
      Bottom row: 0·007, 0·027, 0·045, 0·081
   b Top row: 4·314, 4·332, 4·349, 4·375, 4·389
      Bottom row: 4·303, 4·361, 4·397

4

Counting on in steps of...						
0·001	0·057	0·058	0·059	0·06	0·061	0·062
0·002	0·293	0·295	0·297	0·299	0·301	0·303
0·003	-0·526	-0·523	–0·52	–0·517	–0·514	–0·511
0·005	–3·287	–3·282	–3·277	–3·272	–3·267	–3·262

Counting back in steps of...						
0·001	2·474	2·473	2·472	2·471	2·47	2·469
0·002	0·007	0·005	0·003	0·001	-0·001	-0·003
0·003	–0·535	–0·538	–0·541	–0·544	–0·547	–0·55
0·005	–6·748	–6·753	–6·758	–6·763	–6·768	–6·773

Challenge ❸

5 Answers will vary

Lesson 2: Composing and decomposing decimals

Challenge ❶

1 (3 + 0·7 + 0·09 + 0·003) — 3·793;
   (3 ones + 9 tenths + 7 thousandths) — 3·907;
   (3 ones + 9 hundredths + 7 thousandths) — 3·097;
   (3 + 0·7 + 0·009) — 3·709

Challenge ❷

2 a 0·5 + 0·02 + 0·006
   b 0·9 + 0·02 + 0·008
   c 0·3 + 0·003
   d 0·06 + 0·006
   e 5 + 0·5 + 0·05 + 0·005
   f 8 + 0·008
3 a 1, 0·2, 0·03, 0·004
   b 6, 0·01, 0·009
   c 10, 1, 0·7, 0·04, 0·006
   d 50, 4, 0·8, 0·002
   e 400, 70, 6, 0·3, 0·03, 0·003
   f 700, 70, 7, 0·7, 0·07, 0·007
   g 432·057 = 400, 30, 2, 0·05, 0·007

Challenge ❸

4 Answers will vary – ensure that they follow the example given

Challenge ❷

2 a 5 tenths (0·5)   b 6 hundredths (0·06)
   c 1 thousandth (0·001)   d –1 hundredth (–0·01)
   e 9 tenths (0·9)   f 4 thousandths (0·004)
   g 0 hundredths (0)
   h –2 thousandths (–0·002)
3 a 0·003   b 0·007   c 0·019
   d 0·068   e 0·484   f 0·987

Lesson 3: Regrouping decimals

Challenge ❶

1 a 2 + 0·2 + 0·08 + 0·001
   b 8 + 0·4 + 0·03 + 0·002
   c 20 + 7 + 0·4 + 0·03 + 0·007
   d 60 + 3 + 0·05 + 0·006
   e 80 + 1 + 0·2 + 0·008
   f 500 + 40 + 8 + 0·4 + 0·05 + 0·003

Challenge ❷

2 a 4·265   b 5·819   c 404·187   d 7·082
3 Answers will vary, e.g. 8 + 0·1 + 0·02 + 0·006
4 a i   −26 = −10 + −10 + −6
      ii −51 = −10 + −10 + −10 + −10 + −10 + −1
   b i   −4 + −0·1 + −0·07 + −0·003
      ii −10 + −3 + −0·2 + −0·08 + −0·004

Challenge ❸

5 a 4·225   b 21·922

Lesson 4: Comparing and ordering decimals

Challenge ❶

1 a 2·34 < 2·43   b 5·21 < 5·23
   c 17·03 > 17·02   d 65·46 > 64·56

Challenge ❷

2 a 0·61 > 0·16   b 0·98 > 0·89
   c 2·67 < 2·76   d 5·43 > 5·41
   e 8·17 < 8·71   f 6·04 > 6·03
   g 12·23 < 12·32   h 15·26 > 12·56
   i 22·21 > 21·22   j 37·73 > 37·37
   k 45·04 > 45·03   l 82·23 < 82·32

3 Use the place value charts to help you order the numbers.

a $6{\cdot}24 < 6{\cdot}36 < 6{\cdot}37 < 6{\cdot}41 < 6{\cdot}42$

b $31{\cdot}68 < 32{\cdot}67 < 32{\cdot}68 < 32{\cdot}76 < 32{\cdot}86$

c $4 < 4{\cdot}1 < 4{\cdot}11 < 4{\cdot}13 < 4{\cdot}31$

d $55{\cdot}66 < 56 < 56{\cdot}5 < 56{\cdot}56 < 56{\cdot}65$

Challenge 3

4 Answers will vary, e.g. **a** $2{\cdot}4 < 2{\cdot}2$

Unit 10

Lesson 1: Multiplying whole numbers and decimals by 10, 100 and 1000

Challenge 1

1 a 2800 **b** 327 000 **c** 790 **d** 45 680

Challenge 2

2

Number	× 10	× 100	× 1000
25	250	2500	25 000
98	980	9800	98 000
303	3030	30 300	303 000
2·9	29	290	2900
8·06	80·6	806	8060
55·55	555·5	5555	55 550

3 a 100 **b** 10 **c** 1000
d 100 **e** 1000 **f** 10

4 a 8800, 90 900, 330, 4506
b 16 000, 217 000, 5500, 74 120

Challenge 3

5 a $85{\cdot}47 \times 100 = 8547$

b $912{\cdot}38 \times 10 = 9123{\cdot}8$

c $743{\cdot}08 \times 1000 = 743\,080$

d $734{\cdot}15 \times 100 = 73\,415$

e $5465{\cdot}32 \times 1000 = 5\,465\,320$

f $3942{\cdot}09 \times 10 = 39\,420{\cdot}9$

Lesson 2: Dividing whole numbers and decimals by 10, 100 and 1000

Challenge 1

1 a $4820 \div 10 = 482$

b $5870 \div 1000 = 5{\cdot}870$

c $645{\cdot}1 \div 100 = 6{\cdot}451$

d $76\,069 \div 1000 = 76{\cdot}069$

Challenge 2

2

Number	÷ 10	÷ 100	÷ 1000
99 900	9990	999	99·9
2121·6	212·16	21·216	
46 775	4677·5	467·75	46·775
101	10·1	1·01	0·101
56·14	5·614		
889 865	88 986·5	8898·65	889·865

3 a 100 **b** 10 **c** 10 **d** 1000
e 100 **f** 1000

4 a 462, 532·1, 9·21, 0·883
b 5678, 24·9, 3·82, 0·907

Challenge 3

5 a $4123 \div 100 = 41{\cdot}23$

b $8193 \div 10 = 819{\cdot}3$

c $3485 \div 1000 = 3{\cdot}485$

d $247\,000 \div 100 = 2470$

e $7{\cdot}63 \div 10 = 0{\cdot}763$

f $805 \div 1000 = 0{\cdot}805$

Lesson 3: Rounding decimals to the nearest tenth

Challenge 1

1 a 6·6, 6·6, 6·7, 6·7, 6·7
b 13·2, 13·2, 13·3, 13·3, 13·3

Challenge 2

2 b 2·8 **2·9** **c** 8·1, **8·2**
d **9·6**, 9·7 **e** 17·9, **18**
f 43·6, **43·7** **g** **74·2**, 74·3
h 96·5, **96·6**

3 a 8·8 **b** 7·4 **c** 3·6 **d** 6
e 12·8 **f** 19·2 **g** 23·5 **h** 50
i 89·9 **j** 92·9 **k** 65·8

4 a $2.30 **b** $5.20 **c** $12.60 **d** $38.00
e $42.00 **f** $65.70 **g** $234.30 **h** $546.90

Challenge 3

5 a 8·3: 8·25, 8·26, 8·27. 8·28, 8·29, 8·31, 8·32, 8·33, 8·34

b 14·7: 14·65, 14·66, 14·67, 14·68, 14·69, 14·71, 14·72, 14·73, 14·74

c 35·5: 35·45, 35·46, 35·47, 35·48, 35·49, 35·51, 35·52, 35·53, 35·54

d 86·1: 86·05, 86·06, 86·07, 86·08, 86·09, 86·11, 86·12, 86·13, 86·14

6 Answers will vary

Lesson 4: Rounding decimals to the nearest whole number

Challenge 1

1 b **6**, 7 **c** **8**, 9 **d** **9**, 10
e 11, **12** **f** 16, **17** **g** **20**, 21
h **46**, 47

Challenge 2

2

Number	6·28	12·35	15·91	40·27	69·94
Rounded to the nearest whole number	6	12	16	40	70

3

Number	16·37 cm	28·45 km	47·18 kg	52·99 l	$735.08
Rounded to the nearest whole number	16 cm	28 km	47 kg	53 l	$735
Rounded to the nearest tenth	16·4 cm	28·5 km	47·2 kg	53 l	$735.10

4 a 18·72 b 81·11 c 233·78

Challenge 3

5 Numbers in the top row may vary, for example:

Number	12·35	17·88	27·42	59·07	92·64
Rounded to the nearest whole number	12	18	27	59	93
Rounded to the nearest tenth	12·4	17·9	27·4	59·1	92·6

Unit 11

Lesson 1: Fractions as division

Challenge 1

1 a $1 \div 4$ b $1 \div 3$ c $1 \div 5$ d $3 \div 5$
   e $1 \div 8$ f $5 \div 8$ g $1 \div 16$ h $11 \div 16$
   i $5 \div 4$ j $8 \div 7$ k $19 \div 17$ l $23 \div 20$

2 a $\frac{2}{3}$ b $\frac{4}{5}$ c $\frac{7}{8}$ d $\frac{3}{10}$
   e $\frac{11}{15}$ f $\frac{7}{20}$ g $\frac{6}{5}$ h $\frac{14}{11}$
   i $\frac{19}{16}$

Challenge 2

3 a $\frac{3}{4}, 3 \div 4$ b $\frac{3}{8}, 3 \div 8$
   c $\frac{7}{20}, 7 \div 20$ d $\frac{9}{20}, 9 \div 20$

Challenge 3

4 a $\frac{5}{4}, 5 \div 4$ b $\frac{5}{3}, 5 \div 3$
   c $\frac{11}{4}, 11 \div 4$ d $\frac{8}{5}, 8 \div 5$
   e $\frac{14}{5}, 14 \div 5$ f $\frac{25}{8}, 25 \div 8$
   g $\frac{35}{8}, 35 \div 8$ h $\frac{57}{10}, 57 \div 10$

Lesson 2: Simplifying fractions

Challenge 1

1 a Left section: 24, 12, 8, 4
     Middle section (intersection): 1, 2, 3, 6
     Right section: 30, 15, 10, 5
     HCF = 6; $\frac{24}{30}$
     $= \frac{4}{5}$
   b Left section: 18, 2, 6
     Middle section (intersection): 1, 9, 3
     Right section: 45, 15, 5
     HCF = 9; $\frac{18}{45}$
     $= \frac{2}{5}$

Challenge 2

2 a $\frac{1}{4}$ b $\frac{3}{4}$ c $\frac{2}{5}$ d $\frac{7}{10}$
   e $\frac{9}{10}$ f $\frac{17}{21}$

3 $\frac{36}{45} - \frac{4}{5}$ $\frac{48}{96} - \frac{1}{2}$ $\frac{12}{48} - \frac{1}{4}$ $\frac{56}{80} - \frac{7}{10}$ $\frac{30}{75} - \frac{2}{5}$ $\frac{27}{36} - \frac{3}{4}$

Challenge 3

4 b $1\frac{1}{2}$ c $1\frac{1}{4}$ d $2\frac{1}{5}$

Lesson 3: Comparing fractions with different denominators

Challenge 1

1 a $\frac{1}{6}$ b $\frac{1}{9}$ c $\frac{1}{3}$ d $\frac{1}{6}$
   e $\frac{4}{9}$ f $\frac{1}{3}$ g $\frac{5}{9}$ h $\frac{5}{6}$
   i $\frac{5}{6}$

Challenge 2

2 b $\frac{7}{8} > \frac{3}{4}$ c $\frac{5}{8} < \frac{11}{16}$ d $\frac{3}{8} < \frac{1}{2}$
   e $\frac{4}{5} > \frac{7}{10}$ f $\frac{2}{3} > \frac{9}{15}$

3 a $\frac{1}{2} < \frac{3}{4}$ b $\frac{2}{10} < \frac{2}{5}$ c $\frac{3}{4} > \frac{5}{8}$
   d $\frac{2}{3} < \frac{7}{9}$ e $\frac{2}{5} > \frac{3}{10}$ f $\frac{2}{3} < \frac{5}{6}$

4 b $0·3 > 0·2$ c $0·75 < 0·8$ d $0·6 < 0·7$
   e $0·4 > 0·3$ f $0·6 > 0·5$

Challenge 3

5 a $\frac{3}{5} > 0·5$ b $0·8 > \frac{7}{10}$ c $\frac{4}{5} < 0·9$
   d $0·75 < \frac{8}{10}$ e $\frac{2}{5} > 0·25$ f $\frac{6}{10} > 0·5$

Lesson 4: Ordering fractions with different denominators

Challenge 1

1 a $\frac{1}{9}, \frac{1}{3}, \frac{2}{3}$ b $\frac{2}{6}, \frac{3}{6}, \frac{2}{3}$
   c $\frac{4}{9}, \frac{5}{9}, \frac{2}{3}$ d $\frac{2}{6}, \frac{2}{3}, \frac{5}{6}$
   e $\frac{2}{3}, \frac{7}{9}, \frac{8}{9}$ f $\frac{1}{6}, \frac{1}{3}, \frac{4}{6}$

Challenge 2

2 a $\frac{1}{4}, \frac{1}{2}, \frac{6}{8}$ b $\frac{1}{3}, \frac{4}{9}, \frac{2}{3}$
   c $\frac{7}{10}, \frac{4}{5}, \frac{9}{10}$ d $\frac{1}{2}, \frac{5}{8}, \frac{3}{4}$

3 a $\frac{1}{4}, \frac{3}{8}, \frac{5}{8}, \frac{3}{4}$ b $\frac{3}{10}, \frac{2}{5}, \frac{5}{10}, \frac{3}{5}$
   c $\frac{1}{2}, \frac{2}{3}, \frac{3}{4}, \frac{5}{6}$ d $\frac{2}{6}, \frac{4}{9}, \frac{2}{3}, \frac{7}{9}$

Challenge 3

4 a $\frac{1}{4} < \frac{1}{2} < \frac{5}{8}$ b $\frac{1}{3} < \frac{4}{9} < \frac{5}{9}$
   c $\frac{6}{10} < \frac{7}{10} < \frac{4}{5}$

Unit 12

Lesson 1: Fractions as operators

Challenge ❶

1 **a** $9 **b** 5 km **c** 20 g **d** 30 m
e 60 ml **f** 30 kg **g** $60 **h** 20 l
i 18 cm

Challenge ⚠

2 **a** $21 **b** 39 km **c** 279 g **d** 135
e $240 **f** 250 km **g** 180 **h** $532
i 415 km **j** 304 **k** $616 **l** 264 km
3 **a** $36 **b** 161 km **c** 370 g **d** 80
e $161 **f** 477 km **g** 119 **h** $360
i 715 km **j** 144 **k** $363 **l** 598 km

4. Learners' answer should show 75 x 7 = 525, then 525 ÷ 5 = 105 and converting $\frac{7}{5}$ to $1\frac{2}{5}$ then working out 1 x 75 + $\frac{2}{5}$ x 75 = 75 + 30 = 105.

Challenge ❸

5 **a** 429 ml **b** 644 km

Lesson 2: Adding and subtracting fractions

Challenge ❶

1 **a** $\frac{3}{5}$ **b** $\frac{5}{6}$ **c** $\frac{2}{5}$ **d** $\frac{1}{6}$
e $\frac{7}{8}$ **f** $\frac{5}{8}$ **g** $\frac{11}{10}$ **h** $\frac{5}{4}$
i $\frac{13}{7}$ **j** $\frac{13}{12}$ **k** $\frac{27}{10}$ **l** $\frac{11}{9}$

Challenge ⚠

2 **a** $\frac{27}{20}$ or $1\frac{7}{20}$ **b** $\frac{3}{10}$ **c** $\frac{1}{12}$
d $\frac{49}{30}$ or $1\frac{19}{30}$ **e** $\frac{31}{90}$

Challenge ❸

3 **a** $\frac{203}{117}$ or $1\frac{86}{117}$ **b** $\frac{67}{132}$
c $\frac{51}{20}$ or $2\frac{11}{20}$

Lesson 3: Multiplying fractions by whole numbers

Challenge ❶

1 **a** $\frac{8}{5}$ or $1\frac{3}{5}$ **b** $\frac{25}{6}$ or $4\frac{1}{6}$ **c** $\frac{15}{8}$ or $1\frac{7}{8}$

Challenge ⚠

2 **a** $\frac{9}{10}$ **b** $\frac{20}{8}$ or $2\frac{1}{2}$ **c** $\frac{18}{4}$ or $4\frac{1}{2}$
d $\frac{20}{5}$ or 4
3 **a** $\frac{25}{6}$ or $4\frac{1}{6}$ **b** $\frac{6}{12}$ or $\frac{1}{2}$

Challenge ❸

4 **a** $\frac{15}{7}$ or $2\frac{1}{7}$ **b** $\frac{42}{8}$ or $2\frac{1}{2}$ **c** $\frac{10}{9}$ or $1\frac{1}{9}$
d $\frac{10}{12}$ or $\frac{5}{6}$ **e** $\frac{54}{10}$ or $5\frac{2}{5}$ **f** $\frac{35}{10}$ or $3\frac{1}{2}$
g $\frac{104}{13}$ or 8 **h** $\frac{102}{17}$ or 6

Lesson 4: Dividing fractions by whole numbers

Challenge ❶

1 Learners' diagrams, ensure that diagrams correctly model the fraction

Challenge ⚠

2 **a** $\frac{2}{20}$ or $\frac{1}{10}$ **b** $\frac{3}{40}$ **c** $\frac{5}{36}$ **d** $\frac{3}{35}$
3 **a** $\frac{3}{24}$ or $\frac{1}{8}$ **b** $\frac{5}{28}$ **c** $\frac{3}{20}$ **d** $\frac{5}{18}$

Challenge ❸

4 **a** $\frac{2}{9}$ **b** $\frac{3}{32}$ **c** $\frac{5}{21}$
d $\frac{7}{40}$ **e** $\frac{3}{32}$ **f** $\frac{7}{90}$

Unit 13

Lesson 1: Percentages of shapes

Challenge ❶

1 **a** 60% **b** 20% **c** 50%
d 25% **e** 75% **f** 90%

Challenge ⚠

2 **a** Half of a column shaded
b 3 and a half columns shaded
c 8 and a half columns shaded
3 **a** 95% **b** 65% **c** 15%
4 **a** 65% = $\frac{65}{100}$ = 0·65
b 15% = $\frac{15}{100}$ = 0·15
c 85% = $\frac{85}{100}$ = 0·85
d 35% = $\frac{35}{100}$ = 0·35
e 5% = $\frac{5}{100}$ = 0·05
f 55% = $\frac{55}{100}$ = 0·55

Challenge ❸

5 **a** 9 squares shaded **b** 3 squares shaded
c 17 squares shaded
6 **a** $\frac{9}{20}$ **b** $\frac{3}{20}$ **c** $\frac{17}{20}$

Lesson 2: Percentages of whole numbers (1)

Challenge ❶

1 **a** 200 **b** $220 **c** 340 kg **d** 64 km
e $445 **f** 390 g **g** 41 ml **h** $800
i 210 m **j** $230 **k** 115 g **l** 19 cm
m 49 g **n** 343 l

Challenge ⚠

2

	5% of	25% of	75% of
$20	$1	$5	$15
$120	$6	$30	$90
$4	$0.20	$1	$3

3 28, 280, 140, 1400
4 **a** 420 **b** 1540 **c** 700
d 1260 **e** 2100 **f** 2660

Challenge ❸

5 **a** 30 **b** 91 **c** 31 **d** 18

Lesson 3: Percentages of whole numbers (2)

Challenge ❶

1

	10%	20%	25%	50%	75%
$10	$1	$2	$2.50	$5	$7.50
$20	$2	$4	$5	$10	$15
$60	$6	$12	$15	$30	$45
$140	$14	$28	$35	$70	$105

Challenge ⚠

2

	Rise: 10%	Discount: 20%	Rise: 25%	Discount: 50%	Rise: 75%
$40	$44	$32	$50	$20	$70
$200	$220	$160	$250	$100	$350
$800	$880	$640	$1000	$400	$1400
$2400	$2640	$1920	$3000	$1200	$4200

3 **a** 27 **b** 12 **c** 15 **d** 6

Challenge ❸

4 **a** $380 **b** $6.30 **c** $864 **d** $10560

Lesson 4: Comparing percentages

Challenge ❶

1 **a** 19%, 37%, 48%, 56%
  b 44%, 45%, 51%, 54%
  c 67%, 68%, 74%, 76%
  d 59%, 82%, 85%, 95%
  e 27%, 56%, 57%, 65%, 72%
  f 32%, 33%, 34%, 42%, 43%

Challenge ⚠

2 **a** 50% **b** 60% **c** 0·8
  d 30% **e** 70% **f** 0·2
  g 20% **h** $\frac{3}{4}$ **i** $\frac{3}{10}$
3 **a** $\frac{1}{2}$, 60%, 0·7   **b** 50%, 0·6, $\frac{7}{10}$
  c $\frac{1}{10}$, 20%, 0·3   **d** 0·8, 90%, $\frac{5}{5}$
  e 20%, $\frac{1}{4}$, 0·3   **f** 70%, $\frac{3}{4}$, 0·9

Challenge ❸

4 **a** C, B, A   **b** C, B, A
  c B, C, A   **d** C, B, A
5 Learner's answers, e.g. 20% < 0·3 < $\frac{2}{5}$

Unit 14

Lesson 1: Adding decimals (mental strategies)

Challenge ❶

1 **a** 1·2 **b** 1·1 **c** 1·3 **d** 1·2
  e 1·5 **f** 1·5 **g** 0·92 **h** 0·95
  i 2·02 **j** 2·14 **k** 2·31 **l** 2·1

Challenge ⚠

2 **a** 3·05 **b** 4·23 **c** 1·36
  d 1·31 **e** 5·16 **f** 7·34
3 Lucas should subtract 0·02 to compensate for adding 0·02 too many.
4 **a** 6·14 *l*
  b 8·36 km

Challenge ⚠

5 **a** 0·55 + 0·28 = 0·83
  b 1·4 + 0·72 = 2·12
  c 2·5 + 0·86 = 3·36
  d 0·77 + 0·19 = 0·96
  e 0·67 + 0·49 = 1·16
  f 4·6 + 0·87 = 5·47
  g 7·7 + 0·53 = 8·23
  h 0·87 + 0·49 = 1·36
  i 0·95 + 0·88 = 1·83
  j 8·8 + 0·77 = 9·57

Lesson 2: Adding decimals (written methods)

Challenge ❶

1

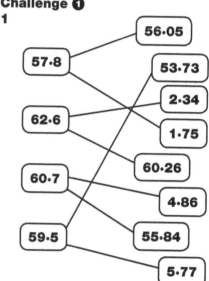

Challenge ⚠

2 **a** 29·245 **b** 53·514 **c** 75·812 **d** 86·518
3 **a** 26·345 **b** 42·224

Challenge ❸

4 **b** 34·472 **c** 82·154 **d** 58·943

Lesson 3: Subtracting decimals (mental strategies)

Challenge ❶

1 a 2·2 b 7·5 c 4·4 d 4·1
   e 4·3 f 3·4 g 2·3 h 0·2
   i 0·7 j 0·1 k 0·2 l 0·2

Challenge ⚠

2 a 3·15 b 7·47 c 0·023
   d 1·37 e 4·31 f 0·014

3 3·54

Challenge ❸

4 a 5·5 – 1·25 = 4·25
   b 7·8 – 4·36 = 3·44
   c 6·28 – 0·9 = 5·38
   d 8·4 – 2·19 = 6·21
   e 4·453 – 4·41 = 0·043
   f 9·678 – 9·63 = 0·048
   g 7·5 – 3·35 = 4·15
   h 3·6 – 1·24 = 2·36
   i 9·57 – 0·9 = 8·67
   j 5·7 – 3·29 = 2·41
   k 6·586 – 6·52 = 0·066
   l 1·645 – 1·61 = 0·035

Lesson 4: Subtracting decimals (written methods)

Challenge ❶

1 a 1·2 b 22·2 c 2·23 d 2·5
   e 26·3 f 5·55 g 4·1 h 84·3

Challenge ⚠

2 a 13·6 b 5·36 c 48·28 d 16·72
3 a 5·04 b 57·625 c 2·22 d 84·826

Challenge ❸

4 a 5·888 b 24·56 c 2·378 d 16·79

Unit 15

Lesson 1: Multiplying decimals by 1-digit whole numbers (1)

Challenge ❶

1 a 10 + 9 + 0·6 + 0·08
   b 30 + 6 + 0·6 + 0·06
   c 90 + 4 + 0·03
   d 50 + 0·9 + 0·01
   e 50 + 7 + 0·8 + 0·02
   f 70 + 3 + 0·4 + 0·08 + 0·005

Challenge ⚠

2 a 218·4 b 12·15
3 a 163·2 b 595·2 c 50·82 d 584·36

Challenge ❸

4 97·44 x 7 — 682·08
   86·3 x 8 — 690·4
   98·57 x 6 — 591·42
   137·5 x 5 — 687·5

Lesson 2: Multiplying decimals by 1-digit whole numbers (2)

Challenge ❶

1 a 172·2 b 59·36
   c 744 d 760·5

Challenge ⚠

2 a 382·2 b 527·52
3 a 46·4 b 2262·8

Challenge ❸

4 a 70 304 b 7628·4

Lesson 3: Multiplying decimals by 2-digit whole numbers (1)

Challenge ❶

1 a 40 b 400 c 4000 d 4 e 0·4
   f 4 g 28 h 280 i 2800 j 2·8
   k 0·28 l 2·8 m 48 n 480 o 4800
   p 4·8 q 0·48 r 4·8 s 54 t 540
   u 5400 v 5·4 w 0·54 x 5·4

Challenge ⚠

2 a 27·2 b 171·6
   c 2509·8 d 18031
   e 491·78 f 1405·32

Challenge ❸

4 8·4 x 38 — 319·2
   6·37 x 56 — 356·72
   47·6 x 24 — 1142·4
   73·44 x 16 — 1175·04

Lesson 4: Multiplying decimals by 2-digit whole numbers (2)

Challenge ❶

1 a 27·2 b 113·9

Challenge ⚠

2 a 5323·5 b 1063·82
3 a 1054·8 b 41·02

Challenge ❸

4 a 75 585·6 b 8711·31
   c 937·02 d 89 083·8

5 Explanations will vary

Unit 16

Lesson 1: Dividing one-place decimals by whole numbers (1)

Challenge ❶

1 a $24 \div 2 = 12$
      $2 \cdot 4 \div 2 = 1 \cdot 2$

   b $33 \div 3 = 11$
      $3 \cdot 3 \div 3 = 1 \cdot 1$

   c $48 \div 4 = 12$
      $4 \cdot 8 \div 4 = 1 \cdot 2$

   d $46 \div 2 = 23$
      $4 \cdot 6 \div 2 = 2 \cdot 3$

   e $66 \div 3 = 22$
      $6 \cdot 6 \div 3 = 2 \cdot 2$

   f $55 \div 5 = 11$
      $5 \cdot 5 \div 5 = 1 \cdot 1$

Challenge ◭

2 a $1 \cdot 7$ b $1 \cdot 4$ c $17 \cdot 8$ d $53 \cdot 9$

Challenge ❸

3 a $2 \cdot 5$ b $1 \cdot 55$ c $1 \cdot 6$ d $18 \cdot 9$ e $14 \cdot 6$
   f $31 \cdot 4$ g $64 \cdot 8$ h $48 \cdot 3$ i $98 \cdot 8$

Lesson 2: Dividing one-place decimals by whole numbers (2)

Challenge ❶

1 a $4 \cdot 1$ b $5 \cdot 2$ c $5 \cdot 1$

Challenge ◭

2 a $1 \cdot 2$ b $1 \cdot 96$ c $18 \cdot 4$ d $12 \cdot 9$
   e $63 \cdot 3$ f $79 \cdot 45$

Challenge ❸

3 a $79 \cdot 2$ b $99 \cdot 55$

Lesson 3: Dividing two-place decimals by whole numbers (1)

Challenge ❶

1 a $846 \div 2 = 423$
      $8 \cdot 46 \div 2 = 4 \cdot 23$

   b $636 \div 3 = 212$
      $6 \cdot 36 \div 3 = 2 \cdot 12$

   c $848 \div 4 = 212$
      $8 \cdot 48 \div 4 = 2 \cdot 12$

   d $505 \div 5 = 101$
      $5 \cdot 05 \div 5 = 1 \cdot 01$

   e $606 \div 6 = 101$
      $6 \cdot 06 \div 6 = 1 \cdot 01$

   f $969 \div 3 = 323$
      $9 \cdot 69 \div 3 = 3 \cdot 23$

Challenge ◭

2 a $6 \cdot 09$ b $7 \cdot 02$ c $4 \cdot 02$ d $5 \cdot 03$

Challenge ❸

3 a $13 \cdot 12$ b $12 \cdot 13$ c $12 \cdot 1$

Lesson 4: Dividing two-place decimals by whole numbers (2)

Challenge ❶

1 a $3 \cdot 05$ b $3 \cdot 09$ c $3 \cdot 07$

Challenge ◭

2 a $3 \cdot 22$ b $1 \cdot 97$ c $6 \cdot 09$
   d $4 \cdot 06$ e $6 \cdot 09$ f $9 \cdot 02$

Challenge ❸

3 a $12 \cdot 13$ b $23 \cdot 16$

Unit 17

Lesson 1: Direct proportion (1)

Challenge ❶

1 a $\frac{1}{10}, \frac{1}{2}, \frac{1}{3}, \frac{1}{2}, \frac{8}{9}$ b Lucy and Brett
   c Satpal. He has the highest proportion scored

Challenge ◭

2 a $1.20   b $1.50   c 4
   Explanations will vary, e.g. 20c × 4 = 80c

3 a 4 times   b $4 \cdot 5$ m

Challenge ❸

4 a $2 \cdot 5$ times   b x: 8 cm, y: $23 \cdot 5$ cm

Lesson 2: Direct proportion (2)

Challenge ❶

1 a $3, \frac{1}{3}$ b $5, \frac{1}{5}$ c $3, \frac{1}{3}$ d $10, \frac{1}{10}$

Challenge ◭

2 a i 8   ii 16   iii 2
   b i 24   ii 40   iii 2
   c i $120   ii $360 iii $5

3 a 440 g   b 900 g   c 8   d 2

Challenge ❸

4 a $9 \cdot 7$ times   b $358 \cdot 9$ cm

Lesson 3: Equivalent ratios (1)

Challenge ❶

a 2:4 = 1:2

b 6:2 = 3:1

c 8:10 = 4:5

d 4:16 = 1:4

Challenge ◭

2 a

| 2:3 | 4:8 | (6:9) | (10:15) | 9:6 | 14:23 | (24:36) |

b

| 5:2 | (10:4) | 20:12 | (20:8) | 6:15 | 70:30 | (45:18) |

c

| 3:8 | 18:53 | 21:49 | (15:40) | (60:160) | 23:72 | (36:96) |

d

| 9:2 | 27:7 | (45:10) | 8:36 | (63:14) | 108:26 | (135:30) |

3 a 15   b 35   c 105

Challenge ❸

4

| orange juice (ml) | 180 | 300 | 30 | 540 | 1620 | 12 | 2760 | 45 |
| pineapple juice (ml) | 390 | 650 | 65 | 1170 | 3510 | 26 | 5980 | $97 \cdot 5$ |

Lesson 4: Equivalent ratios (2)

Challenge ❶

1:3	4:1	6:5	9:7
2:6	8:2	12:10	27:21
3:9	16:4	24:20	45:35
5:15	24:6	42:35	81:63
8:24	40:10	90:75	108:84

Challenge ⚠

2 a i 6   **ii** 30   **iii** 26
   b i 160 g   **ii** 20   **iii** 1040 g

Challenge ❸

4

flour (cups)	$2\frac{1}{2}$	$7\frac{1}{2}$	$12\frac{1}{2}$	20	45
salt (cups)	$\frac{3}{4}$	$2\frac{1}{4}$	$3\frac{3}{4}$	6	13·5

Unit 18

Lesson 1: Quadrilaterals

Challenge ❶

1 a Kite   **b** Rhombus
   c Square   **d** Rhombus
   e Oblong or rectangle   **f** Parallelogram
   g Oblong or rectangle   **h** Trapezium

Challenge ⚠

2

Property	square	oblong	parallelogram	rhombus	kite	trapezium
opposite sides are equal	yes	yes	yes	yes		
opposite sides are parallel	yes	yes	yes	yes	no	One pair
adjacent sides are equal	yes	no	no	yes	Two pairs	no
all angles are 90°	yes	yes	no	no	no	no
opposite angles are equal	yes	yes	yes	yes	One pair	no
diagonals are of equal length	yes	yes	no	no	no	no

3 Learner's choices, ensure that sentences are logical and angles, sides and diagonals are marked correctly.

Challenge ❸

4 A Oblong   **B** Kite
   C Square   **D** Rhombus
   E Parallelogram   **F** Trapezium

Lesson 2: Parts of a circle

Challenge ❶

1

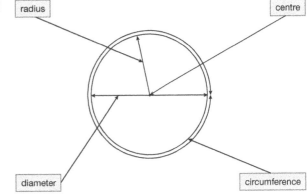

Challenge ⚠

2 A correctly labelled radius drawn on each circle from the centre point to the circumference. A correctly labelled diameter drawn across each circle going through the centre point.

3

a		b	
Radius	**Diameter**	**Diameter**	**Radius**
6 cm	12 cm	36 cm	18 cm
13 cm	26 cm	82 cm	41 cm
47 cm	94 cm	98 cm	49 cm
84 cm	168 cm	146 cm	73 cm
126 cm	252 cm	258 cm	129 cm
458 cm	916 cm	534 cm	267 cm

Challenge ❸

4 64 cm

Lesson 3: Constructing circles

Challenge ❶

1 a Increase the length of the string
   b Decrease the length of the string

Challenge ⚠

2 Learner's drawing, ensure that drawing and labelling are correct

3 Learner's drawing. Diameter: 8 cm

Challenge ❸

4 a (7, 13)
   b (7, 3)
   c (12, 8)
   d (2, 8)

Lesson 4: Rotational symmetry

Challenge ❶

1   a ✓   b ✗   c ✓
     ✗   ✓   ✓

Challenge ▲

2   a 3   b 2   c 6   d 1
    e 4   f 8
3   a 3   b 8   c 6
4   a 2   b 2   c 1   d 4

Challenge ❸

5   Learner's drawings, ensure that grids are completed to create rotational symmetry.

Unit 19

Lesson 1: Identifying and describing compound 3D shapes

Challenge ❶

1   a cube   b pentagonal prism
       hemisphere   pentagonal pyramid

Challenge ▲

2

	Faces	Vertices	Edges
Compound shape	9	9	16
	10	16	24
	9	14	21

Challenge ❸

3   Learner's exercise, e.g. 6 faces

Lesson 2: Sketching compound 3D shapes

Challenge ❶

1   Learner's drawing, ensure throughout this lesson that sketches are done accurately

Challenge ▲

2   Learner's drawings
3   Learner's drawing

Challenge ❸

4   Learner's drawings

Lesson 3: Identifying nets

Challenge ❶

1

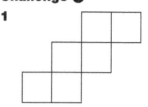

Challenge ▲

2   a four triangles
    b two triangles and three rectangles
    c two pentagons and five rectangles

3

Prism	Pyramid	Does not form a solid shape
A, E, H	B, D, G	C, F

Challenge ❸

4   A, B ticked 'Yes'
    C ticked 'No'
    Explanations may vary

Lesson 4: Sketching nets

Challenge ❶

1   Learner's drawings, ensure throughout this lesson that sketches are accurate

Challenge ▲

2   Learner's drawing
3   Answers will vary, for example:
    a Faces are rectangular, but one is too small.
    b One triangular face is too long.

Challenge ❸

4   Learner's drawing

Unit 20

Lesson 1: Measuring angles

Challenge ❶

1   a 25°   b 57°

Challenge ▲

2   Estimates may vary
    a 60°   b 59°   c 43°   d 130°
3   Estimates may vary
    A: 24° B: 133° C: 66° D: 33° E: 132°

Challenge ❸

4   a 30°, 60°   b 70°
    c 65°,130°, 65°   d 69°, 42°, 69°

Lesson 2: Drawing angles

Challenge ❶

1a–d Angles correctly drawn

Challenge ▲

2a–d Angles correctly drawn
3   Angles correctly drawn

Challenge 3
4a–c Learner's exercise, answers may vary

Lesson 3: Calculating missing angles in a triangle (1)

Challenge ❶
1 39° — second triangle
46° — third triangle
55° — first triangle

Challenge ②
2 **a** 68° **b** 47° **c** 41° **d** 46°, 67°
3

Angles of shape	Is it a triangle?	
	Yes	No
37°, 48°, 95°	✓	
84°, 39°, 57°	✓	
55°, 68°, 58°		✓
29°, 53°, 98°	✓	
17°, 25°, 137°		✓
26°, 66°, 89°		✓

Challenge 3
4 **a** 29° **b** 70°, 96° **c** 46°, 88°

Lesson 4: Calculating missing angles in a triangle (2)

Challenge ❶
1 **a** isosceles **b** scalene **c** equilateral
2 All three angles shaded on equilateral triangle.
Bottom two angles shaded on isosceles triangle.
No angles shaded on scalene triangle.

Challenge ②
3 $a = 65°$ $b = 60°$ $c = 45°$ $d = 54°$ $e = 71.5°$
$f = 78°$ $g = 68°$ $h = 53°$ $i = 64$

Challenge 3
4 $e = 14.5°$ $f = 75.5°$ $g = 14.5°$ $h = 75.5°$

Unit 21

Lesson 1: Converting time intervals (1)

Challenge ❶
1 **a** $\frac{1}{4}$ **b** $\frac{1}{2}$ **c** $\frac{3}{4}$ **d** $1\frac{1}{2}$ **e** $3\frac{3}{4}$
f $2\frac{1}{4}$ **g** $7\frac{1}{4}$ **h** $5\frac{1}{2}$ **i** $9\frac{3}{4}$
2 **a** 60 **b** 180 **c** 300 **d** 30
e 15 **f** 45
3 **a** 60 **b** 240 **c** 360 **d** 30
e 15 **f** 45

Challenge ②
4 **a** 1 min 30 s **b** 3 min 15 s
c 6 min 45 s **d** 8 min 30 s

5 **a** 2 h 30 min **b** 1 h 45 min
c 4 h 15 min **d** 9 h 45 min

Challenge 3
6 **a** 9 min 15 s **b** 2 h 40 min

Lesson 2: Converting time intervals (2)

Challenge ❶
1 **a** $\frac{1}{10}$ **b** $\frac{3}{10}$ **c** $\frac{7}{10}$
d $2\frac{4}{10}$ or $2\frac{2}{5}$ **e** $6\frac{6}{10}$ or $6\frac{3}{5}$ **f** $9\frac{2}{10}$ or $9\frac{1}{5}$
g $11\frac{8}{10}$ or $11\frac{4}{5}$ **h** $16\frac{9}{10}$ **i** $8\frac{1}{2}$
2 **a** 6 **b** 6 **c** 30 **d** 30 **e** 90 **f** 165

Challenge ②
3 **a** 4 min 6 s **b** 2 min 18 s
c 5 min 36 s **d** 8 min 48 s
4 **a** 4 h 6 min **b** 6 h 18 min
c 11 h 42 min **d** 14 h 54 min

Challenge 3
5 **a** 24 min 42 s **b** 3 h 24 min

Lesson 3: Capacity and volume (1)

Challenge ❶
1

Statement	Container
The capacity is 1000 ml.	c
The volume of liquid in the container is 150 ml.	b
The capacity is 500 ml.	a
The volume of liquid in the container is 500 ml.	a
The capacity is 250 ml.	b
The volume of liquid in the container is 800 ml.	c

Challenge ②
2 **a** 300 ml, 240 ml **b** 500 ml, 375 ml
c 1000 ml, 750 ml **d** 500 l, 175 ml
3 **a** Each large increment is 100 ml; each small increment is 25 ml. Shade up to 200 ml.
b Each large increment is 50 ml; each small increment is 12.5 ml. Shade up to 75 ml.

Challenge ❸
4 Explanations will vary

Lesson 4: Capacity and volume (2)

Challenge ❶
1 **a** 3.25 ℓ **b** 4.6 ℓ **c** 4.8 ℓ **d** 8.3 ℓ

Challenge ②
2 **a** 5.666 ℓ **b** 0.851 ℓ **c** 8.17 ℓ **d** 22.269 ℓ
3 **a** 75 ml **b** **i** 0.875 ℓ **ii** 10

Challenge 3
4 **a** 167.06967 ℓ **b** 11.81 ℓ

Unit 22

Lesson 1: Calculating the area of a triangle (1)

Challenge ❶

1   A diagonal line drawn across each rectangle, correctly splitting them into two triangles of equal size.

Challenge ⚠

2   Learner's rectangles drawn correctly with an area of 18 square units. A diagonal line drawn through each rectangle to split it in two creating two triangles with an area of 9 square units.

3   A mirroring triangle drawn next to A, B and C, creating a rectangle.
Area A: 24 square units
Area B: 30 square units
Area C: 44 square units

Challenge ❸

4   a   Area A = 3 cm²
    b   Area B = 8 cm²
    c   Area C = 3 cm²

Lesson 2: Calculating the area of a triangle (2)

Challenge ❶

1   a   8 cm²   b   12 cm²
    c   25 cm²   d   38 cm²

Challenge ⚠

2   a   20 cm²   b   28 m²
    c   90 cm²   d   72 m²
3   a   49·5 m²   b   1914 cm²
    c   944 cm²   d   90 cm²

Challenge ❸

4   x = 8 cm   y = 30 cm

Lesson 3: Surface area (1)

Challenge ❶

1   a   6 squares
    b   6 rectangles
    c   6 squares
    d   6 rectangles

Challenge ⚠

2   Check learner's explanations.
3   a   Check learner's labelling of the net.
    b   Surface area = A1 × 6
    c   Surface area = A1 + A2 + A3 + A4 + A5 + A6
    d   Surface area = A1 × 6

Challenge ❸

4   Ensure that faces are shaded correctly.

Lesson 4: Surface area (2)

Challenge ❶

1   a   6 rectangles: 2 of one size, 2 of another size and 2 of a third size.
    b   6 squares all the same size.

Challenge ⚠

2   Ensure that faces are shaded correctly.

Challenge ❸

3   Check learner's conjectures.
4   Check learner's conjectures.

Unit 23

Lesson 1: Reading and plotting coordinates (1)

Challenge ❶

1

Point	Coordinates
A	(2, 7)
B	(4, 0)
C	(6, 4)
D	(9, 9)

Challenge ⚠

2

Point	Coordinates
A	(−6, 6)
B	(7, 3)
C	(−9, 0)
D	(−5, −5)
E	(2, −4)
F	(−3, 9)
G	(−1, −10)
H	(9, −6)

3

Point	Coordinates (decimals)
P	(4, 3)
Q	(−3·5, 2)
Point	**Coordinates (fractions)**
R	(−2, −4)
S	$(1\frac{1}{2}, -3\frac{1}{2})$

Challenge ❸

4   a   P: (2, 2)   b   M: (−4 , 3)   c   E: (1, 0)
    d   Z: (10, −4)   e   A: (2, 3)   f   C: (7, −10)
    g   X: (−8, 5)   h   G: (1, −4)

Lesson 2: Reading and plotting coordinates (2)

Challenge ❶

1 Learner's plotted points, ensure throughout this lesson that points are plotted correctly

Challenge ⚠

2 Learner's plotted points

3 Learner's plotted points

Challenge ❸

4 **a** Shark correctly plotted

b Boat at (–9, 4)

c Lighthouse correctly plotted

d Hut correctly plotted

Lesson 3: Plotting lines and shapes across all four quadrants (1)

Challenge ❶

1 Learner's plotted points, ensure throughout this lesson that points are plotted correctly

Challenge ⚠

2 Learner's drawings of rectangles

3 **a** Vertex C (2, 0) **b** Vertex H (–9, 3)

Challenge ❸

4 Answers may vary

Lesson 4: Plotting lines and shapes across all four quadrants (2)

Challenge ❶

1 Learner's plotted points joined to form lines

Challenge ⚠

2 Learner's lines and correctly identified points on the lines

3 **a** Point: M: (3, 0); Point: N: (0, –1)

b Point: P: (–2·5, 0); Point: R: (0, –5)

Challenge ❸

4 Answers will vary

Unit 24

Lesson 1: Translating 2D shapes on coordinate grids

Challenge ❶

1 **a** Image vertices at: (–9, 5), (–4, 6), (–9, 1), (–4, 2)

b Image vertices at: (1, –1), (–2, –3), (3, –3), (–1, –5), (2, –5)

Challenge ⚠

2 Image vertices at: A'(–1, 3), B'(2, 3), C'(–3, 0), D'(4, 0)

Image vertices at: A"(–8, 9), B"(–5, 9), C"(–10, 6), D"(–3, 6)

Left 1, up 13

3 Answers for:

Vertex	x-coordinate	y-coordinate
A	–8	8
A'	–4	4
A"	0	0
A'"	4	–4

Challenge ❸

4 **a** (1, 3) **b** (–9, 1)

Lesson 2: Reflecting 2D shapes in a mirror line (1)

Challenge ❶

1

Challenge ⚠

2

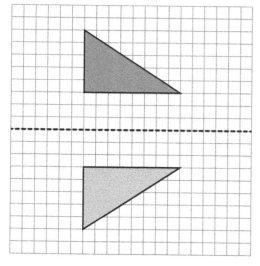

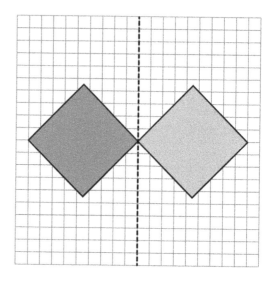

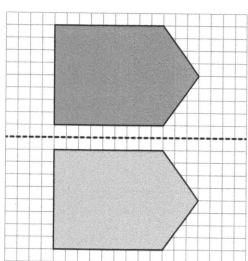

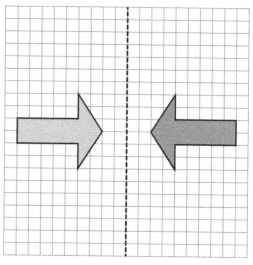

3

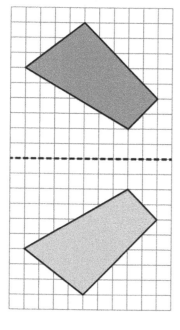

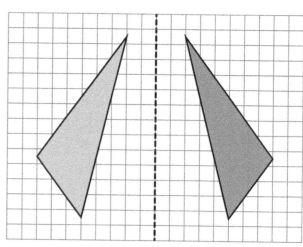

Challenge 3

4

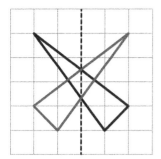 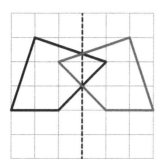 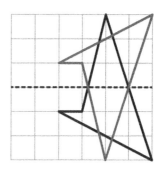 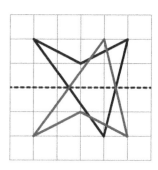

Lesson 3: Reflecting 2D shapes in a mirror line (2)

Challenge 1

1

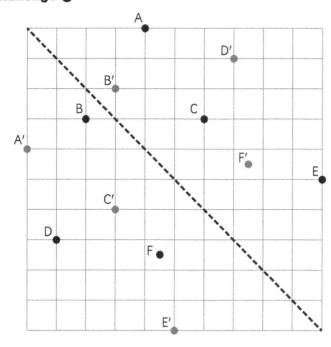

Challenge 2

2

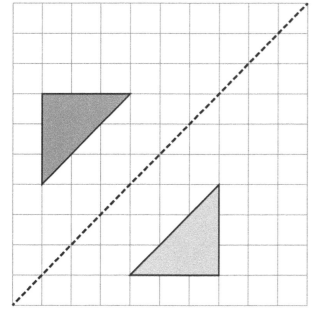 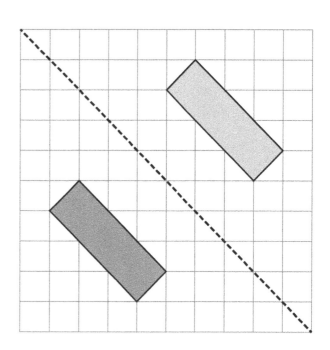

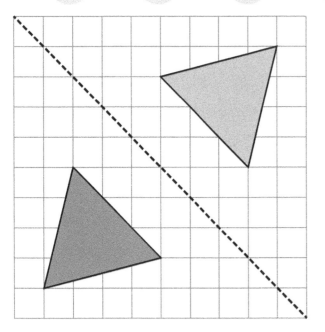

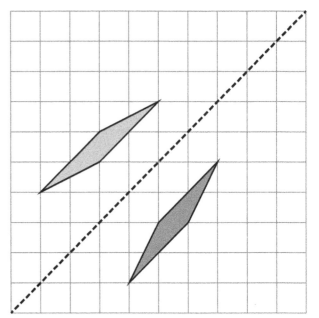

3

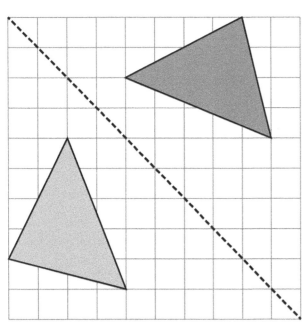

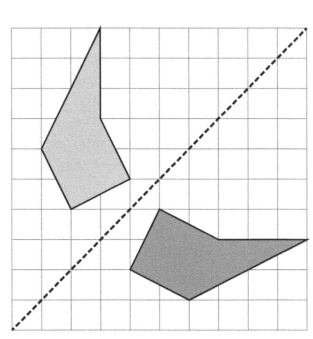

Challenge 3

4

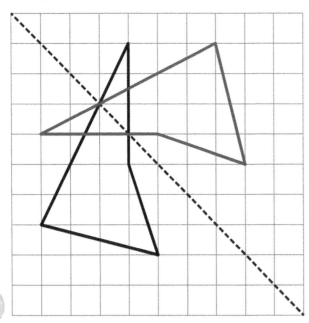

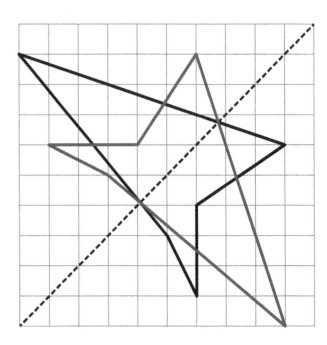

Lesson 4: Rotating shapes 90° around a vertex

Challenge ❶

1 Shape correctly rotated

Challenge ⚠

2 Shape correctly rotated

3 Shape correctly rotated

Challenge ❸

4 Shape correctly transformed

Unit 25

Lesson 1: Venn and Carroll diagrams

Challenge ❶

1 **a** **i** 13   **ii** 9   **iii** 4   **iv** 7

 b Yes. More owners own a cat than a dog.

Challenge ⚠

2 **a**

	Own a car	Do not own a car
Own a bike	xxxxx	xxxxxxxxxxx
Do not own a bike	xxxxxxxxxxxxxxxxxxx	x

 b and c Learner's own response

Challenge ❸

3a–f Learner's own response

Lesson 2: Tally charts, frequency tables and bar charts

Challenge ❶

1 Bar chart with the following bar heights:
Number 3: 7,
Number 4: 14, Number 5: 5, Number 6: 12

Challenge ⚠

2 Learner's own response, e.g. the more occurrences of a number on a spinner, the more frequently it is spun

3 **a–f** Learner's own response

Challenge ❸

4 **a**

Water use (litres per day)	Tally	Frequency
101–110	LH1 I	6
111–120	LH1 II	7
121–130	IIII	4
131–140	LH1 I	6
141–150	LH1 II	7
151–160	LH1 II	7
161–170	IIII	4
171–180	LH1 IIII	9

b A bar graph drawn correctly to match the data in the table provided. Ensure that the axis have been labelled, the graph has been given a title, the axis have been labelled sensibly and the bars are at the correct height.

c Learner's own response

d Learner's own response

Lesson 3: Waffle diagrams and pie charts

Challenge ❶

1

Number of trees counted	Frequency	Percentage
oak	7	35%
birch	6	30%
pine	2	10%
beech	5	25%

Challenge ⚠

2 **a**

Juice	Tally	Frequency	Percentage
blackcurrant	LH1	5	50%
apple	I	1	10%
strawberry	I	1	10%
lemon	III	3	30%

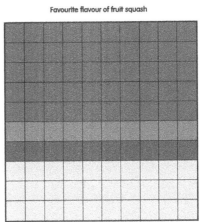

Favourite flavour of fruit squash

Label
- blackcurrant
- apple
- strawberry
- lemon

b

Juice	Frequency	Percentage
blackcurrant	5	50%
apple	1	10%
strawberry	1	10%
lemon	3	30%

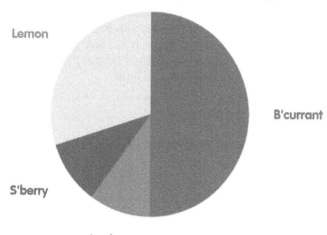

Lemon

B'currant

S'berry

Apple

Challenge 3

3 **a** 5% **b** 30%
c twice **d** 5
e Bananas: 10, oranges: 20, apples: 25, grapes: 40, avocados: 5
f Bananas: 5, oranges: 10, apples: 12·5, grapes: 20, avocados: 2·5
4 Learner's own response

Lesson 4: Mode, median, mean and range

Challenge ❶

1 **a**

Temperature	Frequency
19	1
20	2
21	1
22	5
23	5
24	3
25	8
26	3
27	3

b 25 °C **c** 8 °C

Challenge ⚠

2 **a** Mean: 3 Median: 3 Mode: 2, 4 Range: 4
b Mean: 6 Median: 6 Mode: 6 Range: 5
c Mean: 6·2 Median: 6·5 Mode: 8 Range: 5
d Mean: 63 Median: 62 Mode: 58, 66 Range: 28
3 12, 11, 20, 30, 50

Challenge 3

4 **a** Mean: 8·85 Mode: 8 Median: 8·5 Range: 6
b The mode as it gives an idea of the most popular sizes to stock.
5 **a** Answers may vary, for example: 4, 5, 5, 10 or 5, 5, 5, 9
b 51
c 10, 12, 12, 14, 14, 17, 19
d 7, 7, 8, 14

Unit 26

Lesson 1: Frequency diagrams

Challenge ❶

1 **a** Yes, most learners flicked between 80 and 100 cm.
b Ria's prediction falls in this range.

Challenge ⚠

2 **a** Learner's own response, ensure that the diagram corresponds to the data
b Akihiro's prediction is incorrect. There are more people under the age of 36.
c Learner's own conclusions
3 **a–b** Learner's own response
c Jessica's prediction is a little out. Most days have around 2·5 to 4·5 hours of sunshine.
d Learner's own conclusions

Challenge 3

4 Learner's own response, ensure throughout this question that responses are logical based on their statistical question and data

Lesson 2: Line graphs

Challenge ❶

1 **a** **i** 6000 **ii** 10 000 **iii** 14 000 **iv** 20 000
b 8000 m
c 17 000 m

Challenge ⚠

2 **a** Line graph drawn
b Between 50 and 60 minutes, and between 60 and 70 minutes
c **i** 1500 m **ii** 10 000 m **iii** 3000 m
iv 19 000 m

Challenge 3

3 Learner's own response, ensure throughout this question that responses are logical according to their statistical question and data

Lesson 3: Scatter graphs

Challenge ❶

1 **a**

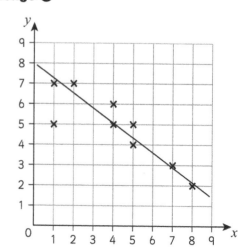

b

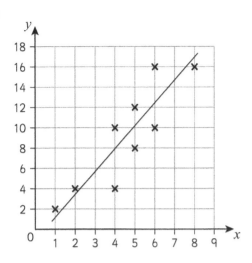

Challenge ⚠
2 **a** Learner's own graph
  b Yes, a line of best fit can be plotted through the points
  c The more customers there are in the shop, the greater the sales
  d Approximately 15
  e Learner's own response

Challenge 🔳
3 Learner's own response

Lesson 4: Dot plots

Challenge ❶
1 **a** 4   **b** 25   **c** 14   **d** 24

Challenge ⚠
a 7; 0, 5, 6 and 12   **b** 10 (11 − 1)
c peaks: 4 and 7
  clusters: 1 to 4, 7 to 11
  gaps: 0; 5 to 6; 12
d Learner's own conclusions

Challenge 🔳
3 **a**

Number of biscuits	Frequency
43	1
44	5
45	6
46	7
47	1
48	1
49	5
50	4

b   **Biscuit count**

43 44 45 46 47 48 49 50
Number of biscuits

c 46
d 43, 47 and 48
e 7
f Peaks: 46, 49; clusters:44–46 and 49–50; gaps: 43 and 47–48
g Yes, as the dot plots peaks around 45 indicating that many packets have around 5 fewer than 50 biscuits

Unit 27

Lesson 1: Describing and comparing outcomes

Challenge ❶
1

Event	Probability		
	Impossible	Even chance	Certain
spin an 11 with a 1–10 spinner	✓		
spin a number less than 11 with a 1–10 spinner			✓
Spin an even number with a 1–10 spinner		✓	
Spin a number less than 6 with a 1–10 spinner		✓	

Challenge ⚠
2 **a** 5 in 10 (1 in 2) or 50%
  b 3 in 10 or 30%
  c 2 in 10 (1 in 5) or 20%

3 **a**

Event	Outcome	Probability
A	spinning a 2	1 in 10 (10%)
B	spinning a 1	3 in 10 (30%)
C	spinning a 4	2 in 10 (20%)
D	spinning a 3	4 in 10 (40%)
E	spinning a number greater than 2	6 in 10 (60%)
F	spinning a number less than 4	8 in 10 (80%)
G	not spinning a 3	6 in 10 (60%)

a **i** A < B   **ii** E > C   **iii** A < F
  iv G = E   **v** F > D   **vi** E > B

Challenge 3

4

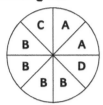

Lesson 2: Independent and mutually exclusive events

Challenge 1

1   Learner's own response, e.g. Tammy can't roll a 1 and a 4 at the same time

Challenge 2

2

Event and outcome	Mutually exclusive (M) or Independent (I)	Reason why
A 1 to 6 is rolled. The number is odd and a multiple of 2.	M	A number cannot both be odd and a multiple of 2
A 4-letter spinner ABCD is spun and a coin is flipped. The letter is C and the coin is 'tails'.	I	One outcome does not affect the other
Winning a cricket match and drawing the same match.	M	A match cannot be both won and drawn
A 50-page book is opened at a random page. The page is 49 or 50.	M	The page can only have one number
Two coins are flipped. The first coin is 'heads' and the second coin is 'heads'.	I	One outcome does not affect the other
On a board numbered 1 to 25, the first dart hits 20 and the second dart hits 20.	I	One outcome does not affect the other

3   Learner's own response

4   Learner's own response

Challenge 3

5   **a**   Rolling a 1 to 6 dice and getting an odd number: 1 in 2 (50%)

Rolling a 1 to 6 dice and getting an even number: 1 in 2 (50%)

   b   Flipping a coin and getting 'heads': 1 in 2 (50%)

Flipping a coin and getting 'tails': 1 in 2 (50%)

   c   Spinning an ABCD spinner and getting a 'D': 1 in 4 (25%)

Spinning an ABCD spinner and **not** getting a 'D': 3 in 4 (75%)

   i   Probabilities in each pair sum to 1.

   ii   Learner's own explanation

Lesson 3: Event probability and the number of trials (1)

Challenge 1

1   **a**   1 in 5 (20%)   **b**   3 in 5 (60%)

   c   1 in 5 (20%)   **d**   2 in 5 (40%)

   e   4 in 5 (80%)

Challenge 2

2   **a**   15   **b**   45

3   **a**   60   **b**   80

Challenge 3

4   **a**

Outcome	Predicted probability (over 80 spins)	Predicted frequency of spins (over 80 spins)
spinning an 'A'	20 %	16
spinning a 'B'	30 %	24
spinning a 'C'	50 %	40

   b   The results differ because experimental probabilities are not the same as theoretical (predicted) probabilities. A theoretical probability is a prediction about the future – and you never know what might happen in the future.

Lesson 4: Event probability and the number of trials (2)

Challenge ❶

1

Colour of sector	Predicted probability	Frequency	Experimental probability
blue	$\frac{1}{4}$ (25%)	30	$\frac{3}{10}$ (30%)
yellow	$\frac{1}{4}$ (25%)	25	$\frac{1}{4}$ (25%)
red	$\frac{1}{4}$ (25%)	20	$\frac{1}{5}$ (20%)
orange	$\frac{1}{4}$ (25%)	25	$\frac{1}{4}$ (25%)

Challenge ⚠

2 a i 1 in 4 (25%)   **ii** 20
     iii 1 in 8   **iv** 10
     v 1 in 2 (50%)   **vi** 40

    b Learner's own response

3 a Harry should expect the probability of spinning each letter to be 20%. That is 20 out of 100 spins. Most of the results are close to 20 out of 100 but not exact.

    b The results differ because experimental probabilities are not the same as theoretical (predicted) probabilities. A theoretical probability is a prediction about the future – and you never know what might happen in the future.

Challenge ❸

4 The gap between experimental and theoretical probability is narrower for 1000 spins compared to 100 spins.

Example:

Spinning letter 'D'	100 spins	1000 spins
Predicted probability	20%	20%
Experimental probability	$\frac{22}{100}$ = 22%	$\frac{190}{1000} = \frac{19}{100}$ = 19%

In general, the greater the number of outcomes you have, the closer a prediction based on probability is likely to be.

Stage 6 Record-keeping

Class: _____ **Year:** _____

KEY

A: Exceeding expectations in this sub-strand	**B:** Meeting expectations in this sub-strand	**C:** Below expectations in this sub-strand

Strand: **Number**
Sub-strand: **Counting and sequences**

Code	Learning objectives
6Nc.01	Count on and count back in steps of constant size, including fractions and decimals, and extend beyond zero to include negative numbers.
6Nc.02	Recognise the use of letters to represent quantities that vary in addition and subtraction calculations.
6Nc.03	Use the relationship between repeated addition of a constant and multiplication to find and use a position-to-term rule.
6Nc.04	Use knowledge of square numbers to generate terms in a sequence, given its position.

A	B	C

Class: _____   **Year:** _____

Strand: **Number**		
Sub-strand: **Integers and powers**		
Code	**Learning objective**	
6Ni.01	Estimate, add and subtract integers.	
6Ni.02	Use knowledge of laws of arithmetic and order of operations to simplify calculations.	
6Ni.03	Understand that brackets can be used to alter the order of operations.	
6Ni.04	Estimate and multiply whole numbers up to 10 000 by 1-digit or 2-digit whole numbers.	
6Ni.05	Estimate and divide whole numbers up to 1000 by 1-digit or 2-digit whole numbers.	
6Ni.06	Understand common multiples and common factors.	
6Ni.07	Use knowledge of factors and multiples to understand tests of divisibility by 3, 6 and 9.	
6Ni.08	Use knowledge of multiplication and square numbers to recognise cube numbers (from 1 to 125).	
A	B	C

Strand: **Number**		
Sub-strand: **Place value, ordering and rounding**		
Code	**Learning objectives**	
6Np.01	Understand and explain the value of each digit in decimals (tenths, hundredths and thousandths).	
6Np.02	Use knowledge of place value to multiply and divide whole numbers and decimals by 10, 100 and 1000.	
6Np.03	Compose, decompose and regroup numbers, including decimals (tenths, hundredths and thousandths).	
6Np.04	Round numbers with 2 decimal places to the nearest tenth or whole number.	
A	B	C

Class: _____ **Year:** _____

Strand: **Number**		
Sub-strand: **Fractions, decimals, percentages, ratio and proportion**		
Code	**Learning objectives**	
6Nf.01	Understand that a fraction can be represented as a division of the numerator by the denominator (proper and improper fractions).	
6Nf.02	Understand that proper and improper fractions can act as operators.	
6Nf.03	Use knowledge of equivalence to write fractions in their simplest form.	
6Nf.04	Recognise that fractions, decimals (one or two decimal places) and percentages can have equivalent values.	
6Nf.05	Estimate, add and subtract fractions with different denominators.	
6Nf.06	Estimate, multiply and divide proper fractions by whole numbers.	
6Nf.07	Recognise percentages (1%, and multiples of 5% up to 100%) of shapes and whole numbers.	
6Nf.08	Understand the relative size of quantities to compare and order numbers with one or two decimal places, proper fractions with different denominators and percentages, using the symbols =, > and <.	
6Nf.09	Estimate, add and subtract numbers with the same or different number of decimal places.	
6Nf.10	Estimate and multiply numbers with one or two decimal places by 1-digit and 2-digit whole numbers.	
6Nf.11	Estimate and divide numbers with one or two decimal places by whole numbers.	
6Nf.12	Understand the relationship between two quantities when they are in direct proportion.	
6Nf.13	Use knowledge of equivalence to understand and use equivalent ratios.	
A	B	C

Strand: **Geometry and Measure**		
Sub-strand: **Time**		
Code	**Learning objectives**	
6Gt.01	Convert between time intervals expressed as a decimal and in mixed units.	
A	B	C

Class: _____ **Year:** _____

Strand: **Geometry and Measure**			
Sub-strand: **Geometrical reasoning, shapes and measurements**			

Code	Learning objectives
6Gg.01	Identify, describe, classify and sketch quadrilaterals, including reference to angles, symmetrical properties, parallel sides and diagonals.
6Gg.02	Know the parts of a circle: - centre - radius - diameter - circumference.
6Gg.03	Use knowledge of area of rectangles to estimate and calculate the area of right-angled triangles.
6Gg.04	Identify, describe and sketch compound 3D shapes.
6Gg.05	Understand the difference between capacity and volume.
6Gg.06	Identify and sketch different nets for cubes, cuboids, prisms and pyramids.
6Gg.07	Understand the relationship between area of 2D shapes and surface area of 3D shapes.
6Gg.08	Identify rotational symmetry in familiar shapes, patterns or images with maximum order 4. Describe rotational symmetry as 'order x'.
6Gg.09	Classify, estimate, measure and draw angles.
6Gg.10	Know that the sum of the angles in a triangle is 180°, and use this to calculate missing angles in a triangle.
6Gg.11	Construct circles of a specified radius or diameter.

A	B	C

Strand: **Geometry and Measure**			
Sub-strand: **Position and transformation**			

Code	Learning objectives
6Gp.01	Read and plot coordinates including integers, fractions and decimals, in all four quadrants (with the aid of a grid).
6Gp.02	Use knowledge of 2D shapes and coordinates to plot points to form lines and shapes in all four quadrants.
6Gp.03	Translate 2D shapes, identifying the corresponding points between the original and the translated image, on coordinate grids.
6Gp.04	Reflect 2D shapes in a given mirror line (vertical, horizontal and diagonal), on square grids.
6Gp.05	Rotate shapes 90° around a vertex (clockwise or anticlockwise).

A	B	C

Class: _____ **Year:** _____

Strand: **Statistics and Probability** Sub-strand: **Statistics**		
Code	Learning objectives	
6Ss.01	Plan and conduct an investigation and make predictions for a set of related statistical questions, considering what data to collect (categorical, discrete and continuous data).	
6Ss.02	Record, organise and represent categorical, discrete and continuous data. Choose and explain which representation to use in a given situation: - Venn and Carroll diagrams - tally charts and frequency tables - bar charts - waffle diagrams and pie charts - frequency diagrams for continuous data - line graphs - scatter graphs - dot plots.	
6Ss.03	Understand that the mode, median, mean and range are ways to describe and summarise data sets. Find and interpret the mode (including bimodal data), median, mean and range, and consider their appropriateness for the context.	
6Ss.04	Interpret data, identifying patterns, within and between data sets, to answer statistical questions. Discuss conclusions, considering the sources of variation and check predictions.	
A	B	C

Strand: **Statistics and Probability** Sub-strand: **Probability**		
Code	Learning objectives	
6Sp.01	Use the language associated with probability and proportion to describe and compare possible outcomes.	
6Sp.02	Identify when two events can happen at the same time and when they cannot, and know that the latter are called 'mutually exclusive'.	
6Sp.03	Recognise that some probabilities can only be modelled through experiments using a large number of trials.	
6Sp.04	Conduct chance experiments or simulations, using small and large numbers of trials. Predict, analyse and describe the frequency of outcomes using the language of probability.	
A	B	C

Class: _____   Year: _____

Thinking and Working Mathematically	
Code	Characteristics
TWM.01	Specialising Choosing *an example* and checking to see if it satisfies or does not satisfy specific mathematical criteria.
TWM.02	Generalising Recognising an underlying pattern by identifying *many* examples that satisfy the same mathematical criteria.
TWM.03	Conjecturing Forming mathematical questions or ideas.
TWM.04	Convincing Presenting evidence to *justify or challenge* a mathematical idea or solution.
TWM.05	Characterising Identifying and describing the mathematical properties of an object.
TWM.06	Classifying Organising objects into groups according to their mathematical properties.
TWM.07	Critiquing Comparing and evaluating mathematical ideas, representations or solutions to identify advantages and disadvantages.
TWM.08	Improving Refining mathematical ideas or representations to develop a more effective approach or solution.

A	B	C

Cambridge Global Perspectives™

Below are some examples of lessons in *Collins International Primary Maths Stage 6* which could be used to develop the Global Perspectives skills. The notes in *italics* suggest how the maths activity can be made more relevant to Global Perspectives.

Please note that the examples below link specifically to the learning objectives in the Cambridge Global Perspectives curriculum framework for Stage 6. However, skills development in a wider sense is embedded throughout this course and teachers are encouraged to promote research, analysis, evaluation, reflection, collaboration and communication as general best practice. For example, the pair work and group activities suggested throughout this Teacher's Guide offer opportunities to develop skills in communication, collaboration and reflection which build towards the specific Global Perspectives learning objectives.

Cambridge Global Perspectives	Learning Objectives for Stage 6	Collins International Primary Maths Stage 6
RESEARCH	Constructing research questions • Begin to construct research questions with support	*There are numerous activities in Units 25 and 26 that give learners the opportunity to write their own statistical questions, which can be used in research. At the end of each activity, encourage learners to reflect on how their original questions could have been better. Examples are:* • Unit 25, Lesson 1: SB, WB Activity 3, TG Teach (Points 6–8) • Unit 26, Lesson 1: SB Pair work activity, WB Activity 4, TG Teach (Points 9–11) • Unit 26, Lesson 2: WB Activity 3, TG Teach (Points 4–5) • Unit 26, Lesson 4: TG Review
	Conducting research • Conduct investigations, using interviews or questionnaires to test a prediction or begin to answer a research question	*There are numerous activities in Units 25 and 26 that give learners the opportunity to conduct investigations to answer statistical questions and/or test predictions. The investigations involve class interviews or group activities. Examples are:* • Unit 25, Lesson 3: SB Pair work activity • Unit 25, Additional practice activities 1 and 2: TG • Unit 26, Lesson 2: WB Activity 3 • Unit 26, Lesson 4: TG Review • Unit 26, Additional practice activities 1 and 2: TG
	Recording findings • Select, organise and record relevant information from sources and findings from research, using an appropriate method	*There are numerous activities in Units 25 and 26 that ask learners to organise and record their findings on a range of tables, charts and graphs. At the end of each Unit, discuss which of the methods used were the most useful and appropriate for recording particular data. Examples are:* • Unit 25, Lesson 1: WB Activity 3 • Unit 25, Lesson 2: SB Pair work activity • Unit 25, Lesson 3: WB Activities 2 and 3 • Unit 25, Additional practice activities 1 and 2: TG • Unit 26, Lesson 3: WB Activity 3 • Unit 26, Additional practice activities 1 and 2: TG

Cambridge Global Perspectives	Learning Objectives for Stage 6	Collins International Primary Maths Stage 6
ANALYSIS	Interpreting data • Find and interpret simple patterns in graphical or numerical data	*There are numerous activities in Units 25 and 26 that ask learners to analyse and interpret the data they have collected to find answers to statistical research questions and/or to compare to predictions. Examples are:* • Unit 25, Lesson 2: WB Activity 2 • Unit 25, Additional practice activities 1 and 2: TG • Unit 26, Lesson 2: WB Activity 3 • Unit 26, Lesson 3: SB Pair work activity, TG Teach (Point 7) • Unit 26, Lesson 3: WB Activity 3 • Unit 26, Lesson 3: WB Activity 3 • Unit 26, Lesson 4: TG Review